Hydrocarbon Exploration to Exploitation West of Shetlands

The Geological Society of London
Books Editorial Committee

Chief Editor
RICK LAW (USA)

Society Books Editors
JIM GRIFFITHS (UK)
DAVE HODGSON (UK)
HOWARD JOHNSON (UK)
PHIL LEAT (UK)
DANIELA SCHMIDT (UK)
RANDELL STEPHENSON (UK)
ROB STRACHAN (UK)
MARK WHITEMAN (UK)

Society Books Advisors
GHULAM BHAT (India)
MARIE-FRANÇOISE BRUNET (France)
JAMES GOFF (Australia)
MARIO PARISE (Italy)
SATISH-KUMAR (Japan)
MARCO VECOLI (Saudi Arabia)
GONZALO VEIGA (Argentina)
MAARTEN DE WIT (South Africa)

Geological Society books refereeing procedures

The Society makes every effort to ensure that the scientific and production quality of its books matches that of its journals. Since 1997, all book proposals have been refereed by specialist reviewers as well as by the Society's Books Editorial Committee. If the referees identify weaknesses in the proposal, these must be addressed before the proposal is accepted.

Once the book is accepted, the Society Book Editors ensure that the volume editors follow strict guidelines on refereeing and quality control. We insist that individual papers can only be accepted after satisfactory review by two independent referees. The questions on the review forms are similar to those for *Journal of the Geological Society*. The referees' forms and comments must be available to the Society's Book Editors on request.

Although many of the books result from meetings, the editors are expected to commission papers that were not presented at the meeting to ensure that the book provides a balanced coverage of the subject. Being accepted for presentation at the meeting does not guarantee inclusion in the book.

More information about submitting a proposal and producing a book for the Society can be found on its website: www.geolsoc.org.uk.

It is recommended that reference to all or part of this book should be made in one of the following ways:

CANNON, S. J. C. & ELLIS, D. (eds) 2014. *Hydrocarbon Exploration to Exploitation West of Shetlands*. Geological Society, London, Special Publications, **397**.

HORSEMAN, C., ROSS, A. & CANNON, S. 2014. The discovery and appraisal of Glenlivet: a West of Shetlands success story. *In:* CANNON, S. J. C. & ELLIS, D. (eds) *Hydrocarbon Exploration to Exploitation West of Shetlands*. Geological Society, London, Special Publications, **397**, 131–144. First published online March 4, 2014, http://dx.doi.org/10.1144/SP397.10

GEOLOGICAL SOCIETY SPECIAL PUBLICATION NO. 397

Hydrocarbon Exploration to Exploitation West of Shetlands

EDITED BY

S. J. C. CANNON

DONG Exploration & Production (UK) Ltd

and

D. ELLIS

Statoil (UK) Ltd

2014
Published by
The Geological Society
London

THE GEOLOGICAL SOCIETY

The Geological Society of London (GSL) was founded in 1807. It is the oldest national geological society in the world and the largest in Europe. It was incorporated under Royal Charter in 1825 and is Registered Charity 210161.

The Society is the UK national learned and professional society for geology with a worldwide Fellowship (FGS) of over 10 000. The Society has the power to confer Chartered status on suitably qualified Fellows, and about 2000 of the Fellowship carry the title (CGeol). Chartered Geologists may also obtain the equivalent European title, European Geologist (EurGeol). One fifth of the Society's fellowship resides outside the UK. To find out more about the Society, log on to www.geolsoc.org.uk.

The Geological Society Publishing House (Bath, UK) produces the Society's international journals and books, and acts as European distributor for selected publications of the American Association of Petroleum Geologists (AAPG), the Indonesian Petroleum Association (IPA), the Geological Society of America (GSA), the Society for Sedimentary Geology (SEPM) and the Geologists' Association (GA). Joint marketing agreements ensure that GSL Fellows may purchase these societies' publications at a discount. The Society's online bookshop (accessible from www.geolsoc.org.uk) offers secure book purchasing with your credit or debit card.

To find out about joining the Society and benefiting from substantial discounts on publications of GSL and other societies worldwide, consult www.geolsoc.org.uk, or contact the Fellowship Department at: The Geological Society, Burlington House, Piccadilly, London W1J 0BG: Tel. +44 (0)20 7434 9944; Fax +44 (0)20 7439 8975; E-mail: enquiries@geolsoc.org.uk.

For information about the Society's meetings, consult *Events* on www.geolsoc.org.uk. To find out more about the Society's Corporate Affiliates Scheme, write to enquiries@geolsoc.org.uk.

Published by The Geological Society from:
The Geological Society Publishing House, Unit 7, Brassmill Enterprise Centre, Brassmill Lane, Bath BA1 3JN, UK

The Lyell Collection: www.lyellcollection.org
Online bookshop: www.geolsoc.org.uk/bookshop
Orders: Tel. +44 (0)1225 445046, Fax +44 (0)1225 442836

The publishers make no representation, express or implied, with regard to the accuracy of the information contained in this book and cannot accept any legal responsibility for any errors or omissions that may be made.

© The Geological Society of London 2014. No reproduction, copy or transmission of all or part of this publication may be made without the prior written permission of the publisher. In the UK, users may clear copying permissions and make payment to The Copyright Licensing Agency Ltd, Saffron House, 6–10 Kirby Street, London EC1N 8TS UK, and in the USA to the Copyright Clearance Center, 222 Rosewood Drive, Danvers, MA 01923, USA. Other countries may have a local reproduction rights agency for such payments. Full information on the Society's permissions policy can be found at: www.geolsoc.org.uk/permissions

British Library Cataloguing in Publication Data

A catalogue record for this book is available from the British Library.
ISBN 978-1-86239-652-4
ISSN 0305-8719

Distributors

For details of international agents and distributors see:
www.geolsoc.org.uk/agentsdistributors

Typeset by Techset Composition India (P) Ltd., Bangalore and Chennai, India.
Printed by Berforts Information Press Ltd, Oxford, UK

Contents

Introduction

AUSTIN, J. A., CANNON, S. J. C. & ELLIS, D. Hydrocarbon exploration and exploitation West of Shetlands — 1

Regional Reviews

ELLIS, D. & STOKER, M. S. The Faroe–Shetland Basin: a regional perspective from the Paleocene to the present day and its relationship to the opening of the North Atlantic Ocean — 11

EGBENI, S., MCCLAY, K., JIAN-KUI FU, J. & BRUCE, D. Influence of igneous sills on Paleocene turbidite deposition in the Faroe–Shetland Basin: a case study in Flett and Muckle sub-basin and its implication for hydrocarbon exploration — 33

Exploration

LOIZOU, N. Success in exploring for reliable, robust Paleocene traps west of Shetland — 59

TRICE, R. Basement exploration, West of Shetlands: progress in opening a new play on the UKCS — 81

Exploitation

KHEIDRI, L. H., CAILLY, F., JONES, K., DE SENNEVILLE, A. & GRAY, R. Laggan Field 3D geological modelling: case study — 107

JONES, K., MACKINNON, D., DE SENNEVILLE, A. & WATT, K. Laggan-Tormore: reservoir to sales production forecasting and optimization using an integrated modelling approach — 123

HORSEMAN, C., ROSS, A. & CANNON, S. The discovery and appraisal of Glenlivet: a West of Shetlands success story — 131

FIELDING, K. D., BURNETT, D., CRABTREE, N. J., LADEGAARD, H. & LAWTON, L. C. Exploration and appraisal of a 120 km^2 four-way dip closure: what could possibly go wrong? — 145

Technologies

WOODBURN, N., HARDWICK, A., MASOOMZADEH, H. & TRAVIS, T. Improved signal processing for sub-basalt imaging — 163

WATTON, T. J., CANNON, S., BROWN, R. J., JERRAM, D. A. & WAICHEL, B. L. Using formation micro-imaging, wireline logs and onshore analogues to distinguish volcanic lithofacies in boreholes: examples from Palaeogene successions in the Faroe–Shetland Basin, NE Atlantic — 173

MILLETT, J. M., HOLE, M. J. & JOLLEY, D. W. A fresh approach to ditch cutting analysis as an aid to exploration in areas affected by large igneous province (LIP) volcanism — 193

MORTON, A., FREI, D., STOKER, M. & ELLIS, D. Detrital zircon age constraints on basement history on the margins of the northern Rockall Basin — 209

Index — 225

Hydrocarbon exploration and exploitation West of Shetlands

J. A. AUSTIN[1], S. J. C. CANNON[2,3]* & D. ELLIS[4]

[1]*OMV (UK) Ltd, 14, Ryder Street, London, SW1Y 6QB, UK*
[2]*DONG E&P (UK) Ltd, 33, Grosvenor Place, London, SW1X 7HY, UK*
[3]*Present Address: Steve Cannon Geoscience Limited, 63, Box Lane, Wrexham, LL12 8BY, UK*
[4]*Statoil (UK) Limited, One Kingdom Street, London, W2 6BD, UK*
*Corresponding author (e-mail: steve@stevecannongeoscience.co.uk)

Abstract: The oil and gas industry has been active in the West of Shetlands area for the last 40 years, but less than 200 wells have been drilled and only four fields have been developed. In the last ten years activity has picked up significantly with very active licensing rounds and increased drilling; however, the challenges of the complex geology and deep-water location, environmental constraints and commercial considerations have also increased at the same time. To fully develop the region successfully an attitude of compromise and collaboration between all the stakeholders must be nurtured, but with the scale of investment required this will remain a challenge for the future.

The Petroleum Exploration Society of Great Britain (PESGB) West of Shetlands Conference convened in October 2011 brought together all the players in this important petroleum province at a significant time in the evolution of the exploration for and development of important new hydrocarbon discoveries in what is now termed by many as the final frontier around the United Kingdom Continental Shelf (UKCS).

It is a frontier area because there are still many challenges facing all companies active in the area. West of Shetlands is at a crossroads at which it is not certain in which direction or at what speed progress is being made to bring new oil and gas fields into production. It is also a crossroads where the lights seem to have been at red (or at least on amber) for far too long for various reasons; it is important to understand why this may be and, more importantly, what can the oil and gas industry do about it? The difficulties and challenges affecting progress are a result of a combination of several factors relating to technical, commercial and political considerations: technical considerations are those due to the nature of West of Shetlands geology and present-day environmental conditions; commercial considerations are those related to high costs and a current lack of infrastructure; and political considerations are those caused by the impact of global events such as Macondo and local events such as fiscal instability.

The impact and the interplay of these considerations provide a relevant overall introduction to the more technical themes of the conference as a whole, taking into context the West of Shetlands with the history of UKCS exploration. A number of specific examples are used to illustrate the nature of various challenges before summarizing how those challenges are being addressed and presenting a view of the future for West of Shetlands, which – notwithstanding the serious challenges – is seen as positive.

Exploration history

Although the Central North Sea sector is considered by the Department of Energy and Climate Change (DECC) to contain the largest yet-to-find potential on the UK Continental Shelf, the West of Shetlands area offers arguably the largest remaining exploration potential for *significant* new finds, by virtue of its relative exploration immaturity (Ellis & Stoker 2014). According to the DECC database, during the 39 years until the end of 2012 over 2400 exploration wells were drilled in the UKCS, of which less than 7% (168) were drilled West of Shetlands. Overall 571 'significant discoveries' have been made, of which 20 (3.5%) are located in the area (Loizou 2014). Figure 1 clearly shows the contrasts in well density, number of fields and infrastructure between West of Shetlands and other UKCS provinces.

It is clear that drilling in the North Sea has predominated, particularly in the early years of UK exploration in the 1970s and 1980s, when a massive growth of exploration success in the Northern and

From: CANNON, S. J. C. & ELLIS, D. (eds) 2014. *Hydrocarbon Exploration to Exploitation West of Shetlands*.
Geological Society, London, Special Publications, **397**, 1–10.
First published online March 19, 2014, http://dx.doi.org/10.1144/SP397.13
© The Geological Society of London 2014. Publishing disclaimer: www.geolsoc.org.uk/pub_ethics

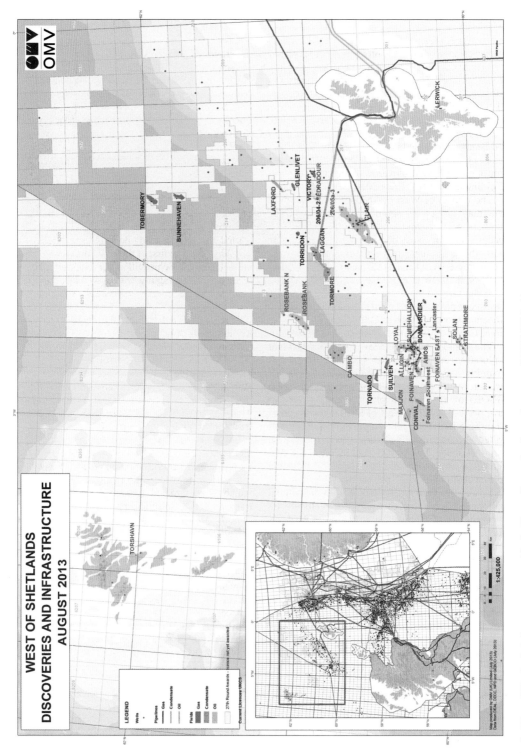

Fig. 1. West of Shetlands infrastructure map with significant discoveries shown (courtesy of OMV UK Ltd.)

Central North Seas provided more than enough resources to boost UK production towards self-sufficiency.

There has been success in the area however and, as a percentage of total, the broad West of Shetlands area (including the Rockall Trough) has certainly grown in influence over the last 20 years. Discoveries seem to have come in batches with Clair and Victory in the late 1970s, Laxford and Laggan in the mid-1980s, Strathmore, Solan, Foinaven, Schiehallion and Suilven in the early 1990s, Tobermory and Benbecula in the late 1990s, Cambo and Rosebank in the early part of the last decade and a stream of around 10 more discoveries in the last few years (Fielding *et al.* 2014; Trice 2014).

The impact of West of Shetland exploration to date may not be as significant in terms of the numbers of discoveries made. While the overall success rate for West of Shetlands is lower than most other areas, there has been a gradual but continued increase over the last 20 years. More significantly still, there is evidence of a rapid increase since the turn of the century, with more than half the discoveries made in the last 12 years including significant discoveries such as Cambo, Rosebank, Tormore, Lancaster, Glenlivet, Tornado and Edradour.

The number of discoveries is only a proxy for exploration success, so for a wider context the creaming curve should also be examined (Fig. 2). Overall, over 3 billions of barrels of oil equivalent (BBOE) of 2P ('proven plus probable') reserves have been discovered West of Shetlands, or almost 20 million barrels of oil equivalent (MMBOE) per drilled exploration well. Historic cost information is not available, but on the assumption of US $50 million per well this would equate to a reasonable finding cost of US $2–3 per barrel.

However, after over 30 years of West of Shetland exploration, it is a fact that only three oilfields are in production (Foinaven, Schiehallion and Clair). The first gas field development (Laggan–Tormore) was sanctioned very recently in 2010. Is this slow pace of activity indicative of particular challenges peculiar to the West of Shetlands or is it just that the industry has concentrated on the more easily attained value in the shallower and more accessible waters of the North Sea?

Technical considerations

The subsurface technical challenges need to be considered first and foremost, of which there are many. West of Shetlands is predominantly a deep-water environment with depths greater than 1500 m in places, and extreme metocean conditions exist. These factors conspire to increase costs and generate delays to exploration programmes. Well control is sparse (which of course provides an opportunity for creative explorers; Egbeni *et al.* 2014) and basalts are extensive and thick across many of the prospective parts of the basin. These factors widen uncertainties and increase risks, as well as costs. In fact, the largest undrilled prospects are probably lurking below or within basaltic sequences. The main Rosebank reservoirs are actually interbedded with basalts, making it much more difficult to seismically map and characterize the reservoir and its distribution.

While geological challenges are constant, technological changes have moved the industry forward. The first West of Shetland well was drilled in 1972, the first hydrocarbon discovery (Clair) was made in 1977, but it was the advent of new seismic technologies combined with better trap definition and other innovative techniques that started a small but steady flow of discoveries in the last 20 years. The first of these – Foinaven (Cooper *et al.* 1999) and Schiehallion (Leach *et al.* 1999) – were developed rapidly with first oil flowing less than seven years after discovery. Clair moved rather slower, with the need to delineate the complex reservoirs and optimize hydrocarbon recovery techniques in what was seen for many years as a marginal project, leading to a delay of more than 27 years from discovery to first oil (Witt *et al.* 2010).

The most recent gas development project, Laggan (planned to come on-stream mid-2014), was discovered in 1986. It was not until the drilling of Tormore in 2007 21 years later, and with improvements in technology (Merlet & Campo 2011; Jones *et al.* 2014), that the potential for a combined and therefore economic project became viable.

Exploration in the West of Shetlands has had a history of unsuccessful drilling results, many of which were targeted specifically at stratigraphic traps related to seismic amplitude anomalies, and these have been well documented in the literature (Smallwood & Kirk 2005; Loizou *et al.* 2006; Waters 2011). However, it was found that by drilling such anomalies using an integrated geophysical and geological approach, including the acquisition of a controlled source electromagnetic survey (CSEM), the risk can be significantly reduced from high to moderate (Rodriguez *et al.* 2010). The result in this instance was the drilling of the Tornado seismic amplitude anomaly, which proved to be successful in identifying the presence of a gas and oil accumulation (Pickering & Rodriguez 2011).

A big challenge to seismic imaging in the West of Shetlands is the presence of a thick Palaeogene basalt pile lying predominantly above the main zone of prospectivity. Exploration and appraisal drilling has identified the presence of hydrocarbon accumulations within these basalt layers, such as

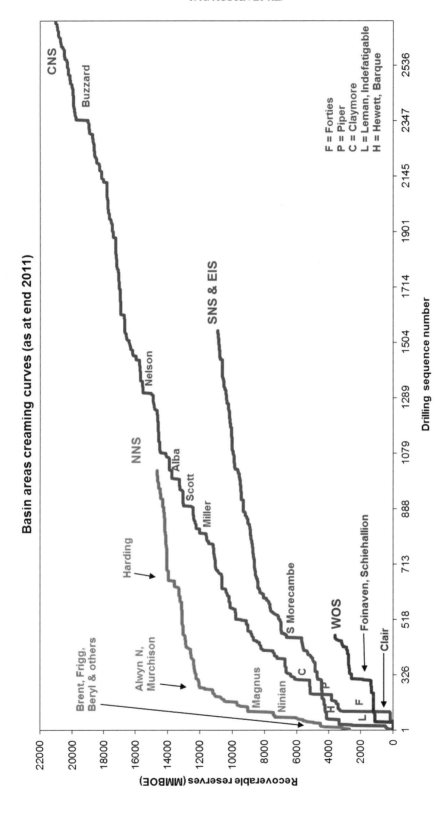

Fig. 2. United Kingdom Continental Shelf creaming curves by basin showing the under-explored West of Shetland (WoS) Basin in comparison (source DECC).

in Rosebank (Jarvie et al. 2011) which lies a few kilometres to the east of the Faroe–UK boundary. The emphasis in the West of Shetlands is now to continue exploring both within and below this basalt section. The latest innovations in low-frequency signal processing, the acquisition of long-offset seismic data and dual-sensor streamer technology for sub-basalt imaging (Whittaker & Dinkleman 2011), focus the need to provide a better-quality seismic response to image the Paleocene section and older events below the thick Tertiary basalt flows. In the case of Rosebank, the use of ocean bottom seismic nodes aims to improve seismic quality of the reservoir intra-basalt imaging (Mathar et al. 2011). The problems of high-frequency signal loss and the impact of multiple noise within the seismic dataset due the presence of the basalts is a challenge which companies are continuing to address.

Commercial considerations

Commercial challenges have been fairly consistent as West of Shetland exploration progressed throughout the 1970s and 1980s as discoveries were made, but these challenges and uncertainties have tended to increase with the passage of time. Fluctuating oil price, low gas price and high costs in a severe environment all conspire to increase economic thresholds and deter and delay investment. A lack of infrastructure also continues to hold back the development of many stranded oil and gas discoveries.

To take an example from current operators' experience, it is clear that commercial challenges have delayed progress on the Rosebank, Cambo and Tornado discoveries. Rosebank, the largest of the three and probably the largest undeveloped discovery West of Shetland (c. 240 MMBOE), was discovered in 2004. At the time there was an expectation (perhaps somewhat naïve) that it would be developed in around seven years (taking Schiehallion as an analogue). Seven years later, at the time of the West of Shetlands conference, it was still uncertain when (or even if) the project would be sanctioned. Cambo and Tornado are smaller, and both will probably depend on developing jointly with other fields or using an existing or planned hub to tie into.

Indeed, nearly all of the unsanctioned projects West of Shetlands are being held back due to questions and concerns about the export route for gas and the cost of new infrastructure.

Political considerations

What has changed dramatically since 2010, and it is still a very dynamic environment, are the political challenges to oil and gas companies active West of Shetlands: firstly global Macondo-related issues and secondly fiscal instability generated by recent UK governments (of all political persuasions), the most recent example of which was a tax increase in 2011.

While the 2010 Macondo oil spill in the Gulf of Mexico was a great human tragedy and environmental catastrophe, it is the political aftershocks that are still rumbling around the world. For the UK, these have been most keenly felt in the West of Shetlands deep-water area. The whole oil industry has spent considerable time and effort in the aftermath of Macondo seeking to understand better the impacts of the tragedy on their businesses, and how to prevent and mitigate such incidents in the future. Initial reactions to Macondo from the European Union (EU) were such that it looked like there might have been a moratorium on deep-water drilling in the UK; it is to the credit of DECC, the UK Government and industry bodies such as Oil & Gas UK and the International Oil & Gas Producers Association (OGP) that this never happened.

The EU Commission also concluded initially that a new EU Regulation should be introduced to improve offshore safety. The UK government and various lobby and pressure groups argued successfully, with strong support from Norway, that an EU Regulation would actually have a damaging effect on safety in countries with a well-developed system already in place (such as UK and Norway). The EU is now moving towards a Directive, with many elements in common with the UK and Norway systems.

Nevertheless, the first post-Macondo deep-water wells in the UK suffered much longer delays to drilling permissions due to the need for all stakeholders to tread very carefully. The first well, drilled to explore the Lagavulin prospect in one of the furthest-north and deepest water areas of the UKCS in 2011–2012, was delayed for almost a month (at full cost to the oil companies awaiting permission to drill).

The increased cost to offshore operators following Macondo is substantial, but clearly necessary and justifiable. What was more difficult to understand was the increased tax imposed on the industry by the UK government in the 2011 budget. That this came only a year after the introduction of the Field Allowance for West of Shetland gas developments that ushered in the sanction of the Laggan–Tormore projects demonstrated that the government did not understand the true value of the oil industry to the UK economy or the implications of penalizing investment in the sector. The announcement by the Chancellor in the 2011 budget eroded at least a quarter of the value from the entire UKCS portfolio, and set back further the most marginal projects (many of which are located West of

Shetlands). The economy needed a lift and the oil industry was identified as one that could afford to pay, but there was a serious lack of joined-up thinking by the UK government at this time.

The lack of consultation with industry, the apparent lack of understanding by the government and the lack of consistency of approach does not serve the UKCS well, and has the potential to lead to a reduction of exploration activity, delays to projects and potential cancellation of many development projects.

Addressing technical and commercial challenges

So, what can the industry say it is doing about these three key challenges to West of Shetlands oil and gas exploration and developments?

From a technical and commercial point of view the oil and gas industry is doing what it is very good at. It is employing the latest technologies in seismic and drilling to reduce uncertainties and improve safety, while hopefully driving down costs. Conferences such as the PESGB West of Shetlands meeting in October 2011 are a great example of the industry sharing knowledge. Geologists and engineers came together and produced a stimulating set of technical presentations in an open networking environment, helping all move forward collaboratively for the common good.

Indeed, the real strength of the petroleum business will be found by working together much more than working against each other. Seeking mutually beneficial commercial arrangements aims to bring oil and gas fields forward, although there is much more to be done in this regard as it sometimes still takes far too long to make progress. If there is a need, DECC initiatives such as the Commercial Code of Practice are there to help reach agreement.

Nevertheless, it is still frustrating that new West of Shetlands infrastructure seems to be project-related rather than strategic. Until new gas lines are oversized enough to encourage the development of currently stranded discoveries, new producing fields will be limited to those large enough to be standalone. The industry needs to be more collaborative in this regard, and the DECC and HM Treasury need to ensure the fiscal and regulatory framework encourages new infrastructure more than at present.

Addressing political considerations

The industry has ways of addressing the technical challenges of exploring and drilling in the West of Shetlands. It has ways of seeking commercially viable solutions to maximize value and ultimate recovery. However, are there ways of addressing the political challenges? The oil and gas industry is global, it is inherently political and it has the scale to influence global decision-makers, but this power is often misunderstood by the public and single-issue groups or the industry does not portray itself well to these stakeholders.

To consider first the Macondo effect on these stakeholders, the initial impact of the incident nearly bankrupted BP. Every oil company in the world has had to look at its exposure to offshore risk in a different way. Every company has had to understand what it needs to do globally and in its specific countries of operations. Every company has had to react to governments, stakeholders and non-governmental organizations' (NGO) demands for improved working practices. It is also clear that all companies active offshore have had to do more to satisfy their managements and shareholders that risks are understood and mitigated as much as possible. The licence to operate has suddenly become higher in the priorities of oil companies, and rightly so. Within the UK, many smaller players also gained from participation in the Oil & Gas UK forum, which was able to marshal views from all companies, provide advice and lobby UK and EU institutions.

The relationship between oil companies, contractors and Oil & Gas UK has also proved helpful in addressing the second political challenge: that of fiscal instability. As soon as the Budget 2011 was announced, the shock and surprise in the oil industry was quickly transformed into task forces and workgroups under the guidance of Oil & Gas UK. These workgroups, including one focused on West of Shetlands, made real progress over the ensuing months. Introductions of a number of new Field Allowances has allowed a more positive view to be taken on many important new projects, resulting in new field approvals arising directly from the Brown Field and Southern North Sea Allowances (DECC/HM Treasury). The new West of Shetlands deep-water oil allowance was a direct result of the industry engagement with HM Treasury and allowed the Rosebank project to move into Front-End Engineering Design (FEED) in mid-2012, and will hopefully give enough fiscal stimulus to allow this project of great importance to the West of Shetlands to be sanctioned.

While the new era of positive engagement between the oil industry and UK government is welcome, the Field Allowances add further complications to an already complicated tax system. The UK Government's objective is to ensure that no economically developable hydrocarbons are left stranded, which is to be lauded by all. However, the process for approvals will become more time-consuming, placing an undue load on an already stretched DECC, and will likely introduce more inefficiencies and potentially delays.

Not only does the industry need to ensure HM Treasury is informed, it could also do much better to inform and educate the public as to the benefits the industry brings to society in terms of taxes, jobs and great economic wealth for the United Kingdom.

Questions posed at the time of the West of Shetlands conference in October 2011 were:

- Will we be stuck in a traffic jam of projects with few infrastructure alternatives and long delays?
- Will we become involved in expansive and expensive projects that may be too complex and uneconomic?
- Or will we end up with a fit-for-purpose collaborative approach between all stakeholders to ensure ready access, no delays and appropriate shares of revenues and profits?

At the time, it was not clear which direction would be taken. The realization by all concerned that the last option is the right one and that all are moving closer to this outcome is however encouraging . It is up to all concerned to create the right framework for the best decisions – technically, commercially and politically.

Conference review and papers

The contents of this volume represent about half of the papers (24) and posters (12) presented at the conference held in October 2011. Conference delegates numbered 250, filling the venue. The conveners' objectives were to capture the challenges of operating West of Shetlands and the state of available technology to meet these challenges. The conveners invited a number of speakers to discuss specific topics such as drilling (D. Reynolds) and environmental (M. Borewell) and commercial (G. Dempster) issues to broaden the overall content of the conference and to introduce many of the delegates to what may be peripheral to their daily task as explorers and developers. Other speakers came from academia, operators and the service sector. The majority of papers presented fell into the categories of regional reviews, exploration history, field case studies and technology advances, all of which are covered in this volume.

Regional reviews

Ellis & Stoker (2014) underscore the importance of better understanding the intricate development of the rift basins in the Faroe–Shetland Basin by challenging the 'classic', and possibly oversimplistic, views on the North Atlantic opening in order to improve our approach to exploration. This concept inevitably extends to the understanding of the tectono-stratigraphic interactions that affected and controlled deposition of the WoS reservoirs (Egbeni *et al.* 2014).

Exploration

Only 162 exploration wells have been drilled West of Shetlands in the last 40 years, and it was highlighted that the majority of failed wells were drilled on poorly defined or invalid traps and on prospects that lacked reservoir or poor top seal (Loizou 2014). Complex rifting and magmatism may also lead to erroneous timing of source maturation and migration. Misinterpretation of high-amplitude features has also contributed to the failure of several wells, especially when the amplitude anomalies were inferred in proximity of updip limit/pinch-out edges of reservoirs.

Trice (2014) passionately describes the exploration for fractured basement plays in the Faroes– Shetland Basin and the challenges associated with drilling and evaluating the results, especially when you are a small player with funding issues.

Exploitation

A number of case studies of fields in development were presented at the conference, but unfortunately not all made it into the volume. A key project that is changing the face of exploitation West of Shetlands is the Laggan–Tormore development operated by Total (TEPUK). Khedri *et al.* (2014) look at the extensive use of seismic imaging and sedimentology to characterize and model the Laggan Field, and Jones *et al.* (2014) discuss the process of optimizing gas production using integrated production modelling. Another gas field recently discovered and waiting development is Glenlivet, where again seismic interpretation and a robust depositional model have been fundamental in establishing an economically viable development project after drilling one well and two sidetracks (Horseman *et al.* 2014). Fielding *et al.* (2103) present an intriguing story of the discovery and appraisal of a 120 km^2 four-way dip-closed structure, the Cambo oilfield. Together with Rosebank, these two oilfields represent the largest undeveloped accumulations in the UKCS.

Technologies

Four papers present some of the many technology solutions being used in the exploration and exploitation of the West of Shetlands, especially in understanding the impact of Paleocene volcanism on acquired data. Extensive thick sequences of basalt dominate the north-western flank of the West of

Shetland region. The key to providing improved images beneath basalt flows is to generate, retain and enhance as much low-frequency energy as possible (Ziolkowski *et al.* 2003). Woodburn *et al.* (2014) describe a project to reprocess 20 000 km of 2D seismic data from the West of Shetlands and how careful signal processing can be applied to improve sub-basalt imaging. Watton *et al.* (2014) demonstrate the value of formation microscanner images in unravelling different basaltic lithofacies, comparing the subsurface images with outcrop analogues in a classic calibration study. A fresh approach to ditch-cuttings analysis is presented by Millet *et al.* (2014) who describe how a robust volcanic stratigraphy can aid the construction of accurate basin models, especially where core data is limited or absent. Morton *et al.* (2014) also address basement history through the use of detrital zircon age dating in the northern Rockall Basin and conclude that the Hebridean Platform is a westward extension of the Lewisian Complex, with age dates that can be directly correlated with events identified in the Outer Hebrides and NW Scotland.

The future

It is unfortunate that a number of the papers presented at the conference did not make it into print and that some of the success stories were not given a wider airing. Two cases in particular spring to mind, that of the Rosebank Field and Schiehallion Field. In the case of the former, although a number of technical papers have been presented at various conferences, these have not yet been published. Since 2011 Rosebank has moved further towards development with a FEED project underway, announced by the operator Chevron on 9 July 2012. In the announcement, Chevron state 'The Rosebank project is another important step forward in our strategy to grow profitably in core areas of our upstream business. Chevron has extensive deepwater capabilities, and the Rosebank project fits well in our portfolio.' Rosebank is located c. 150 km NW of the Shetland Islands in water depths of c. 1100 m and was discovered in 2004. An efficient appraisal campaign including ocean bottom node seismic acquisition has established potential recoverable resources amounting to 240 MMBOE. The project will include a floating production, storage and offloading vessel, production and water injection wells, subsea facilities and a gas export pipeline. As described in Section 'Technical Considerations', the Paleocene reservoirs are interbedded with sub-aerial basaltic lavas making characterization challenging; however, the prize of commercializing the largest remaining known oilfield in the UKCS is an attractive one.

The story of the Schiehallion Field is in many ways even more intriguing. The following is an extract from the Schiehallion and Loyal Decommissioning Programme (BP 2012).

> The Schiehallion and Loyal fields lie within Quadrants 204 and 205 of the UKCS, approximately 130 km west of Shetland in water depths of 350–500 m on the slope of the Faroe-Shetland channel. These fields have been in production since 1998 through the Schiehallion FPSO with production to date totalling over 300 MMBOE of oil and associated gas. The production history and experience obtained from the existing wells and recent reservoir studies have confirmed that a significant oil potential still remains to be exploited from these reservoirs. Additionally, a number of new oil and gas discoveries have been made that could potentially be developed in the future by subsea tie-backs to the Schiehallion and Loyal infrastructure. To fully exploit the remaining Schiehallion and Loyal reserves would require the existing FPSO to remain on-station for a further period in excess of 35 years. In recent years operating challenges on the FPSO have resulted in a deterioration of the production operation efficiency and the existing FPSO is unable to fulfill the processing requirements of the anticipated economic field life. Therefore redevelopment of the surface facilities was deemed necessary for the optimal exploitation of hydrocarbon reserves in the area. Following consideration of a number of options it was determined that the Schiehallion FPSO should be replaced.

In April 2013 the last tanker-load of oil was received at Sullom Voe Terminal on the Shetland Islands before the field was shut-in for a £3 billion redevelopment project. It is expected that oil will start to be produced again in 2016 when the redevelopment is complete. These are only two examples of the importance of the West of Shetlands to the oil industry in the UK and how much investment is needed to operate successfully in the area. However, there are always companies keen to be a part of this exciting arena of exploration and exploitation.

One further topic of interest to the future of oil and particularly gas development in the West of Shetlands is the need for a more permanent infrastructure. A second gas export route to complement the Laggan–Tormore pipeline and the Shetland Islands Regional Gas Export (SIRGE) line will be needed if the UK is to be able to successfully maximize the benefits of current exploration activity and sidelined development projects (Hannke *et al.* 2011).

Conclusions

The West of Shetlands remains an attractive underdeveloped part of the UKCS. That challenges remain cannot be denied, but these can be met by an industry which is commercially driven, environmentally aware and technically advanced. This paper has highlighted the first two of these issues and the volume as a whole addresses many of the technical considerations of exploration and exploitation West of Shetlands.

The conference conveners and volume editors would like to thank the staff and Council of the Petroleum Exploration Society of Great Britain for their hard work and support organizing the conference and sponsoring this volume. All of the contributors to the conference and the volume and their companies/organizations are gratefully acknowledged. Jean Rodriguez of OMV also provided invaluable additions to this introductory paper.

References

COOPER, M. M., EVANS, A. C., LYNCH, D. J., NEVILLE, G. & NEWLEY, T. 1999. The Foinaven Field: managing reservoir development uncertainty prior to start-up. *In*: FLEET, A. J. & BOLDY, S. A. R. (eds) *Petroleum Geology of Northwest Europe: Proceedings of the 5th Conference*. Geological Society, London, 675–682.

EGBENI, S., MCCLAY, K., JIAN-KUI FU, J. & BRUCE, D. 2014. Influence of igneous sills on Paleocene turbidite deposition in the Faeroe-Shetland Basin: a case study Flett and Muckle Sub-basin and its implication for hydrocarbon exploration. *In*: CANNON, S. J. C. & ELLIS, D. (eds) *Hydrocarbon Exploration to Exploitation West of Shetlands*. Geological Society, London, Special Publications, **397**. First published online March 3, 2014, http://dx.doi.org/10.1144/SP397.8

ELLIS, D. & STOKER, M. S. 2014. The Faroe–Shetland Basin: a regional perspective from the Palaeocene to the present day and its relationship to the opening of the North Atlantic Ocean. *In*: CANNON, S. J. C. & ELLIS, D. (eds) *Hydrocarbon Exploration to Exploitation West of Shetlands*. Geological Society, London, Special Publications, **397**. First published online February 20, 2014, http://dx.doi.org/10.1144/SP397.1

FIELDING, K. D., BURNETT, D., CRABTREE, N. J., LADEGAARD, H. & LAWTON, L. C. 2014. Exploration and appraisal on a 120 km^2 four-way dip closure: what could possibly go wrong? *In*: CANNON, S. J. C. & ELLIS, D. (eds) *Hydrocarbon Exploration to Exploitation West of Shetlands*. Geological Society, London, Special Publications, **397**. First published online February 25, 2014, http://dx.doi.org/10.1144/SP397.11

HANNKE, S., SAUNDERS, J. & AUSTIN, J. A. 2011. Does the West of Shetlands need another gas export system? *(Exploration and Exploitation West of Shetlands, PESGB Conference Abstract)*.

HORSEMAN, C. E., ROSS, A. & CANNON, S. J. C. 2014. The discovery and appraisal of Glenlivet: a West of Shetlands success story. *In*: CANNON, S. J. C. & ELLIS, D. (eds) *Hydrocarbon Exploration to Exploitation West of Shetlands*. Geological Society, London, Special Publications, **397**. First published online March 3, 2014, http://dx.doi.org/10.1144/SP397.10

JARVIE, A., STONARD, S. & DUNCAN, L. 2011. The importance of understanding lithology and depositional environment – Rosebank field, West of Shetlands. *(Exploration and Exploitation West of Shetlands, PESGB Conference Abstract)*.

JONES, K., MACKINNON, D., DE SENNEVILLE, A. & WATT, K. 2014. Laggan-Tormore: reservoir to sales production forecasting using an integrated modelling approach. *In*: CANNON, S. J. C. & ELLIS, D. (eds) *Hydrocarbon Exploration to Exploitation West of Shetlands*. Geological Society, London, Special Publications, **397**. First published online February 20, 2014, http://dx.doi.org/10.1144/SP397.4

KHEIDRI, L. H., CAILLY, F., JONES, K., DE SENNEVILLE, A. & GRAY, R. 2014. Laggan field 3D geological modelling – a case study. *In*: CANNON, S. J. C. & ELLIS, D. (eds) *Hydrocarbon Exploration to Exploitation West of Shetlands*. Geological Society, London, Special Publications, **397**. First published online February 20, 2014, http://dx.doi.org/10.1144/SP397.5

LEACH, H. M., HERBERT, N., LOS, A. & SMITH, R. L. 1999. The Schiehallion Development. *In*: FLEET, A. J. & BOLDY, S. A. R. (eds) *Petroleum Geology of Northwest Europe: Proceedings of the 5th Conference*. Geological Society, London, 683–692.

LOIZOU, N. 2014. Success in exploring for reliable, robust Paleocene traps west of Shetland. *In*: CANNON, S. J. C. & ELLIS, D. (eds) *Hydrocarbon Exploration to Exploitation West of Shetlands*. Geological Society, London, Special Publications, **397**. First published online February 25, 2014, http://dx.doi.org/10.1144/SP397.9

LOIZOU, N., ANDREWS, I., STOKER, S. J. & CAMERON, D. 2006. West of Shetland revisted: the search for stratigraphic traps. *In*: ALLEN, M. R., GOFFEY, G. P., MORGAN, R. K. & WALKER, I. M. (eds) *The Deliberate Search for the Stratigraphic Trap*. Geological Society, London, 225–245.

MATHAR, J., ZASKER, J., DAVIES, K., SIM LEE, K. & ZIMMERMAN, J. 2011. The use of ocean-bottom-seismic nodes to improve reservoir imaging at Rosebank. *(Exploration and Exploitation West of Shetlands, PESGB Conference Abstract)*.

MERLET, F. & CAMPO, R. M. 2011. The role of geoscience in achieving the first gas development West of Shetland. *(Exploration and Exploitation West of Shetlands, PESGB Conference Abstract)*.

MILLET, J. M., HOLE, M. J. & JOLLEY, D. M. 2014. A fresh approach to ditch cutting analysis as an aid to exploration in areas affected by large igneous province volcanism. *In*: CANNON, S. J. C. & ELLIS, D. (eds) *Hydrocarbon Exploration to Exploitation West of Shetlands*. Geological Society, London, Special Publications, **397**. First published online February 21, 2014, http://dx.doi.org/10.1144/SP397.2

MORTON, A., FREI, D., STOKER, M. S. & ELLIS, D. 2014. Detrital zircon age constraints on basement history on the margins of the norther Rockall Basin. *In*: CANNON, S. J. C. & ELLIS, D. (eds) *Hydrocarbon Exploration to Exploitation West of Shetlands*. Geological Society, London, Special Publications, **397**. First published online February 25, 2014, http://dx.doi.org/10.1144/SP397.12

PICKERING, G. & RODRIGUEZ, J. M. 2011. Maximising appraisal information from exploration wells: the planning and results from the Tornado well. *(Exploration and Exploitation West of Shetlands, PESGB Conference Abstract)*.

RODRIGUEZ, J. M., PICKERING, G. & KIRK, W. J. 2010. Prospectivity of the T38 Sequence in the Northern Judd Basin. *In*: VINING, B. A. & PICKERING, S. C. (eds) *Petroleum Geology: From Mature Basins to New Frontiers – Proceedings of the 7th Petroleum Geology Conference*. Geological Society, London, 245–259.

SMALLWOOD, J. R. & KIRK, W. J. 2005. Paleocene exploration in the Faroe-Shetland Challel: disappointments and discoveries. *In*: DORE, A. G. & VINING, B. A. (eds) *Petroleum Geology: North–West Europe and Global Perspectives*. Proceedings of the 6th Petroleum Geology Conference, Geological Society, London, 977–991.

TRICE, R. 2014. Basement exploration, West of Shetlands: progress in opening a new play on the UKCS. *In*: CANNON, S. J. C. & ELLIS, D. (eds) *Hydrocarbon Exploration to Exploitation West of Shetlands*. Geological Society, London, Special Publications, **397**. First published online February 28, 2014, http://dx.doi.org/10.1144/SP397.3

WATERS, K. 2011. The West of Shetlands; a regional rock study. *(Exploration and Exploitation West of Shetlands, PESGB Conference Abstract)*.

WATTON, T. J., CANNON, S., BROWN, R. J., JERRAM, D. A. & WAICHEL, B. L. 2014. Using formation micro-imaging, wireline logs and onshore analogues to distinguish volcanic lithofacies in boreholes: examples from Palaeogene successions in the Faroe–Shetland Basin, NE Atlantic. *In*: CANNON, S. J. C. & ELLIS, D. (eds) *Hydrocarbon Exploration and Exploitation West of Shetlands*. Geological Society, London, Special Publications, **397**. First published online February 25, 2014, http://dx.doi.org/10.1144/SP397.7

WHITTAKER, R. C. & DINKLEMAN, M. G. 2011. Long offset pre-stack depth migration improves the regional understanding of sub-basalt geology of the WoS-Faroes margin. *(Exploration and Exploitation West of Shetlands, PESGB Conference Abstract)*.

WITT, A. J., FOWLER, S. R., KJELSTADLI, R. M., DRAPER, L. F., BARR, D. & MCGARRITY, J. P. 2010. Managing the start-up of a fractured oil reservoir: development of the Clair field, West of Shetland. *In*: VINING, B. A. & PICKERING, S. C. (eds) *Petroleum Geology: From Mature Basins to New Frontiers – Proceedings of the 7th Petroleum Geology Conference*. Geological Society, London, 299–313.

WOODBURN, N., HARDWICK, A., MASOOMZADEH, H. & TRAVIS, T. 2014. Improved signal processing for sub-basalt imaging. *In*: CANNON, S. J. C. & ELLIS, D. (eds) *Hydrocarbon Exploration to Exploitation West of Shetlands*. Geological Society, London, Special Publications, **397**. First published online February 20, 2014, http://dx.doi.org/10.1144/SP397.6

ZIOLKOWSKI, A., HANSSEN, P. ET AL. 2003. Use of low frequencies for sub-basalt imaging. *Geophysical Prospecting*, **51**, 169–182.

The Faroe–Shetland Basin: a regional perspective from the Paleocene to the present day and its relationship to the opening of the North Atlantic Ocean

DAVID ELLIS[1]* & MARTYN S. STOKER[2]

[1]*Statoil (UK) Limited, One Kingdom Street, London, W2 6BD, UK*
[2]*British Geological Survey, Murchison House, West Mains Road, Edinburgh, EH9 3LA, UK*
*Corresponding author (e-mail: ddael@Statoil.com)

Abstract: The Faroe–Shetland Basin is located offshore NW Scotland on the SE margin of the Atlantic Ocean and comprises numerous sub-basins and intra-basin highs that are host to a number of significant hydrocarbon discoveries. The principal hydrocarbon discoveries are in Paleocene–Eocene strata, although earlier strata are known, and their existence is therefore intimately linked to the opening and evolution of the North Atlantic from 54 Ma. The final rifting and separation of Greenland from Eurasia is commonly attributed to the arrival of a mantle plume which impacted beneath Greenland during early Tertiary time. Moreover, the ensuing plate separation is commonly described in terms of instantaneous unzipping of the North Atlantic, whereas in reality proto-plate boundaries were more diffuse during their inception and the linked rift system which we see today, including connections with the Arctic, was not established until Late Palaeogene–Early Neogene time. From a regional analysis of ocean basin development, including the stratigraphic record on the adjacent continental margins, the significance of the Greenland–Iceland–Faroe Ridge and the age and role of Iceland, we propose a dual rift model whereby North Atlantic break-up was only partial until the Oligo-Miocene, with true final break-up only being achieved when the Reykjanes and Kolbeinsey ridges became linked. As final break-up coincides with the appearance of Iceland, this model negates the need for a plume to develop the North Atlantic with rifting reliant on purely plate tectonic mechanisms, lithospheric thinning and variable decompressive upper mantle melt along the rifts.

It is a generally accepted model that North Atlantic seafloor spreading began at 54 Ma and rifted Greenland from Eurasia; this process continues to the present day (Pitman & Talwani 1972; Srivastava & Tapscott 1986; Fig. 1). Prior to this final rift, the North Atlantic margins had also been subjected to several earlier phases of extension between the Devonian collapse of the Caledonian Orogen and the early Tertiary break-up (Ziegler 1988; Doré *et al.* 1999; Roberts *et al.* 1999). These pre-Paleocene rifting phases exploited the collapsed Caledonian fold belt and consequently the conjugate North Atlantic margins show similar basinward stepping rift patterns (Doré *et al.* 1999). The final rifting and separation of Greenland from Eurasia at 54 Ma has also been attributed to the arrival and impact of an upwelling mantle plume which impinged on the base of the lithosphere under Greenland duing early Tertiary time (e.g. White 1989; White & McKenzie 1989; Smallwood *et al.* 1999; Smallwood & White 2002). Its modern-day much-reduced expression is the Iceland plume of many authors (e.g. Courtillot *et al.* 2003). A plume here is defined as a convective upwelling of lower mantle material due to thermal instability near the core–mantle boundary (Morgan 1971).

It is interpreted that the impact of the plume head with early Tertiary rifting led to large volumes of surface and subsurface magmatic activity, including the generation of seawards-dipping reflectors and the linear and abnormally thickened (up to *c.* 30 km) Greenland–Faroes Ridge (the GFR of Holbrook *et al.* 2001; the Greenland–Iceland–Faroe Ridge or GIFR of Gaina *et al.* 2009; Fig. 1). Calculated melt volume is estimated at $5 - 10 \times 10^6$ km^3 and led to the North Atlantic being referred to as the North Atlantic Igneous Province (NAIP) (White 1988; Saunders *et al.* 1997). Production of this melt was not uniform, with initially high production and seafloor spreading at 54 Ma declining to almost zero between 35 and 25 Ma before North Atlantic plate reorganization re-established a slow but steady spreading rate up to the present day (Fig. 2).

This instantaneous unzipping model for the North Atlantic has had a major impact on regional post-Cretaceous North Atlantic plate tectonic and palaeo-geographic reconstructions, basin modelling and, consequently, hydrocarbon exploration in offshore UK, Ireland and Norway (e.g. Larsen & Watt 1985; Holmes *et al.* 1999; Knott *et al.* 1999; Roberts *et al.* 1999; Carr & Scotchman 2003). The instantaneous unzipping of the North Atlantic is

From: CANNON, S. J. C. & ELLIS, D. (eds) 2014. *Hydrocarbon Exploration to Exploitation West of Shetlands.*
Geological Society, London, Special Publications, **397**, 11–31.
First published online February 20, 2014, http://dx.doi.org/10.1144/SP397.1
© The Geological Society of London 2014. Publishing disclaimer: www.geolsoc.org.uk/pub_ethics

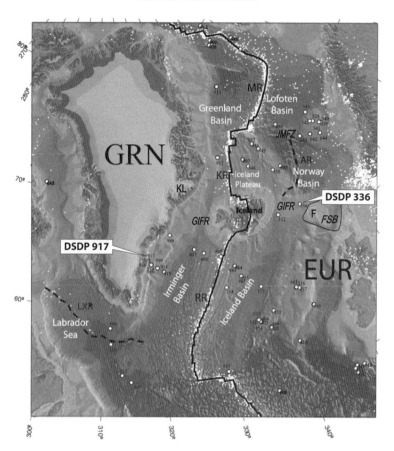

Fig. 1. Topography and bathymetry of the North Atlantic with active plate boundaries (black lines; extinct ridges are dashed black lines). From north to south: MR, Mohns Ridge; KR, Kolbeinsey Ridge; RR, Rekjanes Ridge; AR, extinct Aegir Ridge; LXR, Labrador Sea extinct ridge; GRN, Greenland; EUR, Eurasia; GIFR, Greenland–Iceland–Faroe Ridge; JMFZ, Jan Mayen Fracture Zone; F, Faroe Islands; FSB, Faroe–Shetland Basin; KL, Kangerlussuaq. Open circles indicate sites of Deep Sea Drilling Project (DSDP) or Ocean Drilling Program (ODP) drilling. Small white dots indicate location of recent seismicity (after Gaina *et al.* 2009).

also partially predicated on plate tectonic modelling packages which require 100% defined plate boundaries in order to operate (e.g. Skogseid *et al.* 2000; Gaina *et al.* 2009). In reality, the proto-plate boundaries were probably more diffuse during their inception and more analogous to the NE African Rift data coming out of Ethiopia (Beutal *et al.* 2010; Ferguson *et al.* 2010; Rychert *et al.* 2012).

As authors we use published data and interpretations to challenge the above model and suggest there is more evidence that the North Atlantic did not unzip in one event at 54 Ma or that it was driven by a plume process. The plume engine model has also recently been challenged by others (e.g. Foulger 2002; Foulger & Anderson 2005; Lundin & Doré 2005*a*, *b*; Gaina *et al.* 2009; Gernigon *et al.* 2009; Foulger 2010). On the basis of the evidence presented in this paper, we propose that the North Atlantic opening was the result of the development of opposing rifts, one developing SW from the North Atlantic (between Greenland and Norway) and the other NE from the central Atlantic (between Ireland and SE Greenland). These changed in importance and evolved through time with plate tectonic development of the North Atlantic from Paleocene time to the present.

Initial development and dating of the North Atlantic oceanic crust

Plate reconstructions and their development through time, such as those proposed for the North Atlantic, are constrained by dated sequences of magnetochrons resident in the spreading oceanic crust. It is generally accepted that North Atlantic oceanic

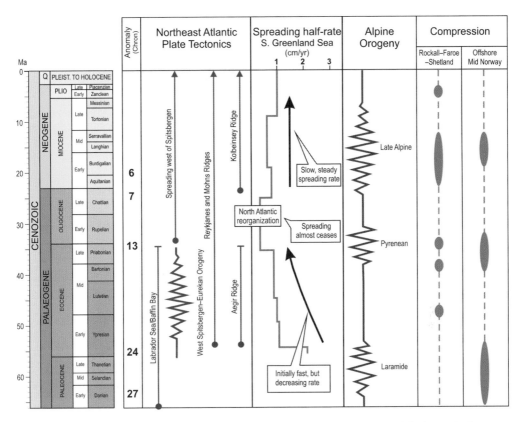

Fig. 2. Regional Palaeogene–Neogene tectonostratigraphic framework of the NW European Atlantic margin. Sources: Doré *et al.* (1999, 2008); Johnson *et al.* (2005); Ritchie *et al.* (2008); Statoil UK Ltd. Timescale from Gradstein *et al.* (2004).

crust first began forming during Chron 24 (54.5 Ma, Fig. 2). Using modern gravity and magnetic data with recent regional seismic and quantitative kinematic analysis, Gaina *et al.* (2009) published a map of North Atlantic magneto-chrons and their confidence in their location and recognition (Fig. 3). There are two key observations which can be made from this data. Firstly, the chrons defining the initial margins of the oceanic crust are not well constrained along the margins of the North Atlantic (Figs 3 & 4). For example, in some areas adjacent to the proposed continent–ocean boundary (COB) they exhibit a patchy magnetic pattern not typical of through-going magnetic seafloor striping (Figs 3 & 4). Similar anomalies have been recognized in the Labrador Sea and have been attributed to highly intruded or extended continental crust (Chalmers & Pulvertaft 2001). This implies that at break-up the crust was rifting, in a manner analogous to the separation of inter-digitating 'fingers', with areas of incipient oceanic crust adjacent to deforming and intruded continental crust. A uniform, instantaneous separation with general full production of oceanic crust along the entire length of the Atlantic rift therefore appears unrealistic. Similar observations have been made along the active Ethiopian and Afar rifts where magmatism is confined to elongate and independent magma chambers with associated active dyke intrusions above. These active magmatic centres are separated by laterally intervening areas of tectonic quiescence (Beutal *et al.* 2010; Pagli *et al.* 2012; Rychert *et al.* 2012).

The first definitive and continuous magnetochron implying extensive oceanic ridge production of basalts is Chron C21 (48 Ma), which is linked to the Aegir Ridge in the Norway Basin and to the Reykjanes Ridge in the Irminger–Iceland basins SE of Greenland (Figs 1 & 3). This chron does not link across the GIFR and leads to the second major observation: the GIFR and the linear interval between the east and west Jan Mayen fracture zones (EJMFZ and WJMFZ) both exhibit patchy magnetic patterns along their entire length. Additionally, they are defined by a thickened and linear

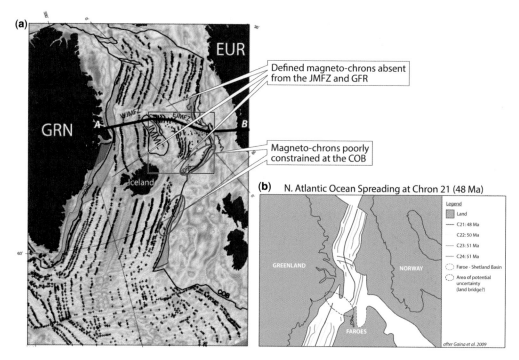

Fig. 3. (a) Magnetic data interpretations in the NW Atlantic. Background image is the free air gravity anomaly (Sandwell & Smith 1997; Forsberg & Kenyon 2004). Seaward-dipping reflectors (SDRS) are shaded transparent light grey. The black line indicates the continent–ocean boundary (COB). JMMC, Jan Mayen micro-continent; EJMFZ and WJMFZ, the east and west Jan Mayen fracture zones, respectively (after Gaina *et al.* 2009). (b) Simplified NW Atlantic magnetic interpretation (after Gaina *et al.* 2009).

crustal signature (Holbrook *et al.* 2001; Gernigon *et al.* 2009) (Fig. 3). Further interpretation of these is discussed in the following.

Evolution and dating of the oceanic ridges separating the North Atlantic province

Evolution of the North Atlantic from Late Paleocene time to the present day involved initial spreading in the Labrador Sea, along the Reykjanes Ridge, along the Aegir Ridge and then finally along the Kolbeinsey Ridge. The Oligo-Miocene linking of the Reykjanes and Kolbeinsey ridges resulted in the separation of the Jan Mayen micro-continent from east Greenland.

In conjunction with early rifting along the Reykjanes Ridge, the spreading in the Labrador Sea created a triple junction to the SE of Greenland which allowed simultaneous spreading along both rift arms (Gaina *et al.* 2002, 2009; Lundin & Doré 2005*a*) (Fig. 1). The Atlantic spreading arm utilized the collapsed Caledonian fold belt with its associated Mesozoic rift system (Lundin & Doré 2005*a*). This dual rifting involving the Labrador Sea and the Reykjanes rift continued between 54 and 33 Ma (Chron 24–Chron 13) when rifting then ceased in the Labrador Sea. Abandonment of the Labrador Sea and Baffin Bay spreading ridge resulted in a North Atlantic plate reorganization and the beginning of the next phase of North Atlantic opening (Røest & Srivastava 1989; Gaina *et al.* 2002; Lundin & Doré 2005*a*; Gaina *et al.* 2009). This coincides with Chron 13 (33 Ma) and represents a significant change in relative motion between Greenland and Eurasia from NW–SE to NE–SW (Gaina *et al.* 2009). Prior to Chron 13 (33 Ma), the NE propagation of the Reykjanes Ridge failed to penetrate any further north than the Kangerlussuaq area in east Greenland. Successive but unproductive attempts to propagate N–NE led to the gradual peeling away of the Jan Mayen micro-continent from east Greenland (Kuvaas & Kodaira 1997; Gaina *et al.* 2009; Fig. 5).

The eastern margin of the Jan Mayen microcontinent is defined by the first appearance of the oldest oceanic crust (magnetic anomaly 24b, 54 Ma) which erupted from the Aegir oceanic spreading ridge (Gaina *et al.* 2009). The western margin is similarly defined by the first appearance of true

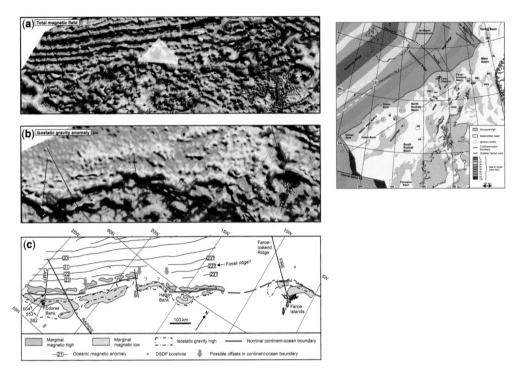

Fig. 4. Geophysical responses across the COB in the region between Edoras Bank and the Faroe–Iceland Ridge. (**a**) Total magnetic field (very poor data in the greyed area); (**b**) isostatically corrected Bouger gravity anomaly; and (**c**) selected magnetic and gravity features. Oceanic magnetic anomaly picks indicated by labels at their northeast ends. Black lines indicate the locations of the CAM77, RAPIDS, N18 and FIRE seismic profiles (Kimbell *et al.* 2005). The rectangular box on the location map locates areas a–c.

oceanic crust represented by magneto-chron 6c/7 (24 Ma) and erupted from the currently active Kolbeinsey oceanic spreading ridge. This indicates abandonment of the Aegir Ridge and the new propagating linkage of the Kolbeinsey and Reykjanes ridges in the Oligo-Miocene, the final separation of Jan Mayen from east Greenland and the destruction of any vestiges of the America–Eurasia land bridge (Gaina *et al.* 2009; Fig. 5). Onshore in east Greenland this final break-up between east Greenland and Jan Mayen is evidenced by alkaline dykes and large syenite intrusives in the Scoresby Sund-Jameson Land and Hold with Hope-Clavering Ø areas (Price *et al.* 1997). Abandonment of the Aegir Ridge in favour of the Kolbeinsey Ridge may have been via ridge 'jump' to the new ridge (Talwani & Eldholm 1977) or by via gradual abandonment of the Aegir Ridge (e.g. Nunns 1983; Muller *et al.* 2001).

The linkage between the final stages of continental rifting and oceanization is uncertain; however, the work of Van Wijk *et al.* (2001) offers a useful insight. They used a dynamic 2D finite element model for the upper and lower crust and the mantle to a depth of 120 km to study melt generation in a rifting environment. Initial unstretched thickened continental crust of 38 km was extended until lithospheric break-up resulted in a zone *c.* 150 km wide and where $\beta \geq 5$. This occurred 15 Myr after stretching was initiated. Outside of this zone transition zones up to 200 km wide (where $\beta = 1.2$–5) were developed, including an outer crustal area where thinning factors were less than 1.2 and extension had minimal effect. The stretching factors mirrored those proposed by Reemst & Cloetingh (2000) and Skogseid *et al.* (2000) for transects described along the North Atlantic volcanic margin. Results showed that decompressional partial melting took place across a 175 km wide zone and at a calculated depth of melt production between 20 and 50 km. The melt initiation began 5 Myr before break-up, with maximum melt just before break-up. Melt volumes fell in the outer edges of the transition zones where lower extension rates existed. They concluded the calculated melt volumes were in agreement with melt volumes 'per unit length of margin' observed along current-day volcanic margins. This did not require the prerequisite of a mantle plume in the North Atlantic, and lithospheric rifting alone could

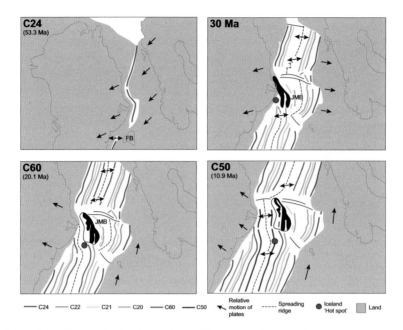

Fig. 5. Evolution of the NW Atlantic plate boundaries and kinematic evolution of the Jan Mayen micro-continent illustrated by a series of tectonic reconstructions in an absolute reference frame (after Gaina *et al.* 2009). JMB, Jan Mayen Block; FB, Faroes Block. Note Jan Mayen peels away from SE Greenland between 20 and 30 Ma, therefore allowing the potential for a long-lived land bridge between America and Eurasia via the Greenland–Faroe Ridge.

produce the enhanced melt volumes observed along the volcanic margins.

Additional evidence from the Afar proto-plate boundary also suggests it involves shallow decompressive melting of the upper mantle with vertical and lateral dyke intrusion within the rift (Ferguson *et al.* 2010; Rychert *et al.* 2012). If continental rifting involves segmented decompressive melting along the rift, the magmatic composition and volume, the eruption site timing and location and rift margin uplift will vary spatially with underlying composition of the upper mantle (see Meyer *et al.* 2007 for documentation of the geochemistry and spatial distribution of varying melts across the entire NAIP, both from enriched and depleted upper mantle sources and with or without crustal contamination).

The Greenland–Iceland–Faroe Ridge (GIFR)

The North Atlantic oceanic province has two anomalously thickened crustal elements: (1) the NW–SE Greenland–Iceland–Faroe Ridge (up to 30+ km of basaltic crust); and (2) the similarly orientated and thickened Vøring Spur located within the Jan Mayen fracture zone (thickened basaltic crust up to ≥15 km) (Fig. 1). Their linearity and non-radial orientation away from any proposed plume centre poses problems for a plume origin.

The GIFR has been drilled by DSDP site 336 which was located on its northern flank (Fig. 1). This site penetrated Middle Eocene basalts dated at 43–40 Ma (K–Ar date) at 515 m below the seabed (Talwani *et al.* 1976). The basalt grades into and is overlain by 8 m of volcanic rubble (conglomerate), which in turn is overlain by 13 m of thick red claystone. The latter is interpreted as a lateritic soil formed *in situ* by sub-aerial weathering of the basalt basement. The palaeosol is overlain by 295 m of Middle Eocene–Upper Oligocene marine mudstones, the bulk of which is probably of Middle–Late Eocene age (Stoker & Varming 2011). Micro-palaeontological evidence indicates that submergence of the GIFR at site 336 from a sub-aerial setting to a marine bathyl environment (shelf to upper slope, occurred during Late Eocene time (Talwani *et al.* 1976; Berggren & Schnitker 1983). The crest of the GIFR, sited about 400 m higher than the sea bed at site 336 (which itself is 463 m above the palaeosol), however suggests that the GIFR remained as either a continuous ridge or a string of closely spaced islands for some considerable time, possibly into Oligo-Miocene time (Talwani *et al.* 1976; Stoker & Varming 2011). Confirmation of this comes from evidence of the continued movement of flora and fauna from

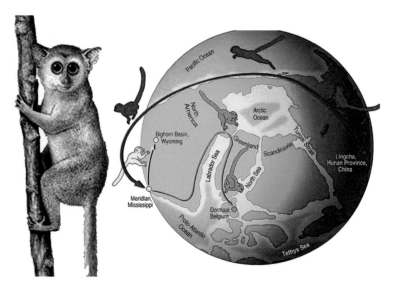

Fig. 6. The Eocene primate *Teilhardina* with fossil locations and their relative stratigraphic age superimposed on an Eocene palaeogeographic map of the globe. A North Atlantic land bridge allowed the primate to migrate from North America to Eurasia during the Eocene (Beard 2008).

America to Eurasia during the Eocene–Miocene via the 'Thulian land bridge'. Beard (2008) records that the minute monkey-like *Teilhardina magnoliana* had migrated from China via the Bering land bridge to the coastal plain of Mississippi by 55 Ma and further migrated to Europe by 47 Ma (as indicated by its discovery in Eocene deposits at Dormaal, Belgium; Fig. 6). Xiang *et al.* (2005) also noted that cornelian cherries (Cornus, cf. dogwoods) were present in North America during Paleocene time, but had later spread to Europe and Africa by Miocene time. While studying Miocene Icelandic oak pollen (Quercus), Denk *et al.* (2010) indicated linkage to the North American white and red oaks and their arrival in Iceland via a land bridge.

A model for the development of the GIFR and JMFZ

So what is the nature and derivation of the GIFR and when did it finally cease to be a land bridge between America and Eurasia? From Figure 3 it can be seen that the GIFR is linear and symmetric about Iceland, which led Morgan (1971) and others (e.g. White 1989; Smallwood *et al.* 1999; Skogseid *et al.* 2000) to suggest it represented the Paleocene–Holocene plume track of the Iceland 'hotspot'. However, Lundin & Doré (2002, 2005*b*) and Foulger (2010) both note that the GIFR is not time-transgressive in one direction as the Iceland 'hotspot' has never been positioned below the Iceland–Faroes side of the GIFR (otherwise the plume head would be located under NW Scotland today). Lundin & Doré (2005*b*) state that lithospheric drift over a fixed mantle plume for the GIFR is untenable, or would require – in the extreme – an early plume capture at the plate boundary and subsequent plume drift to have exactly matched the lithospheric drift. This would allow the 'hotspot' to remain constantly centred on the spreading ridge. Some authors (e.g. Bott 1983; White & McKenzie 1989; Smallwood *et al.* 1999) suggest that the generation of oceanic crust >7 km thick requires anomalously high asthenospheric temperatures (a mantle plume); others advocate decompressive lower-temperature melt of a fertile upper mantle (Foulger & Anderson 2005; Gernigon *et al.* 2009).

The radial plume model also fails to explain the extreme linearity of the GIFR; a clue to its formation might be explained with reference to the recent study of the Vøring Spur, within the Jan Mayen fracture zone, by Gernigon *et al.* 2009 (Fig. 7). Combining Bouguer anomaly analysis with depth to MOHO estimation, they note that the Vøring Spur between the east and west Jan Mayen fracture zones is characterized by a Bouguer 'low' in contrast to adjacent oceanic domains and coincides with thickened (>15 km) oceanic crust. Gernigon *et al.* (2009) propose that the thickened oceanic crust formed within the JMFZ during synrift extension across the NW–SE fractures, which led in turn to lithospheric thinning and subsequent decompressive melting of the mantle during the

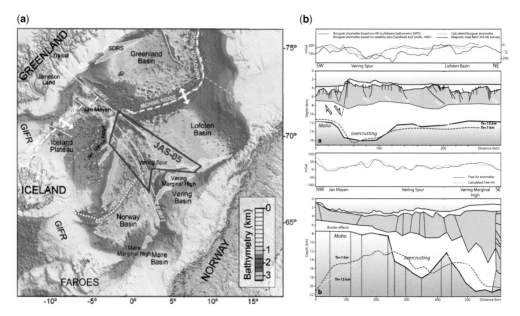

Fig. 7. (**a**) Bathymetric map and main physiographic features of the Norwegian–Greenland Sea. EJMFZ, CJMFZ and WJMFZ, the east, central and west Jan Mayen fracture zones, respectively; SDRS, seawards-dipping reflectors; GIFR, Greenland–Iceland–Faroe Ridge. JAS-05 is the gravity-magnetic survey area designation. (**b**) Gravity forward modelling and crustal model across the Vøring Spur, including gravity and density modelling. The oceanic root, observed beneath the Vøring Spur, is interpreted as a synrift oceanic and mafic feature ('overcrusting') formed during Middle–Late Eocene time (Gernigon *et al.* 2009).

Middle–Late Eocene ('overcrusting'). Similar NW–SE fracture zones have been documented along the entire North Atlantic margin (Rumph *et al.* 1993), and specifically within the Faroe Islands (Ellis *et al.* 2009) and in the Kangerlussuaq area of SE Greenland (Larsen & Whitham 2005; Guarnieri 2011). Plate reconstruction of the North Atlantic prior to 55 Ma places the Faroe Islands some 50–100 km SE of Kangerlussauq during Late Paleocene time (Larsen *et al.* 1999; Skogseid *et al.* 2000; Larsen & Whitham 2005). This allows direct alignment of the NW–SE fracture zones described in the Faroes and Kangerlussuaq area; we therefore propose that the GIFR was formed by a similar process to that described by Gernigon *et al.* (2009) for the Vøring Spur (Fig. 8). We suggest transtensional movement across the Faroes–Kangerlussauq lineaments during Late Paleocene–Early Eocene time caused a zone of linear rifting, lithospheric thinning and decompressive melting of the upper mantle with increased magmatic activity to create the crustally thickened GIFR. Large Late Paleocene–Early Eocene intrusives have also been documented in close association with the Faroes and Kangerlussauq lineaments (Ellis *et al.* 2009; Guarnieri 2011). Walker *et al.* (2011) noted the first phase of faulting in the Faroes involved dip-slip movements along NW–SE- and North–South-orientated faults.

The orientation of these NW–SE lineaments probably represents inheritance from an earlier failed transient North Atlantic rift along a NW–SE axis from Baffin Island to the British Isles (Lundin & Doré 2005*a*). This coincided with Early Paleocene igneous activity (62–58 Ma) in the British NAIP, as represented by igneous centres such as Skye, Rhum and Mull and the NW–SE-orientated dyke swarms which extend across the UK from the Hebrides into the central North Sea and Lundy in the Bristol Channel (e.g. Brown *et al.* 2009; Hansen *et al.* 2009). However, the early Tertiary NAIP igneous activity has also been recently linked with the northern influence of the Large Low Shear Velocity Province (previously called the African Plume) at the core-mantle boundary and located beneath central Africa (Torsvik *et al.* 2006; Ganerød *et al.* 2010).

Eocene development of the Atlantic margin of Britain and Ireland

Recent re-evaluation of the Eocene sequences in the central North Atlantic, involving seismic, wells,

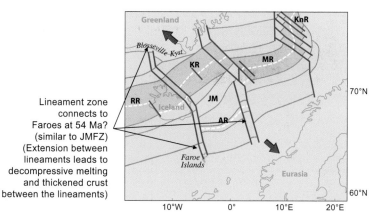

Fig. 8. Interpreted fracture lineaments connecting SE Greenland (Kangerlussuaq) and the Faroe–Shetland Basin (after Lundin & Dóre 2002; see text for additional references). The origin of the Greenland–Iceland–Faroe Ridge is interpreted to be analogous to that proposed for the thickened crust observed between the Jan Mayen Fracture zones by Gernigon et al. 2009 (see Fig. 7).

shallow boreholes, core data, biostratigraphy and onshore geology, has concluded that during Eocene time the northern North Atlantic rift was not connected to the SW North Atlantic rift (Stoker & Varming 2011; Stoker et al. 2013a, b) (Fig. 9). Eocene prograding deltaic (and fan-delta) deposits can be mapped offloading from the Munkagrunnur and Wyville Thompson ridges and the West Shetland margin into the Faroe–Shetland Basin (Robinson 2004; Ólavsdóttir et al. 2010, 2013; Stoker & Varming 2011; Stoker et al. 2013b) (Fig. 9). To the SW, stratigraphically equivalent rocks were deposited on the flanks of the Rockall and Hatton basins (Stoker et al. 2012) and Porcupine Basin (Moore & Shannon 1992; Fig. 9). These deposits are contemporary with similar deposits which accumulated in the North Sea (Jones & Milton 1994; Mudge & Bujak 1994) and onshore east Greenland (Larsen & Whitham 2005; Larsen et al. 2005).

On the eastern flank of the Rockall High, the inter-digitation of sub-aerial volcanic lavas and fan-delta/shallow-marine deposits in BGS borehole 94/3 can be tied to regionally synchronous unconformities across the greater Rockall–Hatton area, as well as in the Faroe–Shetland region and east Greenland (Stoker et al. 2012). Seismic mapping of these units in the Rockall–Hatton area indicates an archipelago of Eocene islands at or near sea level, frequently inundated but capable of supplying sediment into the local area (Stoker et al. 2012, fig. 10). The intra-Eocene unconformities imply a fluctuating response to relatively small sea-level rise and falls associated with the evolving North Atlantic plate tectonic regime (Shannon et al. 1993; Stoker & Varming 2011). This includes the post-depositional (latest Eocene/ Oligocene) tilting of a number of the Eocene prograding units located on the flanks of major compressional folds, such as the Hatton High and the Wyville Thomson Ridge (Stoker et al. 2012).

Paleocene–Eocene sequences offshore and onshore SE and east Greenland

On the opposite side of the rift and offshore SE Greenland, DSDP borehole 917 drilled 775 m of crustally contaminated basalt with sub-aerial weathered horizons (Vallier et al. 1998; Fitton et al. 2000) (Fig. 1). In the borehole, the basalts are underlain by 10 cm of non-metamorphosed quartzitic sandstone (interpreted as being of fluvial origin) and of presumed Paleocene age, which in turn is underlain by 15.9 m of fine-grained metamorphosed sediments (Greenschist grade) including volcaniclastics. These are of presumed Late Cretaceous age and exhibit weak sedimentary structures suggestive of deposition from turbidity currents.

Onshore in SE Greenland, Upper Cretaceous and Early Paleocene shallow-marine sediments in the Kangerlussauq area are overlain by Late Paleocene shallow-marine and fluvial deposits (Fig. 1). These in turn pass upwards and inter-digitate with, and are ultimately inundated by, Upper Paleocene–Lower Eocene flood basalts (Larsen & Whitham 2005). The depositional setting was strongly controlled by large syndepositional NW–SE oblique-slip normal faults trending along the Nansen fjord and the Christian IV glacial valley, as well as the nearby Kangerlussauq fjord (Larsen & Whitham 2005; Guarnieri 2011). These faults also controlled the injection sites and the later deformation of

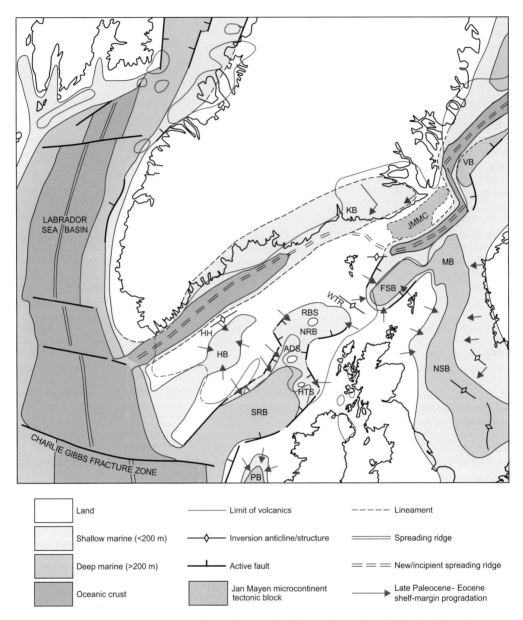

Fig. 9. Generalized palinspastic map for the Late Paleocene–Early Eocene interval (modified after Stoker & Varming 2011). ADS, Anton Dohrn Seamount; KB, Kangerlussuaq Basin; MB, Møre Basin; NRB, North Rockall Basin; NSB, North Sea Basin; PB, Porcupine Basin; RBS, Rosemary Bank Seamount; SRB, South Rockall Basin; WTR, Wyville Thompson Ridge; VB, Vøring Basin.

Early Eocene intrusives (Guarnieri 2011). The subsidiary NE–SW Sortekap fault inland from the coast further constrained the Paleocene–Early Eocene depositional system in the area (Larsen & Whitham 2005). Palaeocurrent evidence from the shallow-marine and fluvial sequences both indicate initial flow to the SE, but then critically to the SW in the centre of the basin (Larsen & Whitham 2005). The SE margin of the basin is not preserved.

Overlying the Lower Paleocene sediments in Kangerlussuaq are a thick succession of Upper Paleocene–Lower Eocene sub-aerial basalts. These were emplaced from east to west and buried a significant Late Paleocene landscape, including deeply incised palaeo-valleys (Larsen et al. 1989; Pedersen et al. 1997). The basalts range up to 6 km thick on the coast, and thin to 2–3 km inland and regionally northwards to c. 1 km in the Trail Ø area (Price et al. 1997). The east Greenland sequence contains age-equivalent basalts of the Beinisfjord, Malinstindur and Enni formations which have been described from the Faroe Islands, and which are now located between Iceland and NW Scotland (Passey & Jolley 2009) (Fig. 1). Additionally, the Kangerlussuaq area has younger Eocene lava formations – the Upper Geikie, Rømer, Skrænterne and Igtertiva formations – which are not present in the Faroe Islands (Pedersen et al. 1997; Larsen et al. 1989, 1999) (Fig. 10). Recent dating of the youngest Kangerlussuaq volcanic sequence, the Igtertiva Formation in the Kap Dalton area, suggests eruption into the late Early Eocene between 49 and 47.9 Ma (Larsen et al. 2005). The significance of this is that offshore of the Faroe Islands the equivalent rocks are represented by NE-seawards-dipping reflectors and oceanic crust, whereas in SE Greenland they comprise east–west prograding sub-aerial basalts. This apparent contradiction is addressed in the following.

In SE Greenland, Middle Eocene fluvial and shallow-marine sediments overlie the Igtertiva basalts in the Kap Dalton area and demonstrate a NE–SW progradational sequence, as identified by palaeocurrent data (Larsen et al. 2005). This is the opposite direction to the Early Eocene landscape developed in the Faroe–Shetland Basin, which was from south to north (Larsen et al. 2005; Stoker & Varming 2011).

The dual rift model

On the basis of the above data, we propose that during Late Paleocene–Early Eocene time two active rift systems were trying to split America and Greenland from Eurasia until 33 Ma (Chron 13) (Fig. 11). The first was developing SW–NE along the SE margin of Greenland and along the

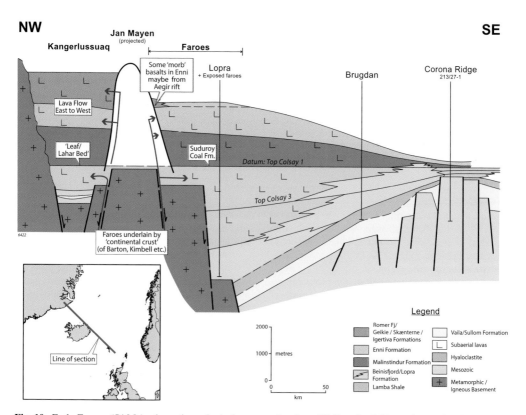

Fig. 10. Early Eocene (54 Ma) schematic geological cross-section from SE Greenland (Kangerlussuaq) and across to the Faroe Islands and the Faroe–Shetland Basin (FSB). Section is balanced on the regionally correlated Colsay 1 Sandstone Formation, based on FSB well penetrations and outcrop studies in the Faroes and SE Greenland (see text for references).

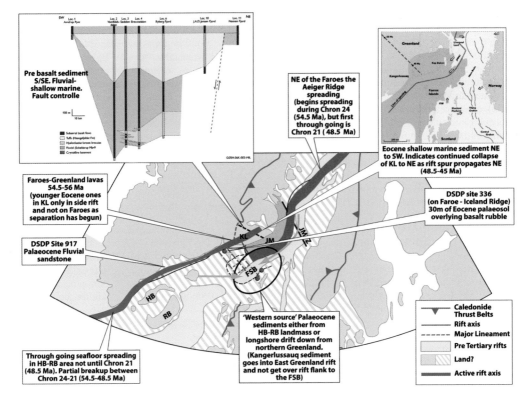

Fig. 11. Summary diagram citing evidence for two NW Atlantic active rift systems attempting to split America and Greenland from Eurasia until 33 Ma (Chron 13). One developed SW–NE along the SE margin of Greenland and formed the line of the proto-Reykjanes Ridge. It penetrated into the Kangerlussauq area where it was unable to penetrate further. A second NE–SW-propagating rift, represented by the Aegir spreading ridge, developed from the Norwegian–Greenland Sea and extended into the area north of the Faroe Islands. The two rifts were independent of each other and did not conjoin, leaving Jan Mayen firmly attached to east Greenland and the Faroes as the centre of a land bridge between east Greenland and Eurasia. The land bridge lasted until at least 33 Ma (and possibly until 25 Ma). KL, Kangerlussuaq; JM, Jan Mayen; JMFZ, Jan Mayen fracture zone; FSB, Faroe–Shetland Basin; HB, Hatton Bank; RB, Rockall Bank.

line of the proto-Reykjanes Ridge. It penetrated into the area of Kangerlussauq and possibly into the Trail Ø area where it was unable to penetrate further northwards (Figs 9 & 11). A second NE–SW propagating rift, represented by the Aegir spreading ridge, developed from the Norwegian–Greenland Sea and extended into the area north of the Faroe Islands. The two rifts were independent of each other and did not conjoin, leaving Jan Mayen firmly attached to east Greenland in the Kangerlussauq and Hold-with-Hope area, with the Faroe region as the centre of a land bridge between East Greenland and Europe. The land bridge lasted until at least 33 Ma (and possibly until 25 Ma). The opposed ridge system has been previously cited by others, although the timing and the ridges involved has varied (see Dewey & Windley 1988; Lundin et al. 2002; Lundin & Doré 2005a; Gaina et al. 2009). The dual ridge system can explain the difference in timing and nature of the flood basalt sequences in the Kangerlussauq rift, which were dominantly sub-aerial and emplaced east–west, against the synchronous development of seawards-dipping reflectors and early oceanic crust from the propagating Aegir rift north of the Faroe Islands (and east of Jan Mayen). The lateral by-passing of the two rift tips would induce severe fracturing in the area of the palaeo-Faroe Island location and anticlockwise rotation of the regional stress field, allowing igneous material to erupt from local volcanic centres (particularly along NW–SE- and NE–SW-orientated fractures). The anticlockwise rotation of the regional stress field has been observed from Paleocene–Early Eocene onshore fault analysis in the Faroe Islands (Walker et al. 2011) and later during the Oligo-Miocene separation of Jan Mayen from Greenland (Gaina et al. 2009) (Fig. 5).

Post-Paleocene to Oligo-Miocene inversion events in the North Atlantic province

The anticlockwise rotation of the regional stress field from 54 to 25 Ma could also partially explain the episodic and gradual inversion of anticlinal and domal structures in the North Atlantic throughout Eocene time, and particularly during the Oligo-Miocene (Johnson et al. 2005; Doré et al. 2008; Ritchie et al. 2008). In the Faroe–Shetland region, the disposition of the Eocene succession is folded about the axes of uplift which form the bathymetric highs of the Fugloy, Munkagrunnur and Wyville Thomson ridges. This implies that a major phase of ocean margin structuring took place during Late Palaeogene–Early Neogene time (Johnson et al. 2005; Stoker et al. 2005b; Ritchie et al. 2008; Ólavsdóttir et al. 2010). The Hatton High (of the Rockall Plateau) was also folded in this interval, as witnessed by the tilting of the Eocene deltaic wedges on this feature (Stoker et al. 2012).

A major consequence of Oligo–Miocene compressional deformation in the Faroe–Rockall region was the formation of the Faroe Conduit (combining the present-day Faroe–Shetland and Faroe Bank channels), which today forms the main passageway for deepwater exchange across the Greenland–Scotland Ridge (Stoker et al. 2005a, b). This Early Neogene instigation of the gateway is consistent with ^{13}C, ^{18}O, taxonomic and sedimentological data from DSDP and ODP sites in the North Atlantic region, which reveal a Miocene instigation for the overflow of North Atlantic deep water into the previously isolated Tethyan (warm water) Reykjanes rift basins to the south (cf. Stoker et al. 2005a, b).

Hot fluid flow evidence coincident with the two major tectonic phases of evolution of the North Atlantic (Eocene and Oligo-Miocene plate rifting and reorganization)

Microthermometry and apatite fission track analysis (AFTA) from the Faroe–Shetland Basin and East Greenland indicates two periods of hot fluid flow through post-Paleocene units until the present: a Late Paleocene–Early Eocene event and another around 20–25 Ma (Parnell et al. 1999; Scotchman et al. 2006; Parnell & Middleton 2009). The first is coincident with the early North Atlantic rifting phase and the second with the major Oligo-Miocene plate reorganization of the North Atlantic, which led to the separation of Jan Mayen from Greenland and North Atlantic regional uplift. Both phases are associated with the highest ocean spreading rates and therefore elevated heat flows (Fig. 2). An omni-present plume from the Paleocene to the present cannot rationally explain these two pulsed hot fluid flows in itself, or the fact that despite these fluid flushes the regional vitrinite reflectance and AFTA data suggest low thermal maturity for most of the Mesozoic–Cenozoic sequences in the Faroe–Shetland Basin and onshore east Greenland (except when in close proximity to intrusive bodies) (Surlyk et al. 1986; Stemmerik et al. 1992; Carr 1999; Scotchman et al. 2006). The AFTA and vitrinite reflectance data could therefore suggest a lower regional thermal gradient more indicative of shallow melts restricted to plate boundaries and not the higher gradient expected from a large-radius mantle-derived plume impacting across the base of the lithosphere.

The dual rift model and Paleocene–Early Eocene sediment provenance studies

The dual rift model may also explain the dichotomy expressed in published North Atlantic Paleocene reservoir provenance studies (e.g. Jolley & Morton 2007; Morton et al. 2012a, b). These studies involved heavy mineral and palynological analysis to identify several Scottish source areas for Paleocene sandstones in the Faroe–Shetland Basin; they also identified a 'westerly' sourced sandstone component (the FSP3 zircon signatures of Jolley & Morton 2007 and Morton et al. 2012a, b).

The palynological data suggest a Greenland affinity but the heavy mineral rutile/zircon data, though distinguishing a non-Scottish source, cannot distinctly type the 'western' source area. In the single rift model the zircons and other heavy minerals from SE Greenland should be observed in the Faroe–Shetland Basin (which was directly SE of Kangerlussuaq) in palaeo-reconstructions (e.g. Larsen et al. 1999; Skogseid et al. 2000; Larsen & Whitham 2005). In particular, the zircon age data profile of westerly FSP3 sandstones does not match surface sampling of Kangerlussuaq basement terrains (Morton et al. 2012b). The Kangerlussuaq low-RuZi sandstones have two main age groups at c. 2700–2750 Ma and c. 2950–3100 Ma, with a subsidiary peak at c. 3200 Ma. Over 50% of the Archaean grains are older than 2950 Ma (Whitham et al. 2004). The c. 2950–3100 Ma and c. 3200 Ma age peaks are either absent or very poorly developed in the FSP3 sandstones, and zircons older than 2950 Ma form only 5–11% of the Archaean population. Derivation of the FSP3 sandstones from the Kangerlussuaq area of east Greenland is therefore dismissed (Morton et al. 2012a, b).

With the currently proposed synchronous dual rift model, any westerly sourced sediments from Kangerlussauq would have entered the northerly progressing rift off SE Greenland and been unavailable to the Faroe–Shetland Basin (Figs 9 & 11). The alternative locations for westerly derived FSP3 sandstones may have been derived from the now-rifted Jan Mayen micro-continent or other areas in NE–central Greenland (which are presently below the icecap). Westerly derived America–Greenland pollen spores can, in contrast, be common to both rifts due to their less-dense silt size and their potential for greater aqueous and regional wind-blown distribution.

The age and role of Iceland

With the progressive dual rift model we now consider the age and role of Iceland. Plume protagonists propose that Iceland is the present signature of a Northern Hemisphere deep mantle plume which impacted the area of NW Greenland during Late Cretaceous–Early Paleocene time, and then migrated via Kangerlussuaq with time and Greenland plate movement to its current Iceland position (e.g. White 1989; White & McKenzie 1989; Smallwood et al. 1999; Smallwood & White 2002). Other authors disagree and propose non-plume origins for Iceland (e.g. Foulger 2002; Foulger & Anderson 2005; Lundin & Doré 2005a, b; Gaina et al. 2009; Gernigon et al. 2009; Foulger 2010).

In the context of the dual propagating rift model and the potential final separation of America–Greenland from Eurasia during Oligo-Miocene time, we note the following pertinent facts regarding Iceland.

- The oldest dated outcrops are 17 Ma (Miocene) in NW Iceland and 13 Ma in east Iceland (Moorbath et al. 1968; Ross & Mussett 1976; Hardarson et al. 1997).
- The time-transgressive V-shaped ridges, extending up to 1000 km along the Reykjanes and Kolbeinsey ridges, are limited to Oligocene–Holocene oceanic crust (Jones et al. 2002). Hey et al. (2010) and Benediktsdóttir et al. (2012) indicate that the V-shaped ridges are not symmetrical about the Reykjanes Ridge axis and this implies formation from either a 'pulsating plume' (the conventional view), or – in their preferred view – the requirement of a simple rift propagation away from central Iceland. They also note the V-shaped ridges have different geographic extent and patterns north and south of Iceland and that this asymmetry is best explained by rift propagation, and not by pulses in a symmetrically radial plume.
- Iceland is dominated by tholeiite basalts (typical Mid-Ocean Ridge Basalts) and not picrites or komatiites which are more distinctive of hotter core–mantle origins (Foulger et al. 2003; Presnall 2003; Søager & Holm 2011). Lower mantle picrites are predicted to have higher MgO concentrations of c. 15–30% (Gill et al. 1992; Lundin & Doré 2005a). Iceland basalts have $\pm 10.5\%$ Mg and are highly depleted with respect to major and trace element composition; they could therefore originate from lower temperatures.
- A larger-than-expected proportion (c. 10%) of compositionally acidic (rhyolite) and intermediate (andesitic) rocks occur in Iceland (Foulger & Anderson 2005; Foulger 2010).
- Mesozoic-age zircons (representing continental crust derivation) have been described from Mount Hvitserker in NE Iceland (Paquette et al. 2006, 2007) and from more widespread Icelandic locations (Bindeman et al. 2012).
- Records of elevated $^{87}Sr/^{86}Sr$ and Pb ratios in rhyolites and basalts in SE Iceland are indicative of an enriched upper mantle source (Prestvik et al. 2001; Søager & Holm 2011). Storey et al. (2004) also describe similar enriched Middle Miocene lavas in east Greenland.
- Tomographic studies show that the low-velocity anomaly below Iceland extends only into the upper mantle (at ± 400 km depth) and not to the core–mantle boundary. It is a vertical and cylindrical anomaly of 200–250 km diameter at depths down to 200 km, but an elongate dyke-like form at greater depths and parallel to the mid-Atlantic Ridge trend (Ritsema et al. 1999; Foulger et al. 2001; Montelli et al. 2003; Hung et al. 2004). The latter suggests a closer relationship to the greater regional morphology, shallow melts and tectonics of the mid-Atlantic rift system rather than a deep mantle plume.

We note the coincidence that the oldest rocks in Iceland post-date the break-away of Jan Mayen in the Oligo-Miocene (33–24 Ma) (Fig. 5) and the dating of the V-shaped ridges beginning in the Oligocene. Similarly, at the point of final break-up, Iceland would have been positioned where the new rift intersected the zone of the Caledonian suture between Scotland and east Greenland, a point noted by Foulger & Anderson (2005). The link to the underlying Caledonian orogen is further strengthened by the fact that the volcanism in Iceland is apparently from a shallow (decompressive) melt and an enriched (eclogitic) upper mantle, and not a deep-rooted plume. We therefore propose that Iceland was initiated during Oligo-Miocene time at the point where the juncture of the Reykjanes and Kolbeinsey ridges crossed the Scotland–Greenland Caledonian orogenic trend. This implies that an earlier proto-Iceland hotspot is not required to explain earlier Paleocene–Eocene rifting and volcanism, and regional plate tectonics can provide a simpler explanation (e.g. van Wijk et al. 2001; Foulger & Anderson 2005; Lundin & Doré 2005a; Gaina et al. 2009). The evidence for an enriched upper mantle source for some Iceland basalts and rhyolites and the Middle Miocene basalts of east Greenland (Storey et al. 2004) can also be attributed to melting in the upper mantle of buoyant subducted Iapetus oceanic crust (Foulger & Anderson 2005; Bindeman et al. 2012), from a residual Caledonian orogenic root (Bott et al. 1974; Ryan & Dewey 1997) or from a metasomatized upper mantle (Storey et al. 2004).

The North Atlantic geoid

Evidence for the plume is often invoked from the observation of the positive geoid (c. 60 m) covering a large sector of the North Atlantic and NW Europe and the resultant low mean Atlantic water depths (Marquart 1991; Köhler 2004; Lundin & Doré 2005b) (Fig. 12). It is interesting to note that the maximum geoid anomaly is non-radial but quasi-linear at its maximum between Iceland and the central Atlantic between Iberia and Newfoundland. This encompasses the modern central Atlantic spreading ridges and the area of the defunct Early Paleocene triple junction SE of Greenland, the time when both the Labrador Sea and proto-Reykjanes spreading ridges were active.

Lundin & Doré (2005a, b) noted that the geoid anomaly has a 3000–4000 km diameter while mantle tele-seismic investigations around Iceland suggest an upper mantle thermal anomaly of only 1000 km radius. Further, this is not centred above the 200–250 km wide low-velocity zone underneath Iceland. If the geoid does represent the thermal effect of a plume from Middle Paleocene time to the present day, then it is also significantly larger than the extent of the North Atlantic igneous province and cannot rationally explain the non-radial and highly disparate distribution of Paleocene–Early Eocene magmatism in space and time (e.g. the west Greenland and Baffin Bay picrites v. the BTIP more normal basalts erupted around 58–62 Ma).

Lundin & Doré (2005a, b) have demonstrated that there is no direct association between geoids and hotspots around the Earth, and suggested that an alternative explanation should be investigated. We speculate that the relevant timings and distribution of the rifting zones in the North Atlantic with their associated melts could have formed the large area of thickened North Atlantic basaltic crust. In particular, the greatest elevation of the geoid coincides with the area of the palaeo-triple junction and the spreading of the central Atlantic ridges into the Labrador Sea and the proto-Reykjanes Ridge (Figs 1 & 12). Synchronous lateral heat flow from both the Labrador Sea and the proto-Reykjanes Ridge would have lifted Greenland at the expense of other areas and generated the 'fossil' centre of the of the geoid (Fig. 12). Transient and synchronous uplift followed by subsidence in west and east Greenland has been

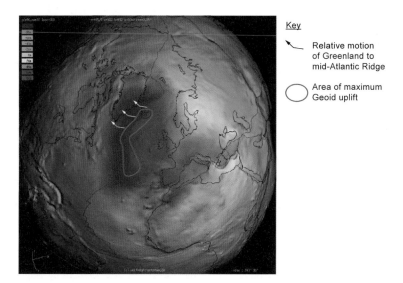

Fig. 12. The North Atlantic geoid anomaly (from Köhler 2004). The geoid anomaly coincides approximately with the extent of the North Atlantic topographic-bathymetric anomaly (Sandwell & Smith 1997); it is not centred on Iceland and is more widespread than the upper mantle low-velocity anomaly (e.g. Ritsema et al. 1999; Foulger et al. 2001).

documented by Dam et al. 1998, who ascribed it to the impact and then the lateral spreading of a plume head; however, it could equally be ascribed to synchronous rift flank uplift in the Labrador Sea and along the Reykjanes Ridge, followed by subsidence as full rifting occurred and magma was depleted from the upper mantle and erupted onto the rift margins. The location of Greenland away from the centre of the plate-tectonic-induced geoid has since been changed by plate migration of Greenland from 62 Ma to the present day. Additionally, the geoid has probably been reinforced by crustal thickening under Europe and north Africa by the Pyrenean and Alpine orogenies between 62 and 14 Ma (Figs 2 & 12).

In our opinion, it was not until the Reykjanes and Kolbeinsey ridges crossed the Scotland–Greenland Caledonian orogenic track that Iceland and its associated smaller upper mantle anomaly came into being, thereby creating the more local geoid around it. The timing and the different interactions of the varying rift zones could also explain the disparate timing of post-Paleocene–Holocene uplift and subsidence of basins and continental margins around the North Atlantic, something not possible with an omni-present and radial plume (Lundin & Doré 2005b).

Conclusions

- On the basis of regional geological and other data, we suggest that the opening of the North Atlantic between Eurasia and North America–Greenland was only partial until the Oligo-Miocene (33–25 Ma). The true final break-up occurred when the Reykjanes and Kolbeinsey ridges conjoined in the area of SE Greenland and offshore Kangerlussuaq.
- Initial attempts at North Atlantic rifting involved two opposing and almost by-passing rifts. These were the proto-Reykjanes Ridge (rift) system propagating NE along the SE Greenland margin and the SW-propagating Aegir Ridge between Norway and offshore east Greenland. Although rifting along the Atlantic margin of Ireland, Britain and the Faroe Islands may have been initiated at 54 Ma, true development of continuous oceanic crust in both rifts did not develop until Chron 21 (48 Ma).
- The two initial propagating rifts failed to join during Paleocene–Early Eocene time in the palaeo-location of Kangerlussuaq and the Faroes Islands. The Jan Mayen micro-continent was still firmly attached to Greenland and continued to form a section of a land bridge between America–Greenland and Eurasia via the volcanic Faroe Islands and into NW Britain. The attempt of the two rifts to by-pass each other in this area caused anticlockwise rotation of the regional stress field.
- Oligo-Miocene North Atlantic plate reorganization, including the separation of Jan Mayen from SE Greenland, destroyed any relict Thulian land bridge and initiated the mixing of cold northern Atlantic waters with warmer southern Tethyan waters.
- Oligo-Miocene plate reorganization in the North Atlantic created the final break between Greenland and Europe and coincided with the appearance of Iceland and the production of the V-shaped seabed ridges to the north and south of the island.
- The dual rift model negates the need for a plume to develop the North Atlantic; the rifting can be wholly explained by plate tectonic mechanisms, lithospheric thinning and variable decompressive upper mantle melting along the rifts. Recent studies from the Afar rift in Ethiopia have shown that decompressive melt generation and dyke swarm propagation are more important than plume influence in the evolution of the proto-plate boundary (Ferguson et al. 2010; Rychert et al. 2012).
- The dual rift model with de-compressive upper mantle melts more closely confined to the plate tectonically induced rifts would imply a lower and more segmented regional heat-flow from 54 Ma to the present. This is in contrast to a higher regional heat-flow evoked by a large radius mantle plume. The implied lower heat-flow with its variable timing and geographical distribution will significantly change results from regional basin modeling studies and the type and timing of hydrocarbon generation around the North Atlantic, in contrast to results from a regional plume model with elevated heat-flow.

We would like to thank Michael Larsen and Gillian Foulger for their thorough reviews and very helpful suggestions for improving the paper, and colleagues in industry, the BGS and academia for stimulating discussions. We would also like to thank drafting staff in both Statoil (UK) Limited and the BGS for help in preparation of the figures. The contribution of MSS is with the permission of the Executive Director, BGS (NERC). This paper reflects the research and views of the authors and not necessarily those of Statoil or the British Geological Survey.

References

BEARD, K. C. 2008. The oldest North American primate and mammalian biogeography during the Paleocene–Eocene Thermal Maximum. *Proceedings of the National Academy of Sciences (USA)*, **105**, 3815–3818.

BENEDIKTSDÓTTIR, Á., HEY, R., MARTINEZ, F. & HÖSKULDSSON, Á. 2012. Detailed tectonic evolution of the Reykjanes Ridge during the past 15 Ma. *Geochemistry, Geophysics, Geosystems*, **13**, 1–27.

BERGGREN, W. A. & SCHNITKER, D. 1983. Cenozoic marine environments in the North Atlantic and Norwegian-Greenland Sea. *In*: BOTT, M. H. P., SAXOV, S., TALWANI, M. & THIEDE, J. (eds) *Structure and Development of the Greeenland–Scotland Ridge: New Methods and Concepts*. Plenum Press, New York, 495–548.

BEUTAL, E., VAN WIJK, J., EBINGER, C., KEIR, D. & AGOSTINI, A. 2010. Formation and stability of magmatic segments in the Main Ethiopian and Afar rifts. *Earth and Planetary Science Letters*, **293**, 225–235.

BINDEMAN, I., GURENKO, A., CARLEY, C., MILLER, C., MARTIN, E. & SIGMARSSON, O. 2012. Silicic magma petrogenesis in Iceland by remelting of hydrothermally altered crust based on oxygen isotope diversity and disequilibria between zircon and magma with implications for MORB. *Terra Nova*, **24**, 1–6.

BOTT, M. H. P. 1983. The crust beneath the Iceland-Faroe Ridge. *In*: BOTT, M. H. P., SAXOV, S., TALWANI, M. & THIEDE, J. (eds) *Structure and Development of the Greeenland–Scotland Ridge: New Methods and Concepts*. Plenum Press, New York, 63–75.

BOTT, M. H. P., SUNDERLAND, J., SMITH, P. J., ASTEN, U. & SAXOV, S. 1974. Evidence of continental crust beneath the Faroe Islands. *Nature*, **248**, 202–204.

BROWN, D. J., HOLOHAN, E. P. & BELL, B. R. 2009. Sedimentary and volcano-tectonic processes in the British Paleocene Igneous Province: a review. *Geology Magazine*, **146**, 326–352.

CARR, A. D. 1999. A vitrinite reflectance kinetic model incorporating overpressure retardation. *Marine and Petroleum Geology*, **16**, 353–377.

CARR, A. D. & SCOTCHMAN, I. C. 2003. Thermal History modelling in the southern Faroe–Shetland Basin. *Petroleum Geoscience*, **9**, 333–345.

CHALMERS, J. A. & PULVERTAFT, T. C. R. 2001. Development of the continental margins of the Labrador Sea: a review. *In*: WILSON, R. C. L., WHITMARSH, R. B. & FROITZHEIM, N. (eds) *Non-Volcanic Rifting of Continental Margins: A Comparison of Evidence from Land and Sea*. Geological Society, London, Special Publications, **187**, 77–105.

COURTILLOT, V., DAVAILLE, A., BESSE, J. & STOCK, J. 2003. Three distinct types of hotspots in the Earth's mantle. *Earth and Science Planetary Letters*, **205**, 295–308.

DAM, G., LARSEN, M. & SØDERHOLM, M. 1998. Sedimentary response to mantle plume: implications from the Palaeocene onshore successions, West and East Greenland. *Geology*, **26**, 207–210.

DENK, T., GRÍMSSON, F. & ZETTER, R. 2010. Episodic migration of oaks to Iceland: Evidence for a North Atlantic 'landbridge' in the latest Miocene. *American Journal of Botany*, **97**, 276–287.

DEWEY, J. F. & WINDLEY, B. F. 1988. Palaeocene-Oligocene tectonics of NW Europe. *In*: MORTON, A. C. & PARSON, L. M. (eds) *Early Tertiary Volcanism and the Opening of the NE Atlantic*. Geological Society, London, Special Publications, **39**, 25–31.

DORÉ, A. G., LUNDIN, E. R., JENSEN, L. N., BIRKELAND, Ø., ELIASSEN, P. E. & FICHLER, C. 1999. Principal tectonic events in the evolution of the northwest European Atlantic margin. *In*: FLEET, A. J. & BOLDY, S. A. R. (eds) *Petroleum Geology of Northwest Europe – Proceedings of the 5th Petroleum Geology Conference*. Geological Society, London, 41–61.

DORÉ, A. G., LUNDIN, E. R., KUSZNIR, N. J. & PASCAL, C. 2008. Potential mechanisms for the genesis of Cenozoic domal structures on the NE Atlantic margin: pros, cons and some new ideas. *In*: JOHNSON, H., DORÉ, A. G., GATLIFF, R. W., HOLDSWORTH, R.W., LUNDIN, E. & RITCHIE, J. D. (eds) *The Nature of Compression in Passive Margins*. Geological Society, London, Special Publications, **306**, 1–26.

ELLIS, D., PASSEY, S. R., JOLLEY, D. W. & BELL, B. R. 2009. Transfer Zones: the application of new geological information from the Faroe Islands applied to the offshore exploration of intra basalt and sub-basalt strata. *In*: *Faroe Island Exploration Conference: Proceedings of the 2nd Conference*. Annales Societatus Scientiarum Færoensis, Supplementum **48**, 198–219.

FERGUSON, D. J., BARNIE, T. D. ET AL. 2010. Recent rift-related volcanism in Afar, Ethiopia. *Earth and Planetary Science Letters*, **292**, 409–418.

FITTON, J. G., SAUNDERS, L. M., HARDARSON, B. S. & KEMPTON, P. D. 2000. Paleogene continental to oceanic magmatism on the SE Greenland continental margin at 63°N: A review of the results of Ocean Drilling Program legs 152 and 163. *Journal of Petrology*, **41**, 951–966.

FORSBERG, R. & KENYON, K. 2004. Gravity and geoid in the Arctic region: The northern polar gap now filled. *In*: *Proceedings of 2nd GOCE User Workshop*. ESA Publication Division, Noordwijk, The Netherlands.

FOULGER, G. R. 2002. Plumes, or plate tectonic processes? *Astronomy and Geophysics*, **43**, 619–623.

FOULGER, G. R. 2010. *Plates vs Plumes: A Geological Controversy*. Wiley-Blackwell, Oxford.

FOULGER, G. R. & ANDERSON, D. L. 2005. A cool model for the Iceland hotspot. *Journal of Volcanology and Geothermal Research*, **141**, 1–22.

FOULGER, G. R., PRITCHARD, M. J. ET AL. 2001. Seismic tomography shows that upwelling beneath Iceland is confined to the upper mantle. *Geophysical Journal International*, **146**, 504–530.

FOULGER, G. R., DU, Z. & JULIAN, B. R. 2003. Icelandic type crust. *Geophysical Journal International*, **155**, 567–590.

GAINA, C., RØEST, W. R. & MULLER, R. D. 2002. Late Cretaceous–Cenozoic deformation of northeast Asia. *Earth and Planetary Science Letters*, **197**, 273–286.

GAINA, C., GERNIGON, L. & BALL, P. 2009. Paleocene–Recent plate boundaries in the northeast Atlantic and formation of the Jan Mayen micro-continent. *Journal of the Geological Society, London*, **166**, 601–616.

GANERØD, M., SMETHURST, M. A. ET AL. 2010. The North Atlantic Igneous Province reconstructed and its relation to the Plume Generation Zone: the Antrim Lava Group Revisited. *Geophysical Journal International*, **182**, 183–202.

GERNIGON, L., OLESEN, O. ET AL. 2009. Geophysical insights and early spreading history in the Jan Mayen fracture zone, Norwegian Greenland Sea. *Tectonophysics*, **468**, 185–205.

GILL, R. C. O., PEDERSEN, A. K. & LARSEN, J. G. 1992. Tertiary picrites in West Greenland: melting at the periphery of a plume? *In*: STOREY, B. C., ALABASTER, T. & PANKHURST, R. J. (eds) *Magmatism and the Causes of Continental Break-Up*. Geological Society, London, Special Publications, **68**, 335–348.

GRADSTEIN, F. M., OGG, J. G. & SMITH, A. G. (eds) 2004. *A Geologic Time Scale 2004*. Cambridge University Press, Cambridge, UK.

GUARNIERI, P. 2011. Analysis of Palaeogene strike-slip tectonics along the southern East Greenland margin (Sødalen area). *Geological Survey of Denmark and Greenland Bulletin*, **23**, 65–68.

HANSEN, J., JERRAM, D. A., MCCAFFREY, K. & PASSEY, S. R. 2009. The onset of the North Atlantic Igneous Province in a rifting perspective. *Geological Magazine*, **146**, 309–325.

HARDARSON, B. S., FITTON, J. G., ELLAM, R. M. & PRINGLE, M. S. 1997. Rift relocation – a geochemical and geochronological investigation of a palaeo-rift in northwest Iceland. *Earth and Planetary Science Letters*, **153**, 181–196.

HEY, R., MARTINEZ, F., HÖSKULDSSON, Á. & BENEDIKTSDÓTTIR, Á. 2010. Propagating rift model for the V-shaped ridges south of Iceland. *Geochemistry, Geophysics, Geosystems*, **11**, doi: 10.1029/2009GC002865.

HOLBROOK, S. W., LARSEN, H. C. & KORENAGA, J. 2001. Mantle structure and active upwelling during continental breakup in the North Atlantic. *Earth and Planetary Science Letters*, **190**, 51–266.

HOLMES, A. J., GRIFFITH, C. E. & SCOTCHMAN, I. C. 1999. The Jurassic petroleum system of the west of Britain Atlantic margin: an integration of tectonics, geochemistry and basin modelling. *In*: FLEET, A. J. & BOLDY, S. A. R. (eds) *Petroleum Geology of Northwest Europe: Proceedings of the 5th Conference*. Geological Society, London, 1351–1365.

HUNG, S. H., SHEN, Y. & CHIAO, L. Y. 2004. Imaging seismic velocity structure beneath the Iceland hotspot: a finite-frequency approach. *Journal of Geophysical Research*, **109**, B08305.

JOHNSON, H., RITCHIE, J. D., HITCHEN, K., MCINROY, D. B. & KIMBELL, G. S. 2005. Aspects of the Cenozoic deformational history of the northeast Faroe-Shetland Basin, Wyville-Thomson Ridge and Hatton Bank areas. *In*: DORÉ, A. G. & VINING, B. (eds) *Petroleum Geology: North-West Europe and Global Perspectives – Proceedings of the 6th Petroleum Geology Conference*. The Geological Society, London, 993–1007.

JOLLEY, D. W. & MORTON, A. C. 2007. Understanding basin sedimentary provenance: evidence from allied phytogeographic and heavy mineral analysis of the Palaeocene of the NE Atlantic. *Journal of the Geological Society, London*, **164**, 553–563.

JONES, R. W. & MILTON, N. J. 1994. Sequence development during uplift: Palaeogene stratigraphy and relative sea-level history of the outer Moray Firth, UK North Sea. *Marine and Petroleum Geology*, **11**, 157–165.

JONES, S. M., WHITE, N. & MACLENNAN, J. 2002. V-shaped ridges around Iceland: implications for spatial and temporal patterns of mantle convection. *Geochemistry, Geophysics, Geosystems*, **3**, 1–23.

KIMBELL, G. S., RITCHIE, J. D., JOHNSON, H. & GATLIFF, R. W. 2005. Intial field imaging and 3D modeling. *In*: DORE, A. G. & VINING, B. A. (eds) *Petroleum Geology: North-West Europe and Global Perspectives – Proceedings of the 6th Petroleum Geology Conference*. Geological Society, London, 933–945.

KNOTT, S. D., BURCHELL, M. T., JOLLEY, E. J. & FRASER, A. J. 1999. Mesozoic to Cenozoic plate reconstructions of the North Atlantic and hydrocarbon plays of the Atlantic margins. *In*: FLEET, A. J. & BOLDY, S. A. R. (eds) *Petroleum Geology of Northwest Europe – Proceedings of the 5th Petroleum Geology Conference*. Geological Society, London, 953–974.

KÖHLER, W. 2004. Section 1.3: Gravity field and Earth models, gravitational geopotential in terms of geoid heights. http://www.gfz-potsdam.de

KUVAAS, B. & KODAIRA, S. 1997. The formation of the Jan Mayen microcontinent: the missing piece in the continental puzzle between the Møre-Vøring Basins and East Greenland. *First Break*, **15**, 239–247.

LARSEN, L. M. & WATT, S. W. 1985. Episodic volcanism during break-up of the North Atlantic: evidence from the East Greenland plateau basalts. *Earth and Planetary Science Letters*, **73**, 116.

LARSEN, L. M., WATT, W. S. & WATT, M. 1989. Geology and petrology of the Lower Tertiary plateau basalts of the Scoresby Sund region, East Greenland. *Grønlands Geologiske Undersøgelse, Bulletin 197*. Danish Geological Survey.

LARSEN, L. M., WAAGSTEIN, R., PEDERSEN, A. K. & STOREY, M. 1999. Trans-Atlantic correlation of the Paleogene volcanic successions in the Faroe Islands and East Greenland. *Journal of the Geological Society, London*, **156**, 1081–1095.

LARSEN, M. & WHITHAM, A. G. 2005. Evidence for a major sediment input point into the Faroe–Shetland basin from southern East Greenland. *In*: DORE, A. G. & VINING, B. A. (eds) *Petroleum Geology: North-West Europe and Global Perspectives – Proceedings of the 6th Petroleum Geology Conference*. Geological Society, London, 913–922.

LARSEN, M., HEILMANN-CLAUSEN, C., PIASECKI, S. & STEMMERICK, L. 2005. At the edge of a new ocean: post-volcanic evolution of the Palaeogene Kap Dalton Group, East Greenland. *In*: DORE, A. G. & VINING, B. A. (eds) *Petroleum Geology: North-West Europe and Global Perspectives – Proceedings of the 6th Petroleum Geology Conference*. Geological Society, London, 923–932.

LUNDIN, E. R. & DORÉ, A. G. 2002. Mid-Cenozoic post break-up deformation in the "passive" margins bordering the Norwegian–Greenland Sea. *Marine and Petroleum Geology*, **9**, 79–93.

LUNDIN, E. R. & DORÉ, A. G. 2005a. NE Atlantic break-up: a re-examination of the Iceland plume model and the Atlantic-Arctic linkage. *In*: DORE, A. G. & VINING, B. A. (eds) *Petroleum Geology: North-West Europe and Global Perspectives – Proceedings of the 6th Petroleum Geology Conference*. Geological Society, London, 739–754.

Lundin, E. R. & Doré, A. G. 2005b. Fixity of the Iceland 'hotspot' on the mid-Atlantic Ridge: observational evidence, mechanisms and implications for Atlantic margins. *In*: Foulger, G. R., Natland, J. H., Presnall, D. C. & Anderson, D. L. (eds) *Plates, Plumes and Paradigms*. Geological Society of America Special Papers, **388**, 627–651.

Lundin, E. R., Torsvik, T. H., Olesen, O. & Roest, W. R. 2002. The Norwegian Sea, a meeting place for opposed and overlapping Arctic and Atlantic rifts, implications for magmatism. *In*: *AAPG Hedberg Conference Hydrocarbon Habitat of Volcanic Rifted Passive Margins*, September 8–11, Stavanger, Norway.

Marquart, G. 1991. Interpretations of geoid anomalies around the Iceland 'hotspot'. *Geophysical Journal International*, **106**, 149–160.

Meyer, R., van Wijk, J. & Gernigon, L. 2007. The North Atlantic Igneous Province: A review of models for its formation. *Geological Society of America Special Papers*, **430**, 525–552.

Montelli, R., Nolet, G., Dahlen, F. A., Masters, G., Engdahl, E. R. & Hung, S. H. 2003. Finite-frequency tomography reveals a variety of plumes in the mantle. *Science*, **203**, 338–343.

Moorbath, S., Sigurdsson, H. & Goodwin, R. 1968. K–Ar ages of the oldest exposed rocks in Iceland. *Earth and Planetary Science Letters*, **4**, 197–205.

Moore, J. G. & Shannon, P. M. 1992. Palaeocene–Eocene deltaic sedimentation, Porcupine Basin, offshore Ireland – a sequence stratigraphic approach. *First Break*, **10**, 461–469.

Morgan, W. J. 1971. Convection plumes in the lower mantle. *Nature*, **230**, 42–43.

Morton, A., Ellis, D., Fanning, M., Jolley, D. & Whitham, A. 2012a. The importance of an integrated approach to provenance studies: a case study from the Paleocene of the Faroe-Shetland Basin, NE Atlantic. *In*: Rasbury, E. T., Hemmings, S. R. & Riggs, N. R. (eds) *Mineralogical and Geochemical Approaches to Provenance*. Geological Society of America, Special Paper, **487**, 1–12.

Morton, A., Ellis, D., Fanning, M., Jolley, D. & Whitham, A. 2012b. Heavy mineral constraints on Paleocene sand transport routes in the Faroe-Shetland Basin *In*: Varming, T. (ed.) *Proceedings of the 3rd Faroe Islands Exploration Conference*. Annales Societatus Scientiarum Færoensis, 59–83.

Mudge, D. C. & Bujak, J. P. 1994. Eocene stratigraphy of the North Sea basin. *Marine and Petroleum Geology*, **11**, 166–181.

Muller, R. D., Gaina, C., Røest, W. & Hansen, D. L. 2001. A recipe for microcontinent formation. *Geology*, **29**, 203–206.

Nunns, A. G. 1983. Plate tectonic evolution in the Greenland–Scotland Ridge and surrounding regions. *In*: Bott, M. H. P., Saxov, S., Talwani, M. & Thiede, J. (eds) *Structure and Development of the Greeenland–Scotland Ridge: New Methods and Concepts*. Plenum Press, New York, 11–30.

Ólavsdóttir, J., Boldreel, L. O. & Andersen, M. S. 2010. Development of a shelf margin delta due to uplift of the Munkagrunnur Ridge at the margin of Faroe-Shetland Basin: a seismic sequence stratigraphic study. *Petroleum Geoscience*, **16**, 91–103.

Ólavsdóttir, J., Andersen, M. S. & Boldreel, L. O. 2013. Seismic stratigraphic analysis of the Cenozoic sediments in the NW Faroe Shetland Basin – implications for inherited structural control of sediment distribution. *Marine and Petroleum Geology*, **46**, 19–35.

Pagli, C., Wright, T. J., Ebinger, C. J., Yun, S., Cann, J. R. & Barnie, T. 2012. Shallow axial magma chamber at the slow spreading Erta Ale Ridge. *Nature Geoscience Letters*, 1–5, March, doi: 10.1038/NGEO1414.

Parnell, J. & Middleton, D. W. J. 2009. Pore fluid and hydrocarbon migration histories in North Atlantic basins intruded by Tertiary sills. *In*: Varming, T. & Ziska, H. (eds) *Faroe Island Exploration Conference. Proceedings of the 2nd Conference*. Annales Societatus Scientiarum Færoensis, Supplementum, **48**, 124–155.

Parnell, J., Carey, P. F., Green, P. & Duncan, W. 1999. Hydrocarbon migration history, West of Shetland: integrated fluid inclusion and fission track studies. *In*: Fleet, A. J. & Boldy, S. A. R. (eds) *Petroleum Geology of Northwest Europe – Proceedings of the 5th Petroleum Geology Conference*. Geological Society, London, 613–626.

Passey, S. R. & Jolley, D. W. 2009. A revised lithostratigraphic nomenclature for the Palaeogene Faroe Islands Basalt Group, NE Atlantic Ocean. *Earth and Environmental Science Transactions of the Royal Society of Edinburgh*, **99**, 127–158.

Paquette, J., Sigmarsson, O. & Tiepolo, M. 2006. Continental basement under Iceland revealed by old zircons. *American Geophysical Union Fall Meeting 2006*, Abstract.

Paquette, J., Sigmarsson, O. & Tiepolo, M. 2007. Mesozoic zircons in Miocne ignimbrite from E-Iceland: a splinter of a continental crust? *Geophysical Research Abstracts*, **9**, 03723.

Pedersen, A. K., Watt, M., Watt, A. S. & Larsen, L. M. 1997. Structure and stratigraphy of the Early Tertiary basalts of the Blosseville Kyst, East Greenland. *Journal of the Geological Society, London*, **154**, 565–570.

Pitman, W. C., III & Talwani, M. 1972. Seafloor spreading in the North Atlantic. *Geological Society of American Bulletin*, **83**, 619–646.

Presnall, D. C. 2003. Petrological constraints on potential temperature. *In*: *The Hotspot Handbook, Proceedings of the Penrose Conference Plume IV: Beyond the Plume Hypothesis*. Geological Society of America, Hveragerdi, Iceland.

Prestvik, T., Goldberg, S., Karlsson, H. *et al.* 2001. Anomalous strontium and lead isotope signatures in the off-rift Oraefajokull central volcano in south-east Iceland. *Earth and Planetary Science Letters*, **190**, 211–220.

Price, S., Brodie, J., Whitham, A. & Kent, R. 1997. Mid-Tetiary rifting and magmatism in the Trail Ø region, East Greenland. *Journal of the Geological Society, London*, **154**, 419–434.

Reemst, P. & Cloetingh, S. A. P. L. 2000. Polyphase rift evolution of the Vøring margin (mid-Norway): constrainst from forward tectonostratigraphic modeling. *Tectonics*, **19**, 225–240.

RITCHIE, J. D., JOHNSON, H., QUINN, M. F. & GATLIFF, R. W. 2008. Cenozoic compressional deformation within the Faroe–Shetland Basin and adjacent areas. In: JOHNSON, H., DORÉ, A. G., HOLDSWORTH, R. E., GATLIFF, R. W., LUNDIN, E. R. & RITCHIE, J. D. (eds) *The Nature and Origin of Compression in Passive Margins*. The Geological Society, London, Special Publications, **306**, 121–136.

RITSEMA, J., VAN HEIJST, H. J. & WOODHOUSE, J. H. 1999. Complex shear wave velocity structure imaged beneath Africa and Iceland. *Science*, **286**, 1925–1928.

ROBERTS, D. G., THOMPSON, M., MITCHENER, B., HOSSACK, J., CARMICHAEL, S. & BJØRNSETH, H.-M. 1999. Paleozoic to Tertiary rift and basin dynamics; mid-Norway to the Bay of Biscay – a new context for hydrocarbon prospectivity in the deep water frontier. In: FLEET, A. J. & BOLDY, S. A. R. (eds) *Petroleum Geology of Northwest Europe – Proceedings of the 5th Petroleum Geology Conference*. Geological Society, London, 7–40.

ROBINSON, A. M. 2004. *Stratigraphic development and controls on the architecture of Eocene depositional systems in the Faroe-Shetland Basin*. PhD thesis (unpublished), University of Cardiff, UK.

RØEST, W. R. & SRIVASTAVA, S. P. 1989. Seafloor spreading in the Labrador Sea: a new reconstruction. *Geology*, **17**, 1000–1004.

ROSS, J. G. & MUSSETT, A. E. 1976. $^{40}Ar/^{39}Ar$ dates for spreading rates in eastern Iceland, *Nature*, **259**, 36–38.

RUMPH, B., REAVES, C. M., ORANGE, V. G. & ROBINSON, D. L. 1993. Structuring and transfer zones in the Faeroe basin in a regional tectonic context. In: PARKER, J. R. (ed.) *Petroleum Geology of Northwest Europe – Proceedings of the 4th Petroleum Geology Conference*. Geological Society, London, 999–1010.

RYAN, P. D. & DEWEY, J. F. 1997. Continental eclogites and the Wilson cycle. *Journal of the Geological Society, London*, **154**, 437–442.

RYCHERT, C. A., HAMMOND, J. O. S. ET AL. 2012. Volcanism in the Afar Rift sustained by decompression melting with minimal plume influence. *Nature Geoscience*, **5**, 406–409.

SANDWELL, D. T. & SMITH, W. H. F. 1997. Marine gravity anomaly from ERS-1 Geosat and satellite altimetry. *Journal of Geophysical Research*, **102**, 10 039–10 054.

SAUNDERS, A. D., FITTON, J. G., KERR, A. C., NORRY, M. J. & KENT, R. W. 1997. The North Atlantic Igneous Province. In: MAHONEY, J. J. & COFFIN, M. L. (eds) *Large Igneous Provinces*. Geophysical Monograph Series, American Geophysical Union, Washington DC, 45–93.

SCOTCHMAN, I. C., CARR, A. D. & PARNELL, J. 2006. Hydrocarbon generation modeling in a multiple rifted and volcanic basin: a case study in the Foinaven Sub-basin, Faroe-Shetland Basin, UK Atlantic margin. *Scottish Journal of Geology*, **42**, 1–19.

SHANNON, P. M., MOORE, J. G., JACOB, A. W. B. & MAKRIS, J. 1993. Cretaceous and Bertiary basin development west of Ireland. In: PARKER, J. R. (ed.) *Petroleum Geology of Northwest Europe – Proceedings of the 4th Petroleum Geology Conference*. Geological Society, London, 1057–1066.

SKOGSEID, J., PLANKE, S., FALEIDE, J. I., PEDERSEN, T., ELDHOLM, Ø. & NEVERDAL, F. 2000. NE Atlantic continental rifting and volcanic margin formation. In: NOTTVELDT, A. ET AL. (eds) *Dynamics of the Norwegian Margin*. Geological Society, London, Special Publications, **167**, 295–326.

SMALLWOOD, J. R. & WHITE, R. S. 2002. Ridge-plume interaction in the North Atlantic and its influence on continental breakup and seafloor spreading. In: JOLLEY, D. W. & BELL, B. R. (eds) *The North Atlantic Igneous Province: Stratigraphy, Tectonic, Volcanic and Magmatic Processes*. Geological Society, London, Special Publications, **197**, 15–37.

SMALLWOOD, J. R., STAPLES, R. K., RICHARDSON, R. K. & WHITE, R. S. 1999. Crust generated above the Iceland mantle plume: from continental rift to ocean spreading centre. *Journal of Geophysical Research*, **104**, 22 885–22 902.

SØAGER, N. & HOLM, P. H. 2011. Changing compositions in the Iceland plume; Isoptic and elemental constraints from the Paleogene Faroe flood basalts. *Chemical Geology*, **280**, 297–313.

SRIVASTAVA, S. P. & TAPSCOTT, C. R. 1986. Plate kinematics of the North Atlantic. In: VOGT, P. R. & TUCHOLKE, B. E. (eds) *The Western North Atlantic Region. The Geology of North America*. Geological Society of America, 379–405.

STEMMERIK, L., CHRISTIANSEN, F. G., PIASECKI, S., JORDT, B., MARCUSSEN, C. & NØHR-HANSEN, H. 1992. Depositional history and petroleum geology of the Carboniferous to Cretaceous sediments in the northern part of East Greenland. In: VORREN, T. O., BERGSAGER, E., DAHL-STAMNES, Ø. A., HOLTER, E., JOHANSEN, B., LIE, E. & LUND, T. B. (eds) *Arctic Geology and Petroleum Potential*. Norwegian Petroleum Society (NPF), Special Publications, **2**, 67–87.

STOKER, M. S. & VARMING, T. 2011. The Cenozoic. In: RITCHIE, J. D., ZISKA, H., JOHNSON, H. & EVANS, D. (eds) *Geology of the Faroe-Shetland Basin and Adjacent Areas*. British Geological Survey Report **RR/11/01**.

STOKER, M. S., PRAEG, D. ET AL. 2005a. Neogene evolution of the Atlantic continental margin of NW Europe (Lofoten Islands to SW Ireland): anything but passive. In: DORÉ, A. G. & VINING, B. (eds) *Petroleum Geology: North-West Europe and Global Perspectives – Proceedings of the 6th Petroleum Geology Conference*. The Geological Society, London, 1057–1076.

STOKER, M. S., HOULT, R. J. ET AL. 2005b. Sedimentary and oceanographic responses to early Neogene compression on the NW European margin. *Marine and Petroleum Geology*, **22**, 1031–1044.

STOKER, M. S., KIMBELL, G. S., MCINROY, D. B. & MORTON, A. C. 2012. Eocene post-rift tectonostratigraphy of the Rockall Plateau, Atlantic margin of NW Britain: linking early spreading tectonics and passive margin response. *Marine and Petroleum Geology*, **30**, 98–125.

STOKER, M. S., LESLIE, A. B., SMITH, K., ÓLAVSDÓTTIR, J., JOHNSON, H. & LABERG, J. S. 2013a. Onset of North Atlantic Deep Water production coincident with inception of the Cenozoic global cooling trend: comment. *Geology*, **41**, e291, http://dx.doi.org/10.1130/G33670C.1

STOKER, M. S., LESLIE, A. B. & SMITH, K. 2013b. A record of Eocene (Stronsay Group) sedimentation in BGS borehole 99/3, offshore NW Britain: implications for the early post-breakup development of the Faroe-Shetland Basin. *Scottish Journal of Geology*, **49**, 133–148.

STOREY, M., PEDERSEN, A. K. ET AL. 2004. Long-lived postbreakup magmatism along the East Greenland margin: Evidence for shallow-mantle metasomatism by the Iceland plume. *Geology*, **32**, 173–176.

SURLYK, F., HURST, J. M., PIASECKI, S., ROLLE, F., SCHOLLE, P. A., STEMMERIK, L. & THOMSEN, E. 1986. The Permian of the Western Margin of the Greenland Sea – a future exploration target. *In*: HALBOUTY, M. T. (ed.) *Future Petroleum Provinces of the World*. American Association of Petroleum Geology, Tulsa, Memoir, **40**, 629–659.

TALWANI, M. & ELDHOLM, O. 1977. Evolution of the Norwegian–Greenland Sea. *Geological Society of America Bulletin*, **88**, 969–999.

TALWANI, M., UDINTSEV, G. ET AL. 1976. Sites 336 and 352. *In*: TALWANI, M., UDINTSEV, G. ET AL. (eds) *Initial Reports of the Deep Sea Drilling Project*. Deep Sea Drilling Project Reports and Publications, **38**, 23–116.

TORSVIK, T. H., SMETHURST, M. A., BURKE, K. & STEINBERGER, B. 2006. Large igneous provinces generated from the margins of the large low-velocity provinces in the deep mantle. *Geophysics Journal International*, **167**, 1447–1460.

VALLIER, T., CALK, L., STAX, R. & DEMANT, A. 1998. Hole 917A, East Greenland margin. *Proceedings of the Ocean Drilling Program, Scientific Results*, **152**.

VAN WIJK, J. W., HUISMANS, R. S., TER VOORDE, M. & CLOETINGH, S. A. P. L. 2001. Melt generation at Volcanic Continental Margins: no need for a mantle plume? *Geophysical Research Letters*, **28**, 3995–3998.

WALKER, R. J., HOLDSWORTH, R. E., IMBER, J. & ELLIS, D. 2011. Onshore evidence for progressive changes in rifting directions during continental break-up in the NE Atlantic. *Journal of the Geological Society, London*, **168**, 27–48.

WHITE, R. S. 1988. A hot-spot model for early Tertiary volcanism in the N. Atlantic. *In*: MORTON, A. C. & PARSON, L. M. (eds) *Early Tertiary Volcanism and the Opening of the NE Atlantic*. Geological Society, London, Special Publications, **39**, 3–13.

WHITE, R. S. 1989. Initiation of the Iceland plume and opening of the North Atlantic. *In*: TANKARD, A. J. & BALKWILL, H. R. (eds) *Extensional Tectonics and Stratigraphy of the North Atlantic Margins*. American Association of Petroleum Geologists, Tulsa, Memoir **46**, 149–154.

WHITE, R. S. & MCKENZIE, D. 1989. Magmatism at rift zones: the generation of volcanic continental margins and flood basalts. *Journal of Geophysical Research*, **94**, 7685–7729.

WHITHAM, A. G., MORTON, A. C. & FANNING, C. M. 2004. Insights into Cretaceous-Paleocene sediment transport paths and basin evolution in the North Atlantic from heavy mineral study of sandstones from southern East Greenland. *Petroleum Geoscience*, **10**, 61–72.

XIANG, Q. Y., MANCHESTER, S. R., THOMAS, D. T., ZHANG, W. & FAN, C. 2005. Phylogeny, biogeography, and molecular dating of cornelian cherries (Cornus, Cornaceae): tracking Tertiary plant migration. *Evolution*, **59**, 1685–1700.

ZIEGLER, P. A. 1988. Evolution of the Arctic-North Atlantic and Western Tethys – A visual presentation of a series of paleogeographic-paleotectonic maps. *American Association of Petroleum Geologists, Memoir*, **43**, 164–196.

# Influence of igneous sills on Paleocene turbidite deposition in the Faroe–Shetland Basin: a case study in Flett and Muckle sub-basin and its implication for hydrocarbon exploration

SYLVESTER EGBENI[1,2]*, KEN McCLAY[2], JACK JIAN-KUI FU[3] & DUNCAN BRUCE[1]

[1]*GDFSUEZ E&P UK Ltd, Holborn Viaduct, London, UK*

[2]*Fault Dynamic Group, Earth Science Dept, Royal Holloway University, Egham, TW20 0EX, UK*

[3]*GE Oil and Gas, Canada*

Corresponding author (e-mail: Sylvester.egbeni@gdfsuezep.co.uk)

Abstract: A new interpretation on three-dimensional seismic data from the Flett and Muckle sub-basins of the Faroe–Shetland Basin has shown dyke and sill emplacement influencing the Paleocene turbidite deposition. Sill and dyke emplacement in the study area created significant inflation anticlines during Paleocene time and affected palaeotopography at the seafloor. The uplift of older Early Paleocene (Sullom) and Middle Paleocene (early Vaila) turbidite deposits results in their erosion and redeposition in adjacent lows or transports them further into the basin by erosive channels. By using public domain well and biostratigraphic data from wells 208/19-1, 206/2a-1 and 206/1-2, seismic data and the use of seismic sequence stratigraphy to map the onlapping seismic reflectors on the flanks of the inflated anticlines, the relative timing of sill emplacement in the study area is shown to be Early–Middle Paleocene. The ability to identify these systems and understand the interaction between palaeotopography and Paleocene turbidite deposition in the basin is key to unlocking the hydrocarbon potential in the UK flank of the Faroe–Shetland Basin.

There are many factors contributing to controls on the distribution of sediments in the deep marine environment of the Faroe–Shetland Basin. One of these is the influence of igneous intrusions on the depositional history of the Paleocene play which has been lightly studied due to a lack of exploration activity. This paper aims to describe the relationship between Paleocene sand deposition and igneous sill emplacement.

Igneous intrusions (dyke and sills) have been studied by a number of authors including Du Toit (1920), Bradley (1965), Pollard (1973), Pollard & Johnson (1973), Burger *et al.* (1981), Francis (1982), Chevalier & Woodford (1999), Smallwood & Maresh (2002), Trude *et al.* (2003), Thomson (2005), Planke *et al.* (2005), Hansen & Cartwright (2006), Lee *et al.* (2006), Cartwright *et al.* (2007), Goulty & Schofield (2008), Smallwood (2008), Smallwood & Harding (2008), Thomson & Schofield (2008) and Schofield *et al.* (2010, 2012*a*, *b*) Without the use of seismic data, Bradley (1965) proposed the concept of topographic compensation formed as a result of sill emplacement. Smallwood & Maresh (2002) were the first to recognize the potential for sills to inflate the seafloor and provide accommodation space influencing the deposition of deep-water turbidites in the Faroe–Shetland Basin. Thomson & Schofield (2008) proposed models for the emplacement of concave-upwards sills formed as a result of lithological control and overpressure, which is important in the evolutionary history of igneous sills.

Radiometric dating, sill emplacement method, evolution, morphologies and qualitative/quantitative analysis of igneous sills has been the main focus of a number of studies in recent years in order to understand the timing of sill emplacement and their influence on sand deposition, with little attention paid to seismic sequence stratigraphy.

A new technique for dating sills using seismic sequence stratigraphy was proposed by Trude *et al.* (2003), with examples from the Faroe–Shetland Basin. This technique was a continuation of the method discussed by Smallwood & Maresh (2002). This demonstrated that shallow level intrusions of igneous sills (<1 km) below the seafloor can deform sediments during the early burial phase by hydraulic elevation of the overburden. These sills forcibly 'jack up' the palaeo-seafloor, thereby creating palaeotopography with small basins that are typically 5 km wide, 250 m deep and >20 km long (Trude *et al.* 2003). By using seismic sequence stratigraphy, Trude *et al.* (2003) dated the subsequent sediment infill to be 54.6–55 Ma. This fitted radiometric dating of the sills at 53–55 Ma (Trude *et al.* 2003). Planke *et al.* (1999, 2000, 2005), Davies *et al.* (2002), Hansen *et al.* (2003) and Trude *et al.* (2003) highlighted the importance of using 3D seismic data in analysing the overall morphology of igneous sills.

From: CANNON, S. J. C. & ELLIS, D. (eds) 2014. *Hydrocarbon Exploration to Exploitation West of Shetlands.*
Geological Society, London, Special Publications, **397**, 33–57.
First published online March 4, 2014, http://dx.doi.org/10.1144/SP397.8
© The Geological Society of London 2014. Publishing disclaimer: www.geolsoc.org.uk/pub_ethics

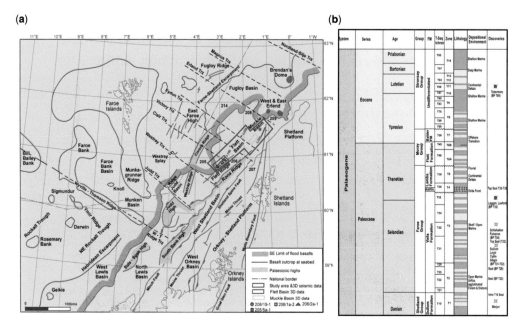

Fig. 1. (a) Regional tectonic map of the Faroe–Shetland Basin showing structural features, basins and location of study. The map is modified from Stoker *et al.* (1993). (b) Tertiary stratigraphy of the Faroe–Shetland Basin, modified from BP *T*-sequence and Ebdon *et al.* (1995).

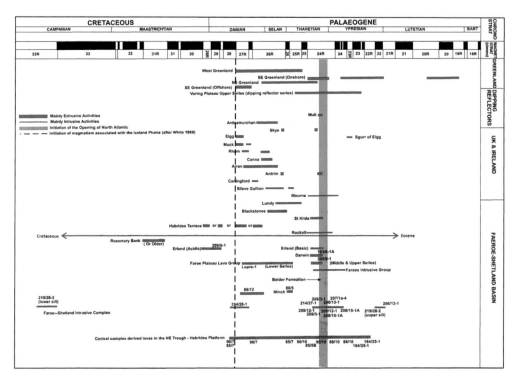

Fig. 2. The age of sill/dyke and basalt of the North Atlantic Igneous Province. Chart modified from Ritchie & Hitchen (1996).

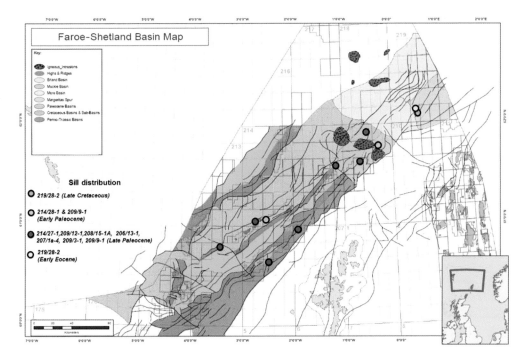

Fig. 3. A regional representation of sill and dyke age and distribution, after Ritchie *et al.* (1999).

The main aims of this paper are: (1) to provide evidence that sill emplacement at shallow levels cause palaeotopography within the Paleocene as earlier proposed by Smallwood & Maresh (2002); (2) to use the methods first proposed by Smallwood and Maresh (2002) and Trude *et al.* (2003) to analyse the timing of sill emplacement in the study area and their relationship with Paleocene Vaila turbidite deposition; and (3) to demonstrate that depositional lows are the preferential location for Paleocene turbidites post sill emplacement with potential onlapping stratigraphic traps.

Geological setting

The Faroe–Shetland Basin is on the north-eastern part of the North Atlantic margin within 8°W–2°E and 58°N–64°N (Fig. 1a). It is bound by the Wyville–Thomson ridge to the SW, the Nordland–Silje transfer fault in the NE and the Shetland Platform to the SE (Fig. 1a). The basin developed during Devonian–Palaeogene time (Dore *et al.* 1997). The two key tectonic events that formed the dominant structure of the Faroe–Shetland Basin were the Early Mesozoic extension that formed the NE–SW-trending faults, and the Late Mesozoic extension which formed the present-day NE–SW fault trend sub-basins (Dean *et al.* 1999). The Late Mesozoic extension led to thick sedimentary sequence deposition within the basin and the formation of a number of important and recognizable highs including the Rona Ridge, Flett Ridge, Shetland Platform, Corona Ridge and Westray Ridge and, to an extent, the Judd, Clair, Erlend and Solan Bank highs (Naylor *et al.* 1999). Geochemical provenance studies using core data indicate that the main source direction during Mesozoic time was from Greenland to the NW (Knott *et al.* 1993). During the Late Cretaceous however, the sediment supply mainly switched to the SE due to regional uplift of the Scottish margin and Faroe–Shetland platform (Knott *et al.* 1993).

A major period of magmatism commenced during Early Paleocene time at *c.* 62 Ma (White & McKenzie 1989; Pearson *et al.* 1996; Ritchie & Hitchen 1996; Jolley 1997; Mussett *et al.* 1988). Large volumes of extruded basaltic rocks marked the onset of seafloor spreading in the NE Atlantic Ocean (Naylor *et al.* 1999). Extension continued during Paleocene time with regional uplift and inversion during Eocene time, which was a result of plume buoyancy, magmatic underplating or probably a combination of these (Brodie & White 1994). Uplift within the Palaeogene stratigraphy is recognized through regressive and transgressive depositional sequences in the basin. Dolerite sill complexes and associated dykes, which are generally

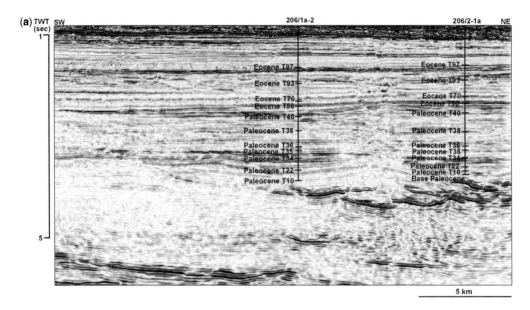

Fig. 4. Seismic to well tie of (**a**) 206/1a-2 and 206/2a-1 and (**b**) 208/19-1.

seen as discontinuous high-amplitude concave-upwards seismic reflectors, intrude the Upper Cretaceous–Paleocene interval (Trude 2004). A study compiled by Ritchie & Hitchen (1987) has provided radiometric dates for the intrusions that range from Late Cretaceous to Eocene in age

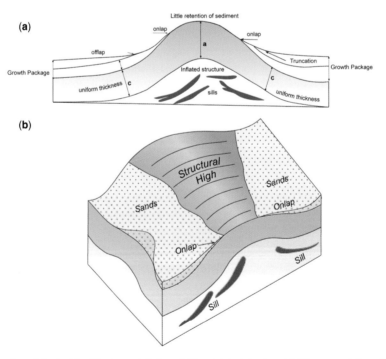

Fig. 5. Approach used for the identification and timing of sill and dyke intrusion. (**a**) 2D representation of the seismic geometries (onlap, offlap, truncation and changes in sediment thickness) observed post-sill and dyke intrusion. (**b**) 3D representation of the geometries post-sill intrusion and control on palaeotopography.

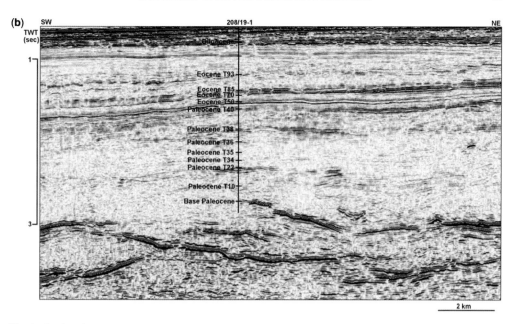

Fig. 4. *Continued.*

(Fig. 2). Although Naylor *et al.* (1999) suggested a more specific age of 55 Ma (Early Eocene) using the published data of Ritchie & Hitchen (1987), the accuracy of this dating is not clearly defined. The work of Ritchie *et al.* (1999) highlighted the compositional difference from core and/or drill sample of the volcanic intrusive across the area (Fig. 3) which could represent different sources at different times.

Dataset and methodology

Dataset

Three-dimensional seismic data across the Flett and Muckle basins was used in this study together with three seismic lines taken from the reprocessed Petroleum Geo-Services (PGS) Mega-merge 3D data (Fig. 1a). The bin spacing of the Flett Basin 3D survey, the Muckle Basin 3D survey and Mega-merge 3D survey are 25 by 25. The polarity of all data presented in this study is North Sea normal with black being soft and red or yellow hard. Data quality in the Flett Basin 3D data and the Mega-merge 3D data is good for two-way time (TWT) >3 s, and becomes fair–poor at TWT 3.5–6.0 s with transparent section and swarms of sills. Seismic reflectors of 2.8–4 s TWT are fairly continuous; there is therefore medium–high confidence on interpretation in the section of interest (Fig. 4a). The Muckle Basin 3D data quality is poor to fair. Between *c.* 2 and 2.7 s TWT, data quality is poor with a thick transparent section and intruding igneous sill (Fig. 4b). Below 3 s TWT are swarm sills and a drastic drop in data quality. The varying quality of this dataset is a limitation to interpretation, especially in the transparent seismic section between *c.* 2 and 2.7 s TWT. The minimum tuning thickness and thickness of a single seismic loop is between *c.* 25 and 60 m. However, the true minimum sill thickness resolvable in the dataset is *c.* 15–50 m with associated metamorphosed clastic as side loops. Evaluations of sill thickness and dimensions, as well as sedimentary thickness, were derived from check-shot velocity function from the wells and are presented in the Results section. Other datasets used include published biostratigraphic tops from three wells within the seismic dataset (206/1a-2, 206/2a-1 and 208/19-1) (Ritchie 2011).

Seismic to well tie and horizon mapping

Released wells in the public domain and biostratigraphic tops from the stratigraphic range chart in the newly published Faroe–Shetland Basin regional study research report (2011) by British Geological Survey formed the basis of the seismic sequence dating using the well 206/2a-1 and 208/19-1. The well was tied to seismic data with check-shot data and had a fair–good tie at Eocene T97, T50 and Paleocene T22, T10 (Ebdon *et al.* 1995). There were misties within the T30 and T40 sequences. The key horizons picked were: the Top Cretaceous,

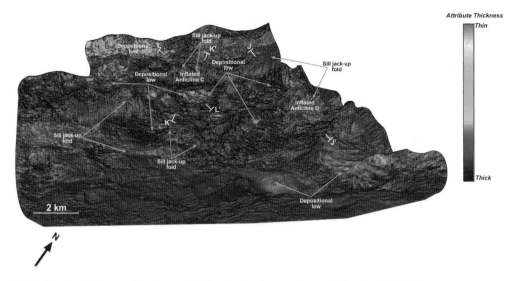

Fig. 6. Seismic thickness attributed to amplitude extraction between the top T10 and top T25 Paleocene sequence, showing depositional low (blue) filled with thick Paleocene Vaila sediments and depositional high (brown) with little presence or retention of Vaila sediments. Data provided by PGS.

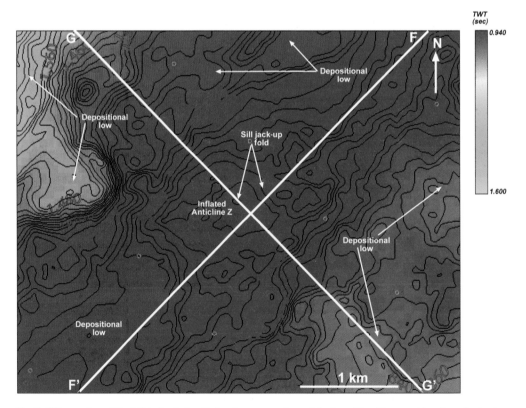

Fig. 7. 2D map view of the inflated anticline Z showing surrounding subtle depositional lows and location of seismic line G–G′ and F–F′. Data provided by PGS.

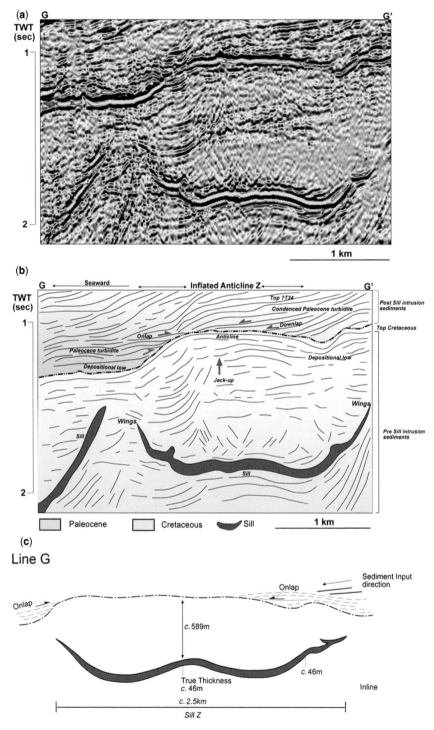

Fig. 8. (**a**) Un-interpreted seismic line G–G′ and (**b**) interpreted seismic line G–G′ showing inflated anticline Z palaeotopography, interpreted to be a result of sill emplacement, onlaps and downlaps in the Muckle Basin. (**c**) A simple sketch of key features and measurements of sill thickness and dimensions. Seismic line located in Figure 7. Vertical exaggeration is *c.* 1.5×. Data provided by PGS.

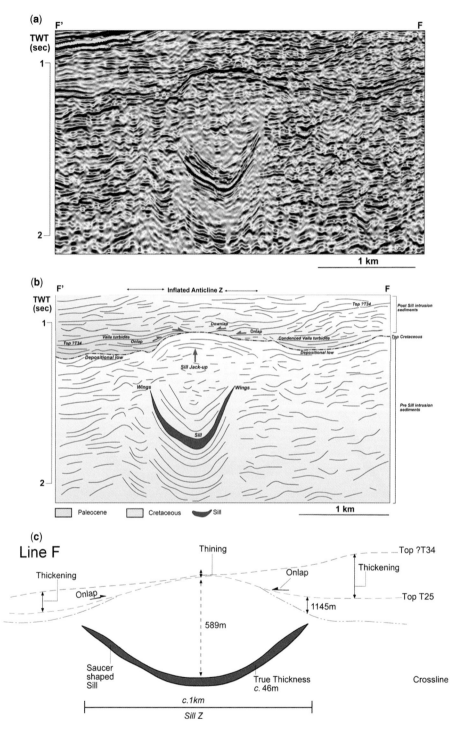

Fig. 9. (a) Un-interpreted seismic line F–F′ and (b) interpreted seismic line F–F′ showing inflated anticline Z and palaeotopography, interpreted to be a result of sill emplacement, onlaps and downlaps in the Muckle Basin. (c) A simple sketch of key features and measurements of sill thickness, dimensions and T21–T25 Vaila thickness within the accommodation space. Seismic line located in Figure 7. Vertical exaggeration is c. 1.5×. Data provided by PGS.

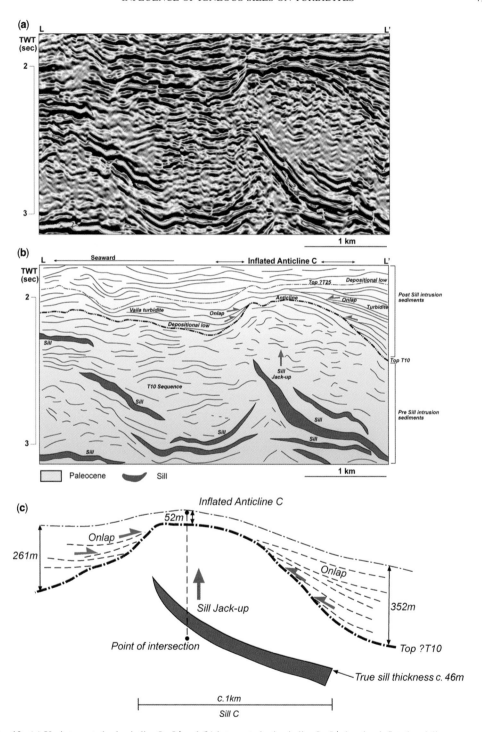

Fig. 10. (a) Un-interpreted seismic line L–L′ and (b) interpreted seismic line L–L′ showing inflated anticline C and palaeotopography interpreted to be a result of sill emplacement probably via faults, onlaps and downlaps in the Muckle Basin. (c) A simple sketch of key features and measurements of sill thickness, dimensions and T21–T25 Vaila thickness within the accommodation space. Seismic line located in Figure 6. Vertical exaggeration is c. 1.5×. Data provided by PGS.

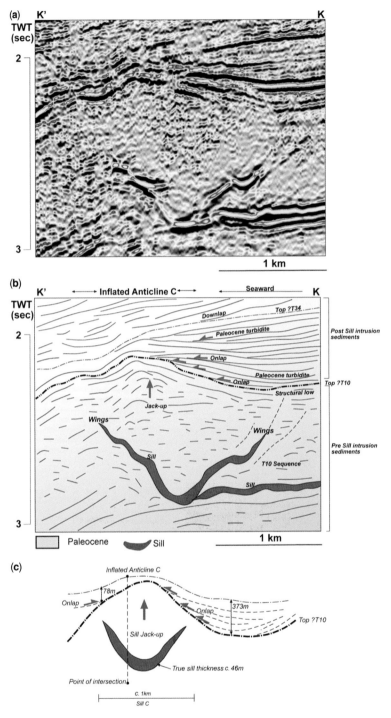

Fig. 11. (a) Un-interpreted seismic line K–K' and (b) interpreted seismic line K–K' showing inflated anticline C and palaeotopography, interpreted to be a result of sill emplacement probably via faults, onlaps and downlaps in the Muckle Basin. (c) A simple sketch of key features and measurements of sill thickness, V-shaped sill and its dimensions and T21–T25 Vaila thickness within the accommodation space. Seismic line located in Figure 6. Vertical exaggeration is c. 1.5×. Data provided by PGS.

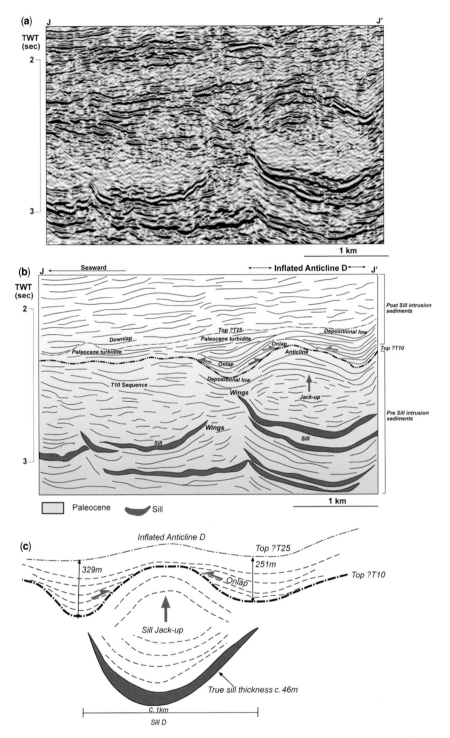

Fig. 12. (a) Un-interpreted seismic line J–J' and (b) interpreted seismic line J–J' showing inflated anticline D and palaeotopography interpreted to be a result of sill emplacement, onlaps and downlaps in the Muckle Basin. (c) A simple sketch of key features and measurements of sill thickness, dimensions and T21–T25 Vaila thickness within the accommodation space. Seismic line located in Figure 6. Vertical exaggeration is c. 1.5×. Data provided by PGS.

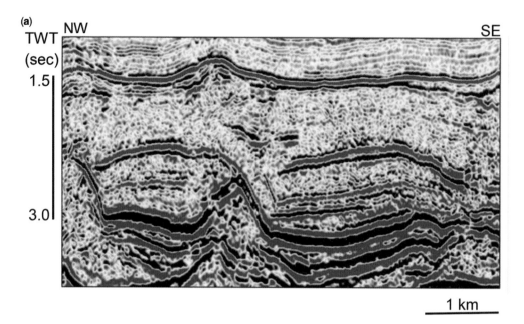

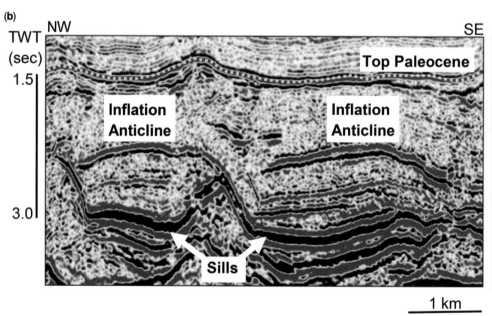

Fig. 13. Line L200 showing close-up of inflation anticline in the Flett Basin. See Figure 1a for location of Line L200. Vertical exaggeration is c. 3×. Data provided by PGS.

Near Top Sullom (T10), Middle Vaila (T25 and T28), Late Vaila (T34/T35), Kettla Tuff, Near Top Lamba (T36) and Balder Tuff (Ebdon et al. 1995) (Fig. 1b). Interpreted horizons and isochron maps enabled the identification of sequences that are key to constraining the timing of sill emplacement across the dataset (Fig. 4a, b).

Seismic geometries as a timing tool (onlap, offlap, truncation and thickness variation)

The seismic sequence stratigraphic method discussed by Trude et al. (2003) was applied to understand the relationship between the timing of emplacement, remobilization of unconsolidated sediments and possible sites where the Paleocene turbidites were deposited. This approach, illustrated in Figure 5, shows that pre-sill sedimentation has uniform thickness while post-sill sedimentation shows onlapping, offlapping and thinning of sedimentary sequences across the inflated anticline (Fig. 5a, b); if biostratigraphically constrained, this help dates the time of sill emplacement. After a good seismic to well tie has been achieved, individual continuous or fairly continuous reflectors are mapped across the dataset. In the section of interest, this becomes difficult in the post-sill intrusion sequence (Paleocene Vaila turbidite section) where reflectors are sometimes non-continuous and therefore difficult to follow. A similar method was used by Burbank & Verges (1994) in studying thrust tectonics.

Results

Paleocene palaeotopography: a response to sill/dyke intrusion

Figures 6–13 shows evidence of inflation anticlines formed as a result of the volcanic sill emplacement. The inflation anticlines become depositional highs and the synclines on either side become potential focus points for turbidite deposition (Fig. 6). These features occur all across the study area. Where interpreted, the sills are saucer-shaped with wings that branch out as dykes (Fig. 8). They forcefully 'jack up' overlying pre-sill Paleocene sediment (T10 Sullom Formation) (Figs 10–13) uplifting large areas, while areas unaffected by their activity accumulate thicker post-sill T21–T25 sediments (Figs 6–12). The T21–T25 sequence is observed to onlap the highs and forms a thin sedimentary sequence over the inflated anticline.

The sill that forms the inflated anticline Z (Figs 8 & 9) is c. 2.5 km × 1 km in size with a thickness of c. 46 m. There is apparent significant thickness change across the anticline (depositional high) with the T21–T25 sediment reaching a gross thickness of c. 145 m in the depositional lows and thinning to c. 0 across the highs.

Sill-inflated anticline C (Figs 10 & 11) shows c. 52 m thick T21–T25 sediment over the anticline with multiple onlaps on either side with T21–T25 thickness of 261–352 m. The sill is c. 1 km × 1 km in size and c. 46 m thick. An alternative interpretation for anticline C is that there was a pre-existing inverted normal fault that was the conduit for sill emplacement which subsequently enhanced the structure. Figure 12 shows sill-inflated anticline D with the sill c. 46 m thick and c. 1 × 1 km in size. The sedimentary thickness of the T21–T25 on either side of the anticline (depositional low) is 251–329 m thick, thinning to <100 m over the crest.

Timing of dyke and sill emplacement

The onlapping and downlapping seismic sequence on the Top Cretaceous and Top T10 in Figures 8–12 and Figure 14 is used to date the timing of sill intrusion as described in the Dataset and methodology section. At the onset of sill emplacement, and with continued injection of magmatic material, there is an instantaneous uplift of the overlying sediments with continuous growth of the inflated anticline. Further uplift coupled with sediment input will result in the formation of onlapping sequences and thinner sedimentary sequences over the inflated anticlines which are timelines for dating the onset of sill emplacement. In the area of study, the first onlapping sequence (T21-T25) on the Top T10 sequence is the onset of sill emplacement or sill arrival (Figs 8–12 & 14).

Interaction of Paleocene Vaila turbidites with the palaeotopography

Along the Flett and Muckle sub-basin margins within the area of study, the Paleocene Vaila turbidite deposition is controlled by the palaeotopography (Figs 6 & 15). Seismic lines through the study area (Figs 7–12 & 15b) show the preferential deposition of T21–?T34 Vaila turbidites (Ebdon et al. 1995) in subtle accommodation space created between inflated anticlines. Figure 15b shows high-amplitude seismic geometries interpreted to be T21–T25 Vaila turbidite with onlapping relationships to the neighbouring highs. Figure 15a, b shows an interpreted channel that is observed to have eroded much of the T10 sequence cutting through the inflated anticline A. Further down-dip, inflated anticline B is observed to divert the channel flow (Fig. 15a, b); much of the high-quality Vaila T21–T25 sands as seen in 208/19-1 are dumped in fault-controlled accommodation space (Fig. 15e) and other depositional lows created by sill emplacement. In Figure 14d, the base T34–T36 syn-post-sill 2 isochron shows a reduction in sediment thickness across the core area, suggesting sill emplacement and therefore inflated anticline and accommodation space creation (Fig. 14d).

Discussions

Age of igneous sills

Observations by Ritchie et al. (1999) show a wide variation in the potential ages of the sills across the basin (Figs 2 & 3). The ages in the

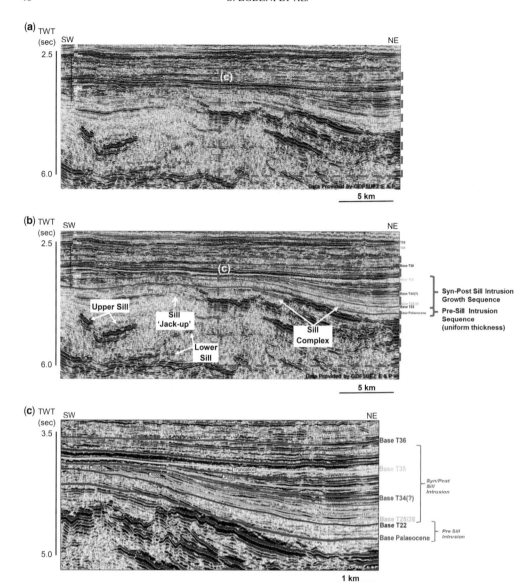

Fig. 14. Line L100 in the SW part of the 3D survey of the study area. (a) Un-interpreted seismic line L100. Vertical exaggeration is c. 2×. (b) Interpreted seismic line L100 showing time relationship of sill intrusion (sill jack-up) and palaeotopography formation. The lower sills jacks up Late Cretaceous and Paleocene sequences, resulting in Paleocene depositional high. Subsequent intrusion in the sill complex zone creates more palaeotopography. Vertical exaggeration is c. 2×. (c) Detailed seismic section showing onlapping on the base T22 sequence, suggesting the onset of sill intrusion (syn-sill intrusion) with subsequent uplift until the early and late T34 sequence with truncation. Pre-sill intrusion is prior to deposition of early T22 sediments. Vertical exaggeration is c. 1.5× (see Fig. 1a for location). (d) Isochron map in the SW part of the study area between Base Paleocene to T22 (pre-sill intrusion), Base T22 to Base T34 and Base T34 to Base T36 (post-sill intrusion) from the Flett Sub-basin area. Syn-post 1 sill isochron map shows thinning round the core area of sill intrusion. Syn-post 2 sill isochron map shows less thinning in the core area of intrusion. The sill influences depositional pattern. Data provided by PGS.

Faroe–Shetland Basin range from Cretaceous to Eocene (Ritchie & Hitchen 1987; Ritchie et al. 1999). Other published work by Trude et al. (2003), Hansen & Cartwright (2006), Thomson (2005), Cartwright et al. (2007), Goulty & Schofield (2008), Thomson & Schofield (2008)

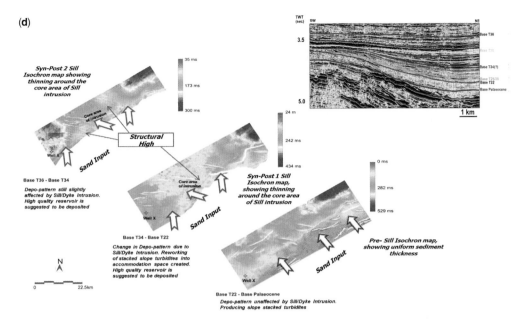

Fig. 14. *Continued.*

and Schofield *et al.* (2010, 2012*a*, *b*) suggest Eocene, Late Paleocene (T36) and Early–Middle Paleocene (T10–T22) ages. Based on geochemical analysis, Ritchie *et al.* (1999) proposed variable composition of the sills from acidic to basic. This could imply multiple magma sources of volcanic materials with different magmatic systems operating from Cretaceous to Eocene in the Faroe–Shetland Basin. The work of Trude *et al.* (2003) proposed a sequence stratigraphic age of 54.6–55 Ma (Eocene) which was comparable with the zircon age dating method of 53–55 Ma in the Flett Basin (Trude *et al.* 2003). This gives some level of confidence to the biostratigraphically tied seismic sequence stratigraphic method used as a relative dating tool. A closer look at the work of Ritchie *et al.* (1999) indicates that the intrusives are older than Eocene in wells 214/28-1, 214/27-1, 209/12-1 and 209/9-1 (Fig. 2). The oldest sill dated by Ritchie *et al.* (1999), found in well 209/9-1 in the Erlend Sub-basin, is Cretaceous in age. They also showed that the ages of the sills in 214/28-1 were Early Paleocene (Danian) and in wells 214/27-1, 208/15-1A, 207/1a-4 and 206/13-1 were Late Paleocene in age. In the area of this study, the interpreted sill age range from top T10 to top ?T25 (late Danian to Selandian) is within the observed age range published by Ritchie *et al.* (1999). The interpreted onlaps of the syn-post-sill sequences (Figs 6–12) represent periods of sill injection which invokes a multiple phase of sill emplacement in the study area. These observations are in line with the work of Svensen *et al.* (2010), who identified multiple-phase injections and different age dating in the upper and lower Utgard sill in the Norwegian 6607/5-2 well.

Sill emplacement and diapric mud models for palaeotopography

Smallwood & Maresh (2002) recognized that intrusions of igneous sill controlled the extent of the sand body. This was demonstrated by the 205/9-1 well and a seismic amplitude extraction that seems to show some deflection of the T35 sands by the already formed structural high caused by sill intrusion similar to that observed in Figure 15a. Smallwood & Maresh (2002) estimated that all the igneous sills were intruded *c*. 1–2 km below the seabed in the Flett Sub-basin. The results presented in this paper support the proposal from Smallwood & Maresh (2002) and Trude *et al.* (2003) that sill emplacement could be one of the mechanisms for palaeotopography formation (Figs 6–12). However, for this method to work it requires that the overburden sedimentary thickness is less than 1 km (Trude *et al.* 2003) or *c*. 1–2 km (Smallwood & Maresh 2002) to physically 'jack up' the overlying sedimentary sequences, forming inflated anticlines in the early burial phase. This mechanism is invoked in this study area and much of the

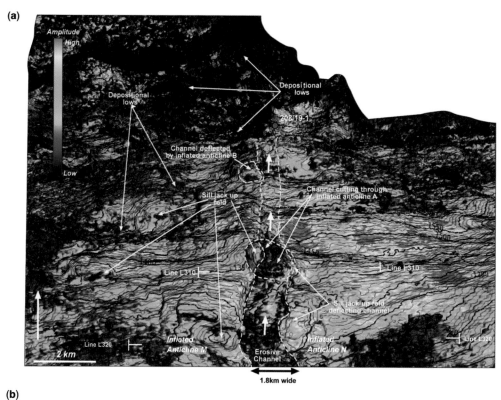

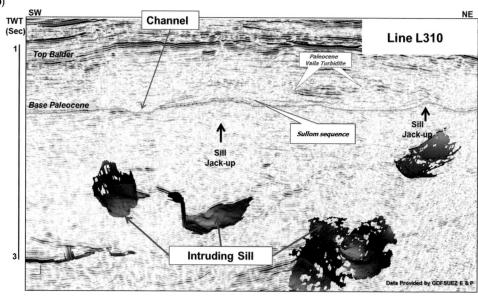

Fig. 15. (a) Seismic amplitude extraction map of top T10 showing channel cutting through inflated anticline A and deflected by anticline B. Also sill jack-up folds and depositional lows which accommodate thick T21–T25 sequence in the Muckle Basin flank. (b) Line L310 across the Muckle Basin flank showing palaeotopography control and the relationship of Paleocene Vaila and Sullom turbidites caused by sill intrusion. See Figure 15a for location of line L310. Vertical exaggeration is c. 2.5×.

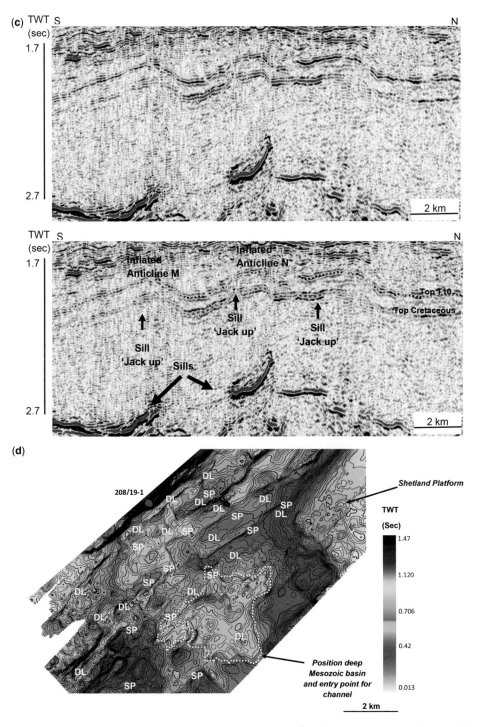

Fig. 15. (*Continued*) (**c**) Line L205 across the Muckle Basin slope to basin-floor showing palaeotopography control, the relationship of Paleocene Vaila and Sullom turbidites caused by sill intrusion (inflated anticline M and N). Paleocene turbidite is deposited in lows. See Figure 1a for location of line L320 Vertical exaggeration is *c*. ×2.5. (**d**) Time thickness map of Base Paleocene and Paleocene T36 Kettla Tuff (Fig. 4b) in the Muckle Basin, showing positions of inflation anticline sediment thinning in the Muckle Sub-basin. DL: depositional lows; SP: sill position.

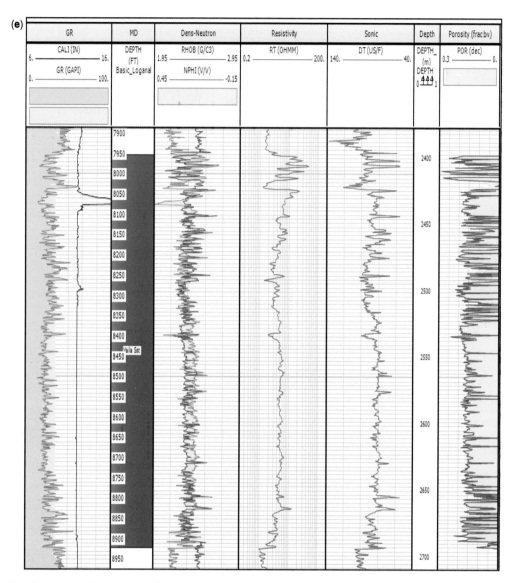

Fig. 15. (*Continued*) (**e**) Well 208/19-1 showing petrophysical interpretation of theT21–T25 turbidite along the Muckle Basin controlled by sill intrusion. Porosity 17–22%, permeability 1598–2674 mD, gross reservoir net reservoir 804 ft and net-to-gross 83%, average gas saturation 38% and net pay 20 ft. The reservoir quality of this sequence is excellent and is approximately what is predicted for the reservoirs deposited in the lows in the study area. Seismic data provided by PGS.

Flett Sub-basin as discussed by Smallwood & Maresh (2002).

Igneous sill intrusion is a complex process controlled by many factors which include magma supply, magma density, viscosity, stress regime, host rock lithology and deformation history (Smallwood & Maresh 2002). It is however possible to form palaeotopography when sills intrude previously reactivated normal faults (Figs 10 & 11), thereby forming inversion anticlines which are expected to be extensive and different geometrically from anticlines formed simply by sedimentary 'jack-up' by igneous sills. In map view, purely sill-induced jacked-up sedimentary sequences form small (2–3 km²) isolated circular to semi-circular anticlines (anticline A and B) underlain by the intruding sills (Fig. 15a, c). On the other hand, a combination of inverted faults and igneous sill

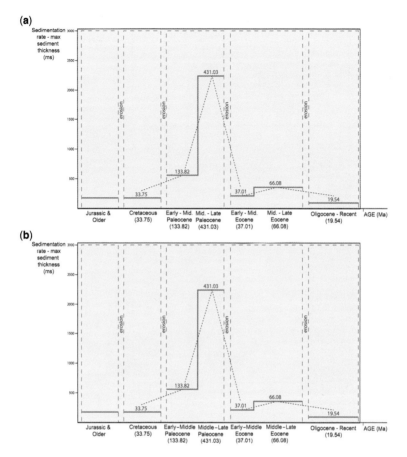

Fig. 16. Sedimentation rate by geological time taken from isochron maps in (**a**) the Muckle Basin and (**b**) the Flett Basin.

intrusions is expected to produce a more complex and elongated structure. It is worth mentioning that differential compaction can also result in apparent palaeotopography. In this case, onlaps, offlaps and downlaps (which are sequence stratigraphic timelines) are not expected because compaction does not occur at the time of deposition, but rather post deposition. The identification of onlaps and offlaps is therefore key to differentiating the method of palaeotopography formation.

The work of Lamers & Carmichael (1999) showed that palaeotopography could be formed by differential loading within the thick, continuously deposited Paleocene mudstones and Cretaceous shale, forming diapric shale ridges and mini-basins. This model for palaeotopographic formation requires a high sedimentation rate, the mudstones or shales to be mobile and unconsolidated and high sediment input along the whole of the basin. However, the sedimentation rate in the Flett and Muckle sub-basins (Fig. 16a, b) shows varying peaks during Cretaceous–Oligocene time. Although there seems to be a significant increase in sedimentation rate during the Paleocene in both areas it is not believed to be sufficient to trigger mud/shale diaprism without significant contraction. An exception is the compressional deformation of unconsolidated sequence during Pliocene–Recent time which resulted in the development of features such as the Pilot Whale Anticline and associated mud volcanoes/mud diaprism (Johnson et al. 2005) in the NW of the basin around the Lagavulin area. Figure 16a, b shows that the Oligocene–Recent sedimentation rate peaked at c. 19 ms thickness per million years compared to c. 6.5 ms thickness per million years during Paleocene–Eocene time, indicating quick sedimentation during the Oligocene–Recent period that could trigger mud volcanoes if contraction is experienced as seen in the NW of the Faroe–Shetland Basin.

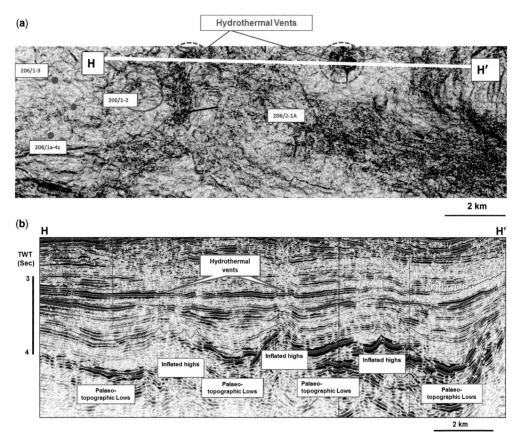

Fig. 17. (a) Dip max similarity of Top Kettla Tuff showing hydrothermal venting system caused by sill intrusion breaching Kettla Tuff top seal. (b) Cross-section (H–H′) across hydrothermal vents in the Flett Sub-basin. Vertical exaggeration is c. 1.5×. Data provided by PGS.

A look at basins that show classic mobile shale or mud diapirs, such as the Niger Delta, show that they require high/quick sedimentation and contraction possibly due to gravity tectonics of under-compacted thick sequences to form mud diapirs or mud ridges (Wiener et al. 2010). The combination of the deformational processes causes overpressure build-up within the thick Eocene shale/mudstone of the Akata Formation (Niger Delta Basin), making it mobile and thereby forming high-relief ridges, shale cored detachment fold, minibasins, diapirs and mud volcanoes (Wiener et al. 2010). Evidence for this mechanism or process has not been clearly observed on seismic data in the Paleocene section of the Faroe–Shetland Basin, nor has it been described by any other researchers in this area. The evidence as presented in this paper supports the sill emplacement model (Figs 6–12 & 15b), as it does not require high sedimentation rates to form palaeotopography as described by Smallwood & Maresh (2002) and Trude et al. (2003).

Implication for hydrocarbon exploration

The timing of sill emplacement with later formation of depositional highs and lows is important for Paleocene Vaila reservoir deposition (Fig. 6). Depositional lows or accommodation space tectonically created by igneous activities (Fig. 15d) become focal points for deposition of T21–T25 Paleocene Vaila reservoir that pinches out against the highs, forming significant pinch-out traps (Figs 8 & 12). Stratigraphic pinch-out trapping mechanism against highs are known as one of the most successful in the basin within the Paleocene play, and one of the most reliable traps in the basin (Loizou et al. 2006). However, intense sill emplacement could be of significant exploration risk to top seal breach, due to hydrothermal vents formed by sill

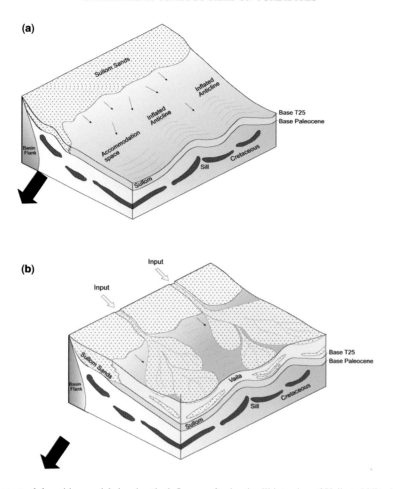

Fig. 18. Conceptual deposition model showing the influence of volcanic sill intrusion of Vaila turbidite deposition in the Flett and Muckle Basin. (**a**) Stage 1: showing Sullom palaeotopography caused by sill intrusion. There is erosion of Early Paleocene shallow marine sands further into the basin. (**b**) Stage 2: creation of depositional lows with palaeotopographic control of early Vaila turbidites. There is reworking and mobilization of sand deposits on the basin flanks. (**c**) Stage 3: more influence of palaeotopography on Vaila turbidite deposition resulting in stacked turbidites in depositional lows which fills and spills into other lows. (**d**) Stage 4: depositional lows become completely filled with subsequent turbidites.

intrusion (Fig. 17a, b) as discussed by Davies *et al.* (2002), Lee *et al.* (2006) and Cartwright *et al.* (2007). This may result in re-migration of trapped hydrocarbons from deeper intervals into younger Eocene traps through the breached Kettla Tuff top seal (Fig. 17b). The palaeotopographic concept is becoming attractive to test future exploration targets in the basin.

Depositional model for Paleocene post-sill intrusion

Figure 18a–d shows a summary depositional model for the area affected by sill emplacement. In Figure 18a, deposition of the T10 Sullom Formation is later followed by the igneous sill emplacement. Post-emplacement, the re-mobilized T10 Sullom sandstones on the shelf edge are transported into palaeotopographic lows downslope (Fig. 18b). This re-mobilization of Paleocene sediment is observed to be transported by the erosive channel downslope in the Muckle study area (Fig. 15a). With later sedimentation (Fig. 18b), there is a preferential accumulation of Middle Paleocene Vaila sandstone (T21–T25) in the lows pinching out or thinning against the highs (Fig 7–12). As sedimentation and burial continues during Middle–Late Paleocene time the effect of differential compaction comes into play (Fig. 18c), impacting the deposition of the stacked Paleocene Vaila turbidite sequence

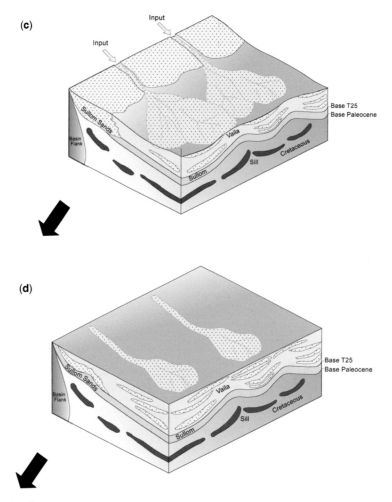

Fig. 18. *Continued.*

which fills and spills into other accommodation space (Figs 7–12). This continues until the palaeotopography formed by sill emplacement (Fig. 18d) is filled and can no longer control the deposition of the turbidites.

Conclusions

The results presented in this paper support the following conclusions in this study area.

(1) Sill/dyke intrusion is one of several causes of Danian–Selandian palaeotopography, and provides alternative explanation to the 'diapric mud model' discussed by Lamers & Carmichael (1999). It is supported by the observations of Smallwood & Maresh (2002) and Trude et al. (2003). The semi-circular to circular intrusion geometry on map view differentiates them from other structures potentially formed by other process.

(2) In the study area, the timing of sill/dyke emplacement is from c. T10 to ?T25 (from growth strata, onlapping and downlapping sequence mapping) in the study area. The result is within the age range of published data and the method supported by the work of Smallwood & Maresh (2002) and Trude et al. (2003).

(3) The emplacement of sills and dykes has a strong influence on post-T10 reservoir deposition and the proposed sill/dyke emplacement model supports the depositional high, lows and inversion concept of Paleocene Sullom and early Vaila turbidites.

(4) This approach is a good tool for identifying Paleocene Vaila turbidites and finding potential prospectivity in the shape of stratigraphic traps further into the basin, and could lead to significant Paleocene discoveries.
(5) Significant volcanic intrusion could breach the Kettla regional top seal and other intra-seals within the Vaila, especially at depositional highs forming hydrothermal vents and gas chimneys. This is one of the risks of trap breaching and must be considered in risk modelling.

The authors wish to thank GDFSUEZ E&P UK Ltd for their support and sponsorship during this project, specifically A. Spencer and C. Johns, together with partners DONG and Dana Petroleum, JX Nippon Exploration and Production UK, Idemitsu Exploration and Production UK and DONG Energy. We are grateful to PGS for the seismic data. Thanks are also due to I. Hubbard and T. Cousins for drafting most of the images included in this paper. The authors are grateful to J. Smallwood for his insightful comments and suggestions during the revision of this paper and the reviewers for their comments that helped to structure this paper.

References

BRADLEY, J. 1965. Intrusion of major dolerite sills. *Transactions of the Royal Society of New Zealand*, **3**, 27–55.

BRODIE, J. & WHITE, N. 1994. Sedimentary basin inversion caused by igneous underplating: Northwest European Continental shelf. *Geology*, **22**, 147–150.

BURBANK, D. W. & VERGES, J. 1994. Reconstruction of topography and related depositional system during active thrusting. *Journal of Geophysical Research*, **99**, 20 281–20 297.

BURGER, C. A. J., HODGSON, F. D. I. & VAN DER LINDE, P. J. 1981. *Hidroliese eienskappe van akwifere in die Suid Vrystaat. Die ontwikkeling en evaluering van tegnieke vir die bepaling van die ontginninspotensiaal van grondwaterbronne in die Suid-Vrystraat en in Noord-Kaapland*. Institute of Groundwater Studies, University of the Orange Free State, Bloemfontein, South Africa, **2**.

CARTWRIGHT, J., HUUSE, M. & APLIN, A. 2007. Seal bypass systems. *AAPG Bulletin*, **91**, 1141–1166.

CHEVALIER, L. & WOODFORD, A. 1999. Morph-tectonics and mechanism of emplacement of the dolerite rings and sills of the western Karoo, South Africa. *South African Journal of Geology*, **102**, 43–54.

DAVIES, R., BELL, B. R., CARTWRIGHT, J. A. & SHOULDERS, S. 2002. Three-dimensional seismic imaging of Palaeogene dike-fed submarine volcanoes from the northeast Atlantic margin. *Geology*, **30**, 223–226.

DEAN, K., MCLACHLAN, K. & CHAMBERS, A. 1999. Rifting and the development of the Faeroe-Shetland Basin. *In*: FLEET, A. J. & BOLDY, S. A. R. (eds) *Petroleum Geology of Northwest Europe: Proceedings of the 5th Conference*. Geological Society, London, 533–544.

DORE, A. G., LUNDIN, E. R., BIRKELAND, O., ELIASSEN, P. E. & JENSEN, L. N. 1997. The NE Atlantic Margin: implications of Late Mesozoic and Cenozoic events for hydrocarbon prospectivity. *Petroleum Geoscience*, **3**, 117–131.

DU TOIT, A. L. 1920. The Karoo dolerite of South Africa: a study of hypabyssal injection. *Transactions of the Geological Society of South Africa*, **23**, 1–42.

EBDON, C. C., GRANGER, P. J., JOHNSON, H. D. & EVANS, A. M. 1995. Early Tertiary evolution and sequence stratigraphy of the Faroe-Shetland Basin: implications for hydrocarbon prospectivity. *In*: SCRUTTON, R. A., STOKER, M. S., SHIMMIELD, G. B. & TUDHOPE, A. W. (eds) *The Tectonics, Sedimentation and Palaeoceanography of the North Atlantic Region*. Geological Society, London, Special Publications, **90**, 51–69.

FRANCIS, E. H. 1982. Magma and sediment – I. Emplacement mechanism of late Carboniferous tholeiite sills in northern Britain. *Journal of the Geological Society, London*, **139**, 1–20.

GOULTY, N. R. & SCHOFIELD, N. 2008. Implications of simple flexure theory for the formation of saucer-shaped sills. *Journal of Structural Geology*, **30**, 812–817.

HANSEN, D. M. & CARTWRIGHT, J. A. 2006. The three dimensional geology and growth of forced fold above saucer-shaped igneous sill. *Journal of Structural Geology*, **28**, 1520–1535.

HANSEN, D. M., CARTWRIGHT, J. A. & THOMAS, D. 2003. 3D seismic analysis of the geometry of igneous sills and sill intersection relationships. *In*: DAVIES, R. ET AL. (eds) *3D Seismic Technology: Application to the Exploration of Sedimentary Basins*. Geological Society, London, Memoirs, **29**, 199–208.

JOHNSON, H., RITCHIE, J. D., HITCHEN, K., MCINROY, D. B. & KIMBELL, G. S. 2005. Aspects of Cenozoic deformational history of the northeast Faeroe-Shetland Basin and the Wyville-Thompson Ridge and Hatton Bank area. *In*: DORE, A. G. & VINING, B. A. (eds) *Petroleum Geology Conference*. Geological Society, London, 993–1007.

JOLLEY, D. W. 1997. Paleosurface palynofloras of the Skye lava field and the age of the British Tertiary volcanic province. *In*: WIDDOWSON, M. (ed.) *Palaeosurfaces: Recognition, Reconstruction and Palaeoenvironmental Interpretation*. Geological Society, London, Special Publications, **120**, 67–94.

KNOTT, S. D., BURCHELL, M. T., JOLLEY, E. J. & FRASER, A. J. 1993. Mesozoic to Cenozoic plate reconstructions of the North Atlantic and hydrocarbon plays of the Atlantic Margins. *In*: PARKER, J. R. (ed.) *Petroleum Geology of Northwest Europe: Proceedings of the 4th Conference*. Geological Society, London, 953–974.

LAMERS, E. & CARMICHAEL, S. M. M. 1999. The Paleocene deep-water sandstone plays West of Shetland. *In*: FLEET, A. J. & BOLDY, S. A. R. (eds) *Petroleum Geology of Northwest Europe: Proceedings of the 5th Conference*. Geological Society, London, 645–659.

LEE, G. H., KWON, Y. I., YOON, C. S., KIM, H. J. & YOO, H. S. 2006. Igenous complexes in the eastern Northern South Yellow Sea Basin and implication for hydrocarbon system. *Marine & Petroleum Geology*, **23**, 631–645.

LOIZOU, N., ANDREWS, I. J., STOKER, S. J. & CAMERON, D. 2006. West of Shetland revisited – the search for stratigraphic traps. *In*: ALLEN, M., GEOFFREY, G., MORGAN, R. & WALKER, I. (eds) *The Deliberate Search for the Stratigraphic Trap: Where Are We Now?* Geological Society, London, Special Publications, **254**, 225–245.

MUSSETT, A. E., DAGLEY, P. & SKELHORN, R. R. 1988. Time and duration of activity in the British Tertiary Igneous Province. *In*: MORTON, A. C. & PARSON, L. M. (eds) *Early Tertiary Volcanism and the Opening of the NE Atlantic*. Geological Society, London, Special Publications, **39**, 337–348.

NAYLOR, P. H., BELL, B. R., JOLLEY, D. W., DURNALL, P. & FREDSTED, R. 1999. Palaeogene magmatism in the Faeroe-Shetland Basin: influences on uplift history and sedimentation. *In*: FLEET, A. J. & BOLDY, S. A. R. (eds) *Petroleum Geology of Northwest Europe: Proceedings of the 5th Conference*. Geological Society, London, 545–559.

PEARSON, D. G., EMELEUS, C. H. & KELLEY, S. P. 1996. Precise $^{40}Ar/^{39}Ar$ age for the initiation of Palaeogene volcanism in the Inner Hebrides and its regional significance. *Journal of the Geological Society, London*, **153**, 815–818.

PLANKE, S., ALVESTAD, E. & ELDHOLM, O. 1999. *Seismic Characteristics of Basaltic Extrusive and Intrusive Rocks*. The Leading Edge, March 1999.

PLANKE, S., SYMONDS, P. A., ALVESTAD, E. & SLOGSEID, J. 2000. Seismic volcanostratigraphy of large-volume basaltic extrusive complexes on rifted margins. *Journal of Geophysical Research-Solid Earth*, **105**, 19 335–19 351.

PLANKE, S., RASMUSSEN, T., REY, S. S. & MYKLEBUST, R. 2005. Seismic characteristics and distribution of volcanic intrusions and hydrothermal vent complexes in the Vøring and Møre basins. *In*: DORE, A. G. & VINING, B. A. (eds) *North-West Europe and Global Perspectives: Proceedings of the 6th Petroleum Geology Conference*. Geological Society, London, 833–844.

POLLARD, D. D. 1973. Derivation and evaluation of a mechanical model for sheet intrusions. *Tectonophysics*, **19**, 233–269.

POLLARD, D. D. & JOHNSON, A. M. 1973. Mechanics of growth of some laccolith intrusions in the Henry Mountains, Utah, Part II. *Tectonophysics*, **18**, 311–354.

RITCHIE, J. D. 2011. *Geology of the Faroe-Shetland Basin and Adjacent Areas*. GBS research report RR/11/01. Published by BGS.

RITCHIE, J. D. & HITCHEN, K. 1987. Geological review of the West Shetland area. *In*: BROOKS, J. & GLENNIE, K. W. (eds) *Petroleum Geology of North West Europe*. Graham and Trotman, London, 737–749.

RITCHIE, J. D. & HITCHEN, K. 1996. Early Palaeogene offshore igneous activity to the northwest of the UK and its relationship to the North Atlantic igneous province. *In*: KNOX, R. B., CORFIELD, M. & DUNNAY, R. E. (eds) *Correlation of the Early Palaeogene in Northwest Europe*. Geological Society, London, Special Publications, **101**, 63–78.

RITCHIE, J. D., GATLIFF, R. W. & RICHARDS, P. C. 1999. Early Tertiary magmatism in the offshore NW UK margin and surrounds. *In*: FLEET, A. J. & BOLDY, S. A. R. (eds) *Petroleum Geology of Northwest Europe: Proceedings of the 5th Conference*. Geological Society, London, 573–584.

SCHOFIELD, N., STEVENSON, C. & RESTON, T. J. 2010. Magma fingers and importance of host rock fluidization in sill emplacement. *Geology*, **38**, 63–66, http://dx.doi.org/10.1130/G30142.1

SCHOFIELD, N., HEATON, L., HOLFORD, S. P., ARCHER, S. G., JACKSON, C. A.-L. & JOLLEY, D. 2012a. Seismic imaging of 'broken bridges': linking seismic to outcrop-scale investigations of intrusive magma lobes. *Journal of the Geological Society, London*, **169**, 421–426.

SCHOFIELD, N., BROWN, D. J., MAGEE, C. & STEVENSON, C. T. 2012b. Sill morphology and comparison of brittle and non-brittle emplacement mechanisms, Geological Society of London. *Journal of the Geological Society, London*, **169**, 127–141.

SMALLWOOD, J. R. 2008. Comment on "Determining magma flow in sills, dykes and laccoliths and their implications for sill emplacement mechanisms by K. Thomson [Bulletin of Volcanology 70: 183–210.]". *Bulletin of Volcanology*, **70**(9), 1139–1142, http://dx.doi.org/10.1007/s00445-008-0203-4

SMALLWOOD, J. R. & HARDING, A. 2008. New seismic imaging methods, dating, intrusion style and effects of sills: a drilled example from the Faroe-Shetland Basin. *In: Faroe Islands Exploration Conference: Proceedings of the 2nd Conference*. Annales Societatis Scientiarum Færoensis, Torshavn, Supplementum, **48**, 96–116.

SMALLWOOD, J. R. & MARESH, J. 2002. The properties, morphology and distribution of igneous sills: Modelling, borehole data and 3D seismic from the Faroe-Shetland area. *In*: JOLLEY, D. W. & BELL, B. R. (eds) *The North Atlantic Igneous Province: Stratigraphy, Tectonic, Volcanic, and Magmatic Processes*. Geological Society, London, Special Publications, **197**, 271–306.

STOKER, M. S., HITCHEN, K. & GRAHAM, C. C. 1993. *The Geology of the Hebrides and West Shetland Shelves, and Adjacent Deep Water Areas*. United Kingdom Offshore Regional Report, British Geological Survey, London.

SVENSEN, H., PLANKE, S. & CORFU, F. 2010. Zircon dating ties NE Atlantic sill emplacement to initial Eocene global warming. *Journal of the Geological Society, London*, **167**, 433–436.

THOMSON, K. 2005. Extrusive and intrusive magmatism in the North Rockall Trough. *In*: DORE, A. G. & VINING, B. A. (eds) *Petroleum Geology: North-West Europe and Global Perspectives – Proceedings of the 6th Petroleum Geology Conference*. Geological Society, London, 1621–1630.

THOMSON, K. & SCHOFIELD, N. 2008. Lithological and structural controls on the emplacement and morphology of sills in sedimentary basins, Structure and emplacement of high-level magmatic systems. *In*: THOMSON, K. & PETFORD, N. (eds) *Structure and*

Emplacement of High-Level Magmatic Systems. Geological Society, London, Special Publications, **302**, 31–44.

TRUDE, J. 2004. Geometry and growth of sill complexes: insights using 3D seismic data from the North Rockall Trough. *Bulletin of Volcanology*, **66**, 364–375.

TRUDE, J., CARTWRIGHT, J., DAVIES, R. J. & SMALLWOOD, J. 2003. New technique for dating igneous sills. *Geology*, **31**, 813–816.

WHITE, R. S. & MCKENZIE, D. P. 1989. Magmatism at rift zones: the generation of volcanic continental margins and flood basalts. *Journal of Geophysical Research*, **94**, 7865–7729.

WIENER, R. W., MANN, M. G., ANGELICH, M. T. & MOLYNEUX, J. B. 2010. Mobile shale in the Niger Delta. Characteristics, structure and evolution. *In*: WOOD, L. J. (ed.) *Shale Tectonics*. AAPG, Memoir, **93**, 145–161.

Success in exploring for reliable, robust Paleocene traps west of Shetland

N. LOIZOU

Department of Energy and Climate Change, 3 Whitehall Place, London, SW1A 2AW, UK
(e-mail: nick.loizou@decc.gsi.gov.uk)

Abstract: This paper incorporates the results of recent exploration wells to provide a re-evaluation and re-validation of earlier publications by the author.

Following the discovery of the Foinaven and Schiehallion fields in the early 1990s, exploration success for Paleocene targets outside the Quadrant 204 area of the Faroe–Shetland Basin was initially rather limited. However, renewed interest in the last 10 years has seen 12 Paleocene exploration wells drilled, of which 9 have encountered notable hydrocarbons.

Since 1972, 151 exploration wells have been drilled west of Shetland with 79 wells (52%) specifically positioned on Paleocene prospects which resulted in 23 discoveries. Analysis of the 56 failed Paleocene wells shows that around 80% were drilled on either a poor or invalid trap, with the remaining 20% failing mainly due to either lack of or poor reservoir or poor top or lateral seal. Intriguingly, only 4 exploration wells have been exclusively positioned on Paleocene 4-way dip (or structural closures) structures, with all of these encountering hydrocarbons.

Fifty-six Paleocene prospects contained a stratigraphic component with 15 notable successes, all of which are within the Upper Paleocene Vaila sequence. Ten key discoveries are located in the Judd Sub-basin or adjacent Westray High (Foinaven, SE Foinaven, SW Foinaven, Schiehallion, Loyal, Alligin, Cuillin, Arkle, Amos and Tornado). A further four are located in the Flett Sub-basin (Laggan, Tormore, Torridon, Laxford) and one is in the Foula Sub-basin (Glenlivet).

Forty-six of the wells were positioned on an amplitude or amplitude-variation-with-offset (AVO) anomaly, of which 16 encountered notable hydrocarbons. Following post-mortem studies, the majority of the remaining 30 failed wells could be shown to have drilled poorly defined amplitude anomalies (various lithologies including igneous), AVO artefacts or spurious direct hydrocarbon indicator (DHIs) (which include multiples). Wells 204/17-1 and 204/18-1 are good examples of poorly interpreted AVO responses, in which high amplitudes are mainly present on the nears (low offset data) and there is significantly decreased amplitude on the highs (far offset data). Moreover, post-drill AVO analyses categorically show that both wells were indeed drilled on Class I AVO anomalies.

Not surprisingly, in recent years all of the wells positioned on sound AVO anomalies have been successful in encountering hydrocarbons. Moreover, these wells show both a clear increase in amplitude with offset and a conformance with structure. Undoubtedly, the recent triumphs firmly demonstrate there is potential for maintaining a high success rate in future Paleocene exploration west of Shetland. Equally, the evaluation and use of proven examples of successful traps such as Foinaven, Schiehallion, Laggan, Tormore, Glenlivet and other analogues can add value to future exploration programmes.

History of West of Shetland Paleocene exploration success

Since 1972, 79 exploration wells have targeted the West of Shetland (WoS) Paleocene succession (Fig. 1), resulting in 23 notable discoveries. Many of these can also be classed as significant discoveries, that is, hydrocarbon accumulations that if tested would flow to surface (Table 1). This does not necessarily indicate the commercial potential of the discovery. The overall technical success rate of 29% is skewed by an improved success rate achieved during the last 10 years through a mini exploration renaissance (Fig. 2). Although hydrocarbons have been found from fractured Lewisian basement to Eocene strata, the Paleocene is by far the most prolific. Not surprisingly, hydrocarbons have been encountered within almost every Paleocene interval (Fig. 3). The majority of the Paleocene discoveries are located in three main geological provinces: the Judd Sub-basin (and adjacent Westray Ridge), the Flett Sub-basin and overlying the Corona Ridge (Fig. 1). For the latter province there have been three key discoveries: Cambo, Rosebank and Bunnehaven.

During the 1980s there were just four discoveries in the Lower–Middle Paleocene Vaila Sandstone play: 206/2-1A, Laxford 214/30-1, Torridon 214/27-1 and Laggan 206/1-2A. The 1980s discoveries are all located in the Flett Sub-basin, and their focus was on Paleocene structural/stratigraphic traps identified on 2D seismic data. Only the

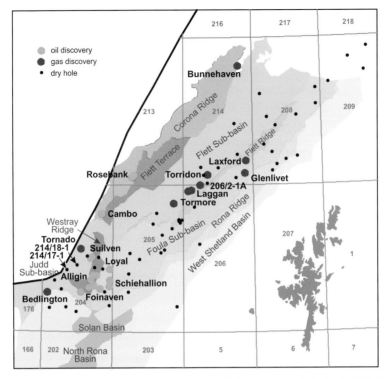

Fig. 1. Locations of Paleocene exploration wells and key structural elements west of Shetland. Unlabelled oil discoveries in the SW include Arkle, Cuillin, SE Foinaven, SW Foinaven and Amos.

Table 1. *Summary of Paleocene discoveries west of Shetland*

Well	Year	Operator	Status	Discovery/Field	Basin/Structure	Reservoir
206/2-1A	1980	Shell	Gas discovery	–	Flett Sub-basin	T34
214/30-1	1984	BG	Gas discovery	Laxford	Flett Sub-basin	T35
214/27-1	1985	Gulf	Gas discovery	Torridon	Flett Sub-basin	T31/T32 & T20
206/1-2A	1986	Shell	Gas discovery	Laggan	Flett Sub-basin	T35
204/24-1A	1990	BP	Oil & Gas discovery	Foinaven	Judd Sub-basin	T31–T34
204/19-2	1991	BP	Oil discovery	Arkle	Judd Sub-basin	T22–T25
204/20-1	1993	BP	Oil & Gas discovery	Schiehallion	Judd Sub-basin	T35
204/19-3A	1994	BP	Oil discovery	Cuillin	Judd Sub-basin	T36
204/20-3	1994	BP	Oil discovery	Loyal	Judd Sub-basin	T35
204/19-6	1995	BP	Oil & Gas discovery	Alligin	Judd Sub-basin	T34
204/25b-5	1995	BP	Oil discovery	SE Foinaven	Judd Sub-basin	T34–T36
204/19-8Z	1996	BP	Gas discovery	Suilven	Judd Sub-basin	T35
214/9-1	2000	Mobil	Gas discovery	Bunnehaven	Corona Ridge	T40
204/10-1	2002	Hess	Oil discovery	Cambo	Corona Ridge	T45
213/27-1Z	2004	Chevron	Oil discovery	Rosebank	Corona Ridge	T45
204/21-1	2004	BG	Gas discovery	Bedlington	Judd Sub-basin	T35
205/5-1	2007	Total	Gas discovery	Tormore	Flett Sub-basin	T35
204/23-2	2007	BP	Oil discovery	SW Foinaven	Judd Sub-basin	T32
204/25a-8Z	2008	Hess	Oil discovery	Amos	Judd Sub-basin	T31 & T34
213/27-4	2009	Chevron	Oil discovery	Rosebank North	Corona Ridge	T45
214/30a-2	2009	DONG	Gas discovery	Glenlivet	Foula Sub-basin	T35
204/13-1	2009	OMV	Gas discovery	Tornado	Judd Sub-basin	T38

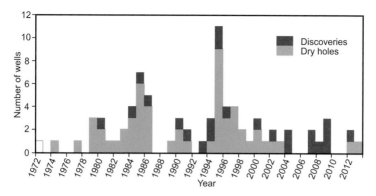

Fig. 2. Histogram of discoveries and dry holes drilled per year west of Shetland.

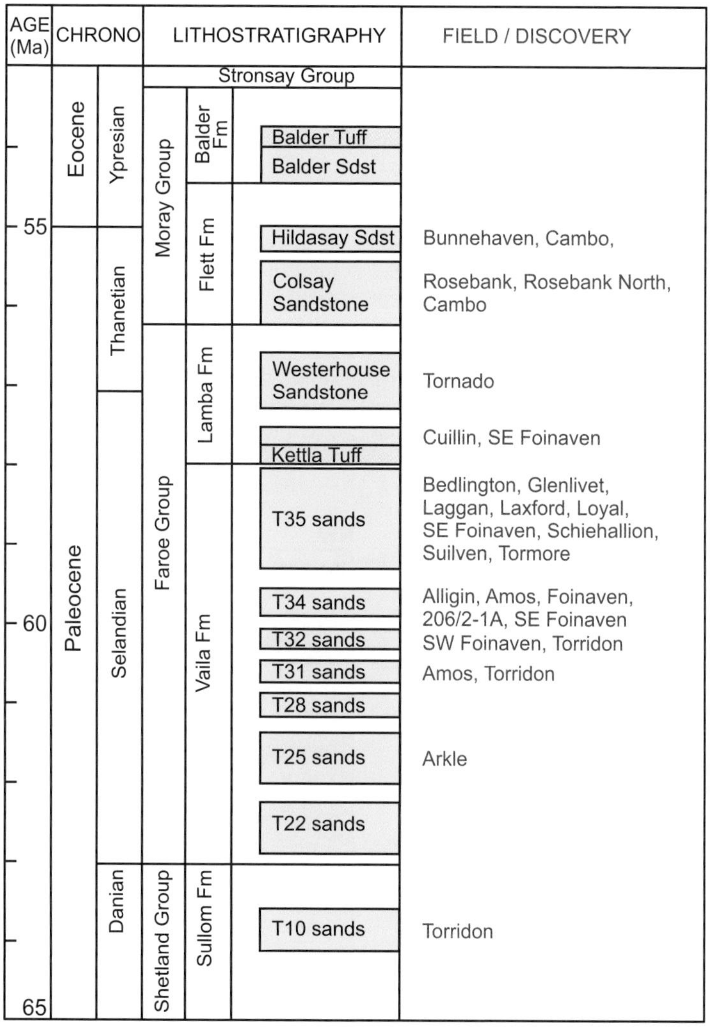

AGE (Ma)	CHRONO		LITHOSTRATIGRAPHY				FIELD / DISCOVERY
	Eocene	Ypresian	Moray Group	Balder Fm	Balder Tuff	Stronsay Group	
55					Balder Sdst		
		Thanetian		Flett Fm	Hildasay Sdst		Bunnehaven, Cambo,
					Colsay Sandstone		Rosebank, Rosebank North, Cambo
	Paleocene			Lamba Fm	Westerhouse Sandstone		Tornado
			Faroe Group		Kettla Tuff		Cuillin, SE Foinaven
		Selandian		Vaila Fm	T35 sands		Bedlington, Glenlivet, Laggan, Laxford, Loyal, SE Foinaven, Schiehallion, Suilven, Tormore
60					T34 sands		Alligin, Amos, Foinaven, 206/2-1A, SE Foinaven
					T32 sands		SW Foinaven, Torridon
					T31 sands		Amos, Torridon
					T28 sands		
					T25 sands		Arkle
					T22 sands		
		Danian	Shetland Group	Sullom Fm	T10 sands		Torridon
65							

Fig. 3. Generalized Paleocene stratigraphy and age of principal reservoirs in all WoS oil and gas fields and discoveries.

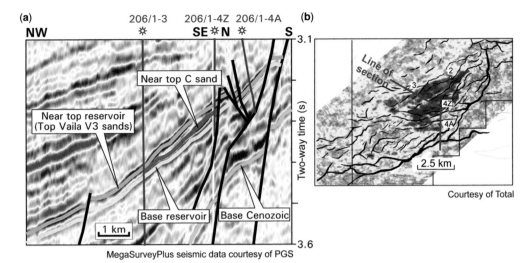

Fig. 4. (a) Example seismic profile across the Laggan Field (interpretation after Gordon *et al.* 2010) and (b) map of depth to top reservoir in the Laggan Field.

Laggan discovery can be counted as a 'significant' discovery from the 1980s phase of exploration. The other successes (204/30-1, 206/02-1A and 214/27-1) are smaller, currently sub-economic, gas pools (<150 billion cubic feet pool size). Laggan (Fig. 4) is a classic combination trap of stratigraphic with updip closure against bounding faults with a T35 reservoir comprising sand-rich turbidite channelized lobes (Gordon *et al.* 2010). Reservoir properties are good (permeability range is 30–300 mD) due to the presence of ubiquitous chlorite grain coating and pre-sorting on the shelf (Sullivan *et al.* 1999). Exploration interest waned through the late 1980s with a notable lack of success. By the end of the 1980s, none of the WoS finds had been successfully brought forward to development sanction.

The 1990 discovery of Foinaven in the Judd Sub-basin, reservoired in Vaila T32–T34 sandstones, marked an important turning point in WoS exploration in proving the presence of oil in what was previously considered to be a gas play. The

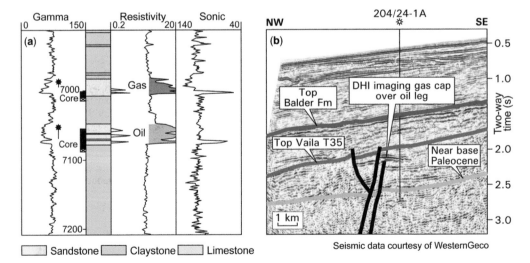

Fig. 5. (a) Log responses of the principal reservoirs in Foinaven discovery well 204/24-1A (depths are in feet below Kelly Brushing). (b) The well was positioned to test the DHI imaged on the illustrated 1978 2D seismic line.

Foinaven discovery well 204/24-1A was primarily positioned on a Cretaceous prospect with a secondary Paleocene target, and it was located on reprocessed 1978 WesternGeco 2D seismic data (Fig. 5). The well encountered 11 m of gas- and oil-bearing Paleocene sandstones but, mainly because of the sparse coverage of 2D seismic data, the potential of the discovery was not fully recognized at the time. In 1992 the Foinaven discovery was successfully appraised, thus confirming the potential of the Vaila Formation play. Following the discovery of Foinaven, the exploration focus was mainly within the Paleocene Vaila Formation, specifically from the T31–T35 sandstones which are the main producing reservoirs in the Foinaven area (Lamers & Carmichael 1999). Nearby discoveries made within the Vaila Formation include Schiehallion in 1993 and Loyal in 1994 (Cooper et al. 1999). Toward the close of the 1990s there was another discovery Suilven, which shares a similar structural character to its southerly Judd Sub-basin neighbours. Both Foinaven and Schiehallion fields have stratigraphic and structural elements and are associated with distinct amplitude anomalies, which demonstrate a Class III AVO effect (a brightening with offset) in response to their fluid content.

Subsequent to the discoveries in the Foinaven area, many companies believed that there was also strong potential for prospectivity in the Flett Sub-basin (Smallwood & Kirk 2005). In the 16th UK Offshore Licence Round 22 firm wells were pledged on Paleocene prospects, which together were associated with huge P50 reserves totalling 8 billion barrels of oil equivalent. Eleven 16th Round Paleocene exploration wells were drilled, all of which were disappointingly dry holes. A 12th Paleocene well 206/1-3 targeted an amplitude shut-off down dip in what is now the Laggan Field (it has since been re-categorized from an exploration to an appraisal well). This well encountered a gas–water contact (GWC) and confirmed that there was no oil rim in Laggan. Parr et al. (1999) wrote that 'the search for mountains of oil was a gross exaggeration built on a poor understanding for the petroleum geology of the area'. It is now recognized that the majority of the 11 failed 16th Round wells were drilled on poorly defined traps. What became a geophysically led exploration emphasis often lost sight of geological reality. It became evident that seismic-driven exploration risk reduction was not a silver bullet solution for the Paleocene WoS play.

Interestingly, the next decade featured the first non-Vaila Formation discovery, 214/9-1. Drilled in 2000, the Bunnehaven discovery was located in a water depth of 1551 m that was more than double that of previous Paleocene wells. The Bunnehaven well primarily targeted a Mesozoic fault block, but encountered a gas column within shallower Paleocene Flett Formation T40 sandstones. Until the discovery of Bunnehaven, and subsequently Cambo in 2002 and Rosebank in 2004, all of the Paleocene discoveries had been made in reservoirs below the Kettla Tuff and underlying T36 shale, which was regarded as a major regionally widespread pressure seal. The reservoir facies in Bunnehaven could either be shallow marine or possibly deeper marine. The Rosebank reservoir sediments are fluvial to shallow marine, but are interbedded with sub-aerial volcanic lava flows derived from the west. The Rosebank structure overlies the Corona Ridge (Fig. 1), and is highly complex with multiple reservoirs within the uppermost Paleocene Colsay Formation. Cambo is located at the junction between the Westray and Corona ridges (Fig. 1) and is reservoired mainly in Late Paleocene Hildasay Sandstone.

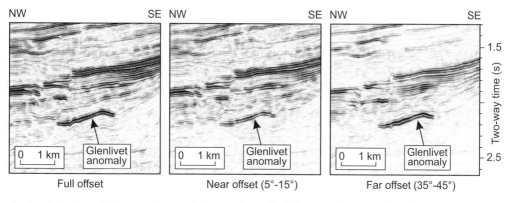

Fig. 6. Glenlivet amplitude anomaly: note the increase in amplitude between the near and far offset data. MegaSurveyPlus seismic data courtesy of PGS.

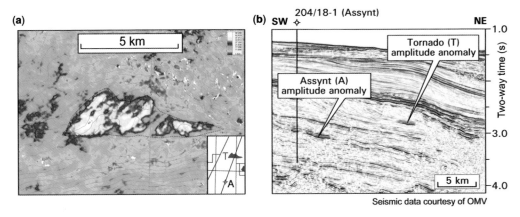

Fig. 7. (a) Seismic amplitude map across a T38 reservoir at the Tornado 2014/13-1 discovery (reproduced with permission from OMV) and (b) example seismic profile across the Tornado discovery.

After Rosebank, the next important find was the 2007 Tormore 205/5a-1 discovery. As with Laggan, its trap is stratigraphic in nature with updip closure against a NE–SW-bounding fault. Reservoir properties are very similar to Laggan with good porosity and permeability (30–300 mD), again mainly due to the presence of chlorite grain coating.

More recently, two interesting gas discoveries were made in 2009: 214/30a-2 (Glenlivet) and 204/13-1 (Tornado). Both Glenlivet and Tornado are largely stratigraphic in nature, and are highlighted on 3D seismic data by strong amplitude and positive AVO responses with strong increases in amplitude with offset (Fig. 6). While Glenlivet is reservoired in T35 Vaila Formation sandstones (Stephensen et al. 2013), Tornado represents the first major discovery within the shallower Lamba Formation T38 sandstones. Until the Tornado well, all of the discoveries in the Judd Sub-basin had been made below the Kettla Tuff and underlying T36 regional shale seal. The Tornado reservoir is located within clinoforms characterized by strong amplitude anomalies (Fig. 7) associated with a toe-of-slope fan that is hydrocarbon filled (Rodriguez et al. 2010).

Key elements for a WoS Paleocene trap

Trap definition

The most important prospect-specific success factor west of Shetland is the presence of a reliable trap model (Loizou 2005), in particular for stratigraphic traps requiring the accurate prediction of the pinch-out of reservoir sands. The ingredients that produce a good stratigraphic trap include the clear identification of reservoir, seal and source. When all these combine favourably with good-quality seismic and other key data, the likelihood is that they will produce a robust trap model with an optimum chance of success. Figure 8 shows a simplified stratigraphic trap model for the Flett

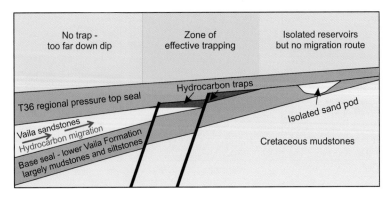

Fig. 8. Simplified model for the WoS Paleocene petroleum system.

Sub-basin. Many of the traps are linked by the deposition of reservoir sands related to pre-existing fault-controlled monoclinal structures, with the sands generally pinching out on the hanging walls of (or at) the NE-trending faults. There are also a number of dry wells such as 204/14-1 and -2 in which the Paleocene sands are somewhat isolated from the main migration fairway.

Aside from discoveries along the Corona Ridge, all the Paleocene discoveries west of Shetland to date have an amplitude anomaly with an associated AVO. However, not all amplitude anomalies are associated with hydrocarbons. Within the Vaila Formation, seismic amplitudes can be seen to vary considerably over short distances and be locally very strong, particularly in updip situations where sands are thinning and an opportunity for stratigraphic trapping of migrating hydrocarbons is possible. This is clearly apparent at the Laggan, Tormore and Glenlivet gas discoveries, where high-amplitude reflectors dominate the T35 Middle Vaila Sandstone reservoir in updip positions. Trap definition can be further improved by enhanced processing techniques or through the availability of new purpose-designed 'high-resolution' long-offset 3D seismic data, particularly if processed to Pre Stack Depth Migration level to define with greater precision the pinch-out edge for these sandstone units. Longer offset data also provide a clearer idea as to whether there is a potential AVO effect that may confirm the presence of hydrocarbons.

The Foinaven trap is partly structural and partly stratigraphic. The structure is a faulted anticline overlying the Westray Ridge intra-basinal high that was formed by compaction and inversion during Late Eocene–Miocene time. Foinaven is dip closed along its western side and relies on stratigraphic pinch-out of reservoir towards its SE corner. The southern limit of the field is fault closed against a footwall block, and its northern closure is partly stratigraphic and partly dip closed.

The structural grain is predominantly east–west in the Judd Sub-basin, which not only influenced sediment transport into the basin but was also fundamental to trap formation of the Foinaven and Schiehallion fields (Smallwood & Kirk 2005). Schiehallion is a good example of a convincing AVO anomaly conforming to mapped structure. This field has a clear amplitude and AVO anomaly displaying depth shut-off in the NW segment of the field, exactly where the sheet/amalgamated channel sands reservoir passes below the oil–water contact (Leach et al. 1999).

Cambo and Rosebank are intra- and sub-basalt low-relief four-way inversion structures that probably formed either due to compactional or tectonic 'drape' over gently inverted Mesozoic fault blocks.

Reservoir presence

With a portfolio of 79 exploration wells supplemented by a number of appraisal wells that have penetrated the Paleocene, the risk associated with reservoir presence and quality has been much reduced. However, understanding the environment of deposition with the aid of good-quality 3D seismic data is a major key to identifying reservoir presence. Despite this, there remain areas where seismic amplitude anomalies have been observed but are at some distance from ideal well control. Such anomalies can potentially be de-risked by using a combination of techniques including analogues, AVO analysis and high-quality electromagnetic (3D) data.

The Judd Sub-basin T10–T40 sands were deposited as a series of amalgamated channel complexes (Cooper et al. 1999; Leach et al. 1999) in a base-of-slope fan environment. The sands were derived from a predominantly sandy shelf to the SE and entered the basin at the junction of the Judd and Rona ridge faults (Ebdon et al. 1995). The distribution of T31–T35 sands is further complicated by continuing structural activity, as evidenced by lobe switches and unconformities in the Foinaven and Schiehallion areas. Hydrocarbon-bearing sands in Foinaven have low impedance relative to shale and stand out on conventional full-stack seismic data (Fig. 9). However, it would not be possible to accurately define the extent of its reservoir without good-quality 3D seismic data, Ocean Bottom Cable or Ocean Bottom Node data.

During the Late Paleocene (T35–T38), the depositional system of the Judd Sub-basin changed from mainly aggradational to progradational. A large prograding shelf system built out from the SE, represented on seismic data as a package of NW-directed amalgamated clinoforms. Eventually, by T50 time, the entire Judd Sub-basin was a non-marine paralic environment. The discoveries at Cambo and Rosebank were made in latest Paleocene Colsay (T40) and Hildasay (T45) sandstone members. Shallow-marine to fluvio-deltaic environments typify the T38–T50 sequences. Basalt lava flows are also present over much of the Cambo–Rosebank area. These lavas and volcaniclastic rocks are interbedded with the Colsay reservoirs, but underlie the Hildasay reservoirs. The T50 Balder Formation records the final infilling of the Paleocene WoS basin, with its progradational phase being succeeded by transgressive flooding of the shelf during latest Balder times.

Reservoir quality

Good-quality reservoir sandstones occur in most of the Paleocene Upper Vaila sequences (T31–T35)

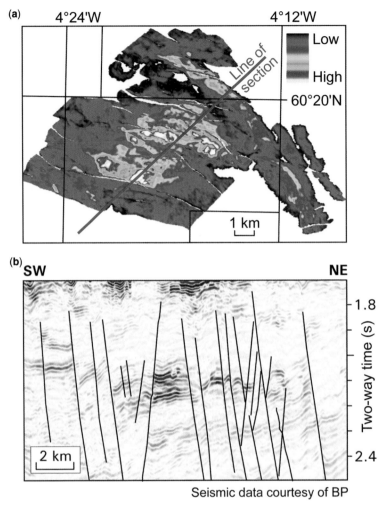

Fig. 9. (a) 3D amplitude inversion of the T32–T34 Vaila Formation reservoir in the Foinaven Field (image courtesy of BP) and (b) seismic profile across the Foinaven Field.

west of Shetland. In the Flett Sub-basin, their porosities range from less than 10% to greater than 30%, and permeabilities range from 0.1 mD to 2 D (Fig. 10). Normally there is an overall reduction of reservoir quality with depth of burial, but some WoS reservoirs have retained high porosities (>20%) and permeabilities (10–100 mD) at burial depths below 3 km (Sullivan et al. 1999). T35 Vaila Formation sandstones in Laggan Field wells 206/1-1, -2, -3 and -4 have porosity and permeability preservation (or enhancement) at depth. Although showing the same detrital composition as the older sandstones, the T35 sandstones are much better sorted with ubiquitous chlorite grain coating (Sullivan et al. 1999; Gordon et al. 2010). The presence of this coating appears to have prevented further quartz diagenesis and led to locally preserved, anomalously high porosities. The chlorite is believed to be derived either as an early clay coat around the sand grains during a period of residence on a shallow marine shelf prior to final redeposition in deep water, or as a result of early diagenesis in situ, with iron perhaps being provided by the degradation of the abundant tuffaceous material deposited with the sandstones.

In the adjacent Torridon area wells 206/2-1 and 214/27-3 are at slightly shallower burial depths to Laggan but have poorer-quality T35 sandstones, which are devoid of chlorite. Furthermore, between 150 and 200 m of tight, non-reservoir quality T25–T28 Lower Vaila Formation sandstones were also penetrated by these

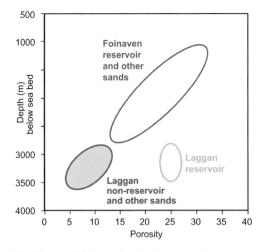

Fig. 10. Simplified porosity–depth trends in Paleocene sandstones west of Shetland (modified after Lamers & Carmichael 1999).

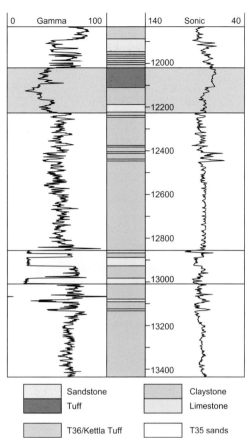

Fig. 11. Log responses of the T35–T36 interval in Laggan 206/1-3 appraisal well. Depths are in feet below Kelly Brushing.

wells. Coincidentally, close to Torridon is a lineament of intrusive rocks interpreted by Doré et al. (1999) to delineate a small NW–SE transfer zone, and this may have some bearing towards explaining the poorer-quality reservoir. Prediction of areas where reservoir quality is best preserved is a major challenge for continuing exploration, particularly in the deeper parts of the Faroe–Shetland Basin. Using the Laggan case, there is a strong relationship between high porosities and high seismic amplitudes; true amplitude preservation is therefore certainly an important element for predicting reservoir quality prior to drilling in some areas west of Shetland.

The T31–T38 Judd Sub-basin sandstones generally show good reservoir properties. They are at shallower burial (less than 2 km below sea bed) and typically over 20% porosity with permeabilities of several hundred milliDarcies. T10 and T20 sandstones in the Judd Sub-basin occur over a wider depth range and generally follow a predictable trend of decreasing porosity with depth.

Seal presence and effectiveness

In the Flett Sub-basin, shales within the T35–T36 Vaila Formation sequence combine with the overlying Kettla Tuff Member to form an effective, basin-wide pressure seal (Lamers & Carmichael 1999). The Kettla Tuff is typically 10–50 m thick, while the underlying shales add up to a further 200 m of seal thickness (Fig. 11). A seal of equivalent age is also present in the Judd Sub-basin, but it is less well defined and much thinner.

In the Flett Sub-basin, the distribution of hydrocarbons within Vaila Formation sandstones strongly relates to the extent of the T35–T36 regional seal. In general, an increase in aquifer pressure of 350–650 psi can be observed across the T31–T35 sequence over most of the Flett Sub-basin, as in well 214/27-2 (Fig. 12). In both the 205/14-1 and -2 wells the Kettla Tuff is absent, and the Paleocene sequence is normally pressured. The composition of the Kettla Tuff Member varies across the area; Linnard & Nelson (2005) interprets this unit as an influx of basalt outwash material rather than an airfall deposit.

A map illustrating the extent of the Kettla Tuff in the Judd and Flett Sub-basins (Fig. 13) has been constructed as a proxy for the T35–T36 regional pressure seal, and superimposed onto the underlying T34–T35 sand play fairway as an aid to understanding whether the wells were optimally located to test a valid trap. This map demonstrates that there

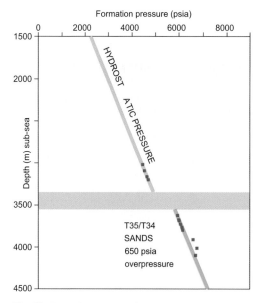

Fig. 12. Formation pressure data above and below the regional T36 pressure seal from Flett Sub-basin well 214/27-2.

is indeed a strong relationship between the extent of this seal and of hydrocarbon occurrences in the Faroe–Shetland Basin. However, this is only one stage in predicting the location of subtle hydrocarbon accumulations below that regional seal. At the Tornado discovery, the top seal is provided instead by basinal shales and overlying mudstones and siltstones of subsequent progradational sequences.

Foinaven has three separate reservoir layers, each comprising stacked slope and basin-floor submarine fan sands sandwiched between mudstone units (Carruth 2003). These mudstones may affect hydraulic communication between the reservoir sands (Cooper et al. 1999). The Schiehallion and Loyal fields are partly sealed by east–west-trending normal faults that juxtapose the reservoirs against mudstones (Leach et al. 1999). To the east, their reservoirs pinch out onto an intrabasinal high. Top and bottom seals are provided by Vaila Formation T35 and T28 mudstones, respectively.

Source rocks and charge

The west of Shetland lies within the UK's passive Atlantic continental margin that formed as a result of multiphase extension. This extension generated

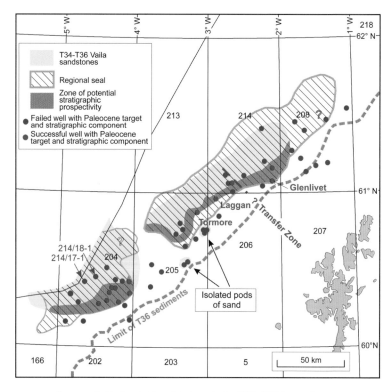

Fig. 13. Summary map of prospectivity for stratigraphic traps in the T34–T36 interval.

a complex assortment of rift basins during Mesozoic and Tertiary time. Because source rocks have been encountered in only a few WoS wells, identifying and mapping the extent of candidate source rock intervals on seismic data remains problematic. As a result, there have been no realistic estimates of the volumes of hydrocarbons potentially generated and expelled in the area prior to the major Cretaceous rifting phase. Nonetheless, the presence of source rocks is not a key geological constraint for the area west of Shetland. From the distribution of hydrocarbon shows it is clear that there is an extensive petroleum system working within the Judd Sub-basin, Flett Sub-basin and Corona Ridge. Studies to date, including analysis of hydrocarbons from the Cambo and Rosebank wells, indicate that the main source rock is the Kimmeridge Clay Formation or equivalent. This has been penetrated by the 213/27-1Z well and is believed to be widely distributed except on the structural highs, where it is likely to be absent either due to non-deposition or erosion.

The Judd Sub-basin is known to be underlain by source rocks of both Middle and Late Jurassic age (Fig. 14). Well data and geochemical modelling suggest that the pre-Tertiary strata here initially reservoired oil and gas, but these traps were subsequently breached by later overpressuring caused by rapid burial during Tertiary time. Fields such as Foinaven and Schiehallion, which directly overlie pre-Tertiary fault blocks and lie on an inversion anticline, experienced multiple phases of charge (Iliffe et al. 1999). The Foinaven hydrocarbon accumulation consists of a fully gas-saturated oil leg of between 24–27° API including a mixed biogenic–thermogenic gas cap (Carruth 2003). Its hydrocarbons are thought to have been sourced mainly from the SW with oil accumulating in its reservoirs by a fill and spill mechanism, resulting in multiple oil–water contacts within the field. In the Flett Sub-basin the understanding of source rock distribution and pre-Tertiary burial history still remains somewhat speculative (Lamers & Carmichael 1999). However, based on the distribution of wells that have encountered gas shows within the Vaila Formation (Fig. 15), gas charge in the Flett Sub-basin appears to be almost ubiquitous.

Suilven is a good example of a hydrocarbon accumulation in a favourable position that could have been sourced from two areas: from either the gas-prone deeper Flett Sub-basin or from the shallower oil-prone Judd Sub-basin. On the contrary, Tornado is located above the main T35-36 regional seal, and was presumably sourced via an adjacent fault zone where the seal's mudstones and tuff were breached, allowing hydrocarbons to migrate from the Upper Jurassic source kitchen into its trap.

DHIs, AAs, AVOs and related geophysical features

Given that true stratigraphic traps have little or no structural control, the location of drilling targets that contain convincingly predicted hydrocarbons west of Shetland has relied heavily on additional geophysical techniques, such as DHIs and AVO technology. AVO analysis forms a particularly valuable tool for exploration. However, an understanding of the quality and the suitability of the seismic data for such analysis before any AVO

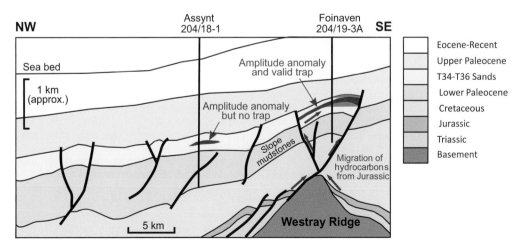

Fig. 14. Geoseismic profile illustrating the merits and demerits of the traps associated with the Foinaven Field and the 204/18-1 Assynt dry hole.

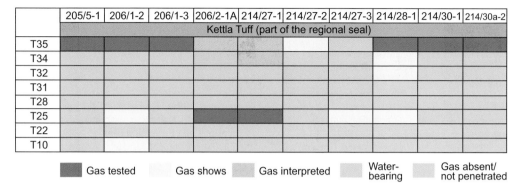

Fig. 15. Summary of principal reservoir intervals for Flett Sub-basin hydrocarbon occurances.

analysis is performed is paramount (Loizou et al. 2008; Whitcombe et al. 2002).

AVO studies to date indicate that conventional DHI anomalies (soft amplitude anomalies conforming with structure) should be represented in typical hydrocarbon-bearing sands west of Shetland at depths shallower than 2000 m sub-sea bed (Smallwood & Kirk 2005; Loizou et al. 2008). The same studies suggest that DHI anomalies should not be expected below 2500 m (sub-sea bed) in typical oil-bearing sandstones, or below 2700 m (sub-sea bed) in typical gas-bearing sandstones. In certain areas such as Laggan and Tormore, where high-porosity sands have been retained at burial depths greater than 3500 m true vertical depth subsea (TVDSS), AVO techniques continue to work well.

When so-called DHI anomalies are seen at depths below 2000 m sub-sea bed, it could indicate particularly favourable conditions (e.g. anomalously high porosity reservoir as in the Laggan gas accumulation, the presence of gas, very good signal-to-noise ratio data or very uniform rock properties), or it could indicate secondary effects associated with the presence of hydrocarbons (e.g. cementation contrasts near the hydrocarbon–water contact). In these circumstances of seismically invisible pay, the trap must be very well defined as the level of risk will be much higher than for amplitude-supported targets. Amplitude anomalies are influenced by other factors, such as lithology, porosity, anisotropy and also fluids. It would be incorrect to infer an infallible direct link between amplitude anomalies and the presence of hydrocarbons.

Controlled Source Electromagnetic technology can also provide a further de-risking tool in the exploration of Paleocene prospects if it provides evidence for the presence and extent of a highly resistive hydrocarbon-bearing sandstone at depth.

Post-drill analysis of 77 WoS Paleocene exploration wells

A total of 50 WoS exploration wells are considered to have targeted Paleocene traps with a significant stratigraphic component. In this post-drill analysis, wells were classed as a success if significant volumes of hydrocarbons were discovered. Analysis shows that almost all of the successful wells were located close to or at the basin margins, with 10 discoveries located in the Judd Sub-basin (including Foinaven, SE Foinaven, Schiehallion, Loyal, Alligin, Cuillin, Arkle, Amos, SW Foinaven and Tornado) and a further 4 located in the Flett Sub-basin (Laggan, Tormore, Torridon and Laxford). The Flett Sub-basin discoveries all lie immediately west of the underlying Flett Ridge (Fig. 1). Most of these discoveries have a north-westerly structural dip and are sealed updip by an east–west or NE–SW fault in combination with stratigraphic pinch-out of the Vaila Sandstones.

Analysis of the key stratigraphic trap elements (i.e. trap, reservoir, seal and charge) shows that the key reason for most failures in both the Judd and Flett Sub-basins has been poor trap definition. However, many wells failed on a combination of geologic components. For the analysis, if the trap constituted more than 50% towards the well failing to find hydrocarbons then trap is assigned as the key element for failure. The majority of unsuccessful wells (80%) are deduced to have failed as a result of a poorly defined trap; 13% of the wells failed as a result of thin or absent reservoir; and 7% failed due to the seal being either thin or absent. Intriguingly, none of the wells specifically failed as a result of source rock absence. However, many poorly defined traps could also have failed due to lack of a viable migration route.

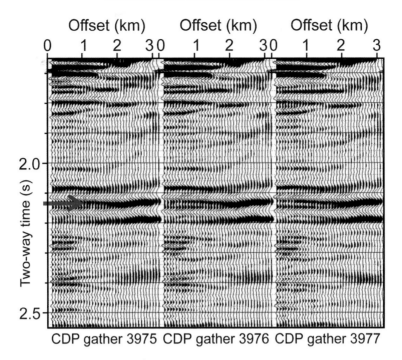

Fig. 16. CDP gathers for Foinaven well 204/24a-2: note increase in amplitude with offset (the target reservoir is indicated by the red arrow).

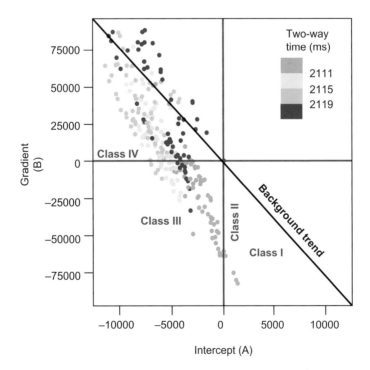

Fig. 17. AVO cross-plot between 2017 and 2123 ms two-way time in Foinaven well 204/24a-2, illustrating a Class III AVO response associated with its thin gas cap within T34 sandstones.

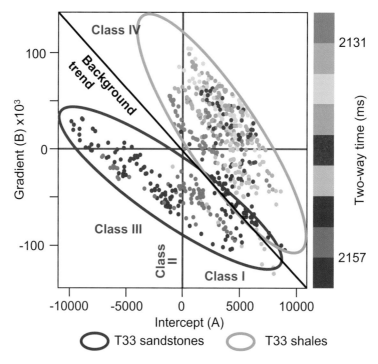

Fig. 18. AVO cross-plot between 2130 and 2160 ms two-way time in Foinaven well 204/24a-2, illustrating a separation in AVO response from its T33 shales (Class I to Class IV) and its deeper T32 sandstones (Class III).

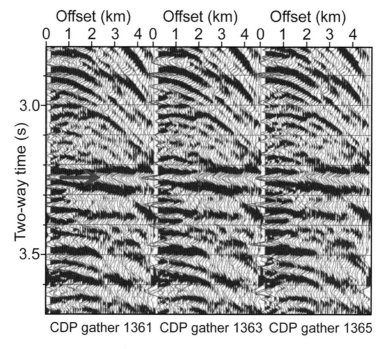

Fig. 19. CDP gathers for Laggan well 206/1-2: note increase in amplitude with offset (the target reservoir is indicated by the red arrow).

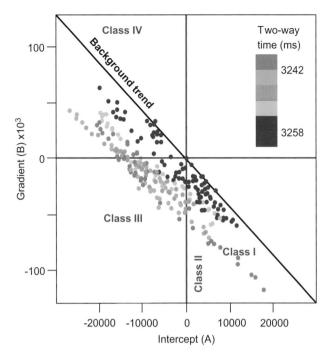

Fig. 20. AVO cross-plot between 3240 and 3260 ms two-way time in Laggan well 206/1-2, illustrating a Class III response associated with its gas-filled reservoir.

Approximately 40 WoS Paleocene wells were positioned on an amplitude or AVO anomaly but only 16 encountered notable hydrocarbons, resulting in a drilling success rate for this group of 40%. A large number of failed wells were positioned on interpreted high-amplitude features believed

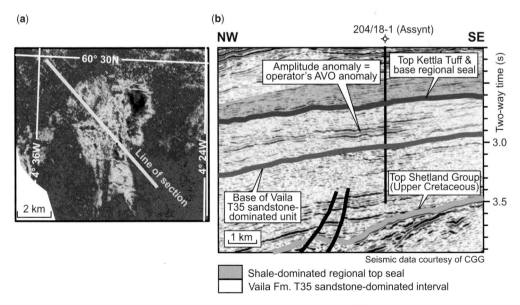

Fig. 21. (a) Seismic amplitude anomaly drilled by (Assynt prospect) dry hole 204/18-1 and (b) interpreted seismic profile across the Assynt prospect.

pre-drill to be coincident with the termination or updip limit/pinch-out edge of a sandstone interval (Loizou et al. 2006, 2008). Moreover, the early work in a number of cases indicated that there was a high chance of hydrocarbons being present, and that the anomaly conformed to a Class III type (for the definition of AVO Classes see Rutherford & Williams 1989). The majority of these failures relate to poor definition of traps that had been mapped on the basis of AVO artefacts and spurious amplitudes, and those wells that encountered hydrocarbons were positioned on robust structural closures or combination traps (Loizou 2005; Loizou et al. 2006, 2008). The problems associated with the AVO analysis in the WoS region include: (1) many of the 1990s 3D seismic datasets were acquired with a 3–3.5 km cable length and these datasets in many cases were not ideal for robust AVO studies; (2) considerable difficulties have been experienced in separating amplitudes associated with contrasting lithologies to those associated with fluids (hydrocarbons); and (3) deeply buried, low-porosity thin sands may contain gas but will have a weak or non-existent AVO response.

Complications in AVO response due to overlying condensed sections or variations in rock properties can significantly reduce or even destroy AVO responses. For example, Margesson & Sondergeld (1999) deduced that dry well 208/17-2 had drilled a manifestation of anisotropy and not a predicted hydrocarbons-related AVO anomaly. Hence, AVO studies cannot be used as the only key measure of prospect risk, but they must be combined with other techniques.

An AVO study was undertaken by Loizou et al. (2008) of four wells, of which 204/24a-2 and 206/1-2 were hydrocarbon-bearing and 204/17-1 and 204-18-1 were dry holes. Well 204/24a-2 was the first appraisal well on the Foinaven Field, and it provides an excellent example of a soft/negative acoustic response that increases with offset angle (Fig. 16). The well actually penetrated multiple reservoirs of Palaeocene T31–T34 sandstones, each with separate hydrocarbon contacts. Cross-plot analysis (cf. Castagna et al. 1998) over a 16 ms two-way time window (2107–2123 ms) through the deeper T31 reservoir sands shows a discrete Class III AVO trend (Fig. 17). An AVO cross-plot analysis (Fig. 18) for a slightly deeper interval between 2130 and 2160 ms shows an impressive and clear separation of sealing intra-formational shales. These exhibit a Class I–IV trend while the underlying T34 oil sands show an apparent Class III AVO response.

The Laggan Field discovery well 206/1-2 encountered gas pay in its T35 Vaila Formation sandstones. The logs show that the gas is contained in three separate reservoirs totalling 62 m, which are equivalent to 33 ms of two-way time (interval velocity of 3730 m s^{-1}) at a depth below 3815 m TVDSS. This interval is represented on 3D seismic data by only one event as a result of seismic tuning. Examination of the CDP gathers at the 206/1-2 well location shows a definite increase in amplitude with offset (Fig. 19). A 20 ms AVO cross-plot extraction from the gas-bearing interval shows a distinct Class III AVO anomaly (Fig. 20).

Well 204/18-1 was positioned on a high-amplitude anomaly referred to as Assynt (Fig. 21). Located in the Judd Sub-basin, it was largely a geophysically driven prospect. The amplitude anomaly was interpreted by a number of operators to be a direct fairway analogue to discoveries such as Foinaven. However, Loizou et al. (2006, 2008) showed that unlike Foinaven there is no evidence of true amplitude conformance with depth. Furthermore, as a predominantly stratigraphic prospect, Assynt relied heavily on a valid sealing mechanism. At the nearby Foinaven and Schiehallion fields, a thick and dominantly mud-prone T35

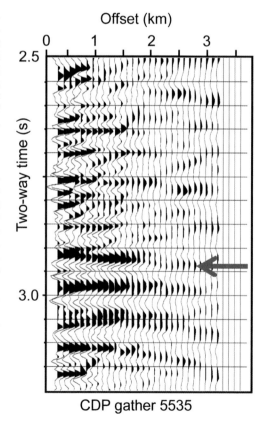

Fig. 22. CDP gathers for Assynt well 204/18-1: note decrease in amplitude with offset (the target reservoir is indicated by the red arrow).

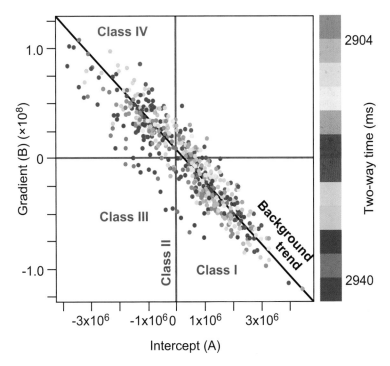

Fig. 23. AVO cross-plot between 2901 and 2943 ms two-way time in Assynt well 204/18-1, illustrating a Class I response associated with its non-hydrocarbon bearing reservoir.

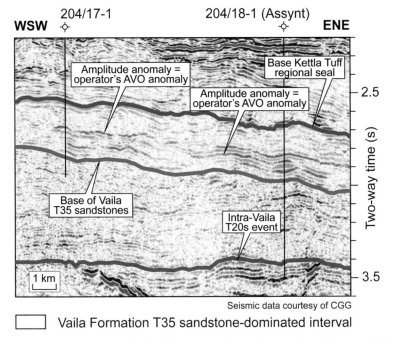

Fig. 24. Interpreted seismic profile across the 204/17-1 prospect, located 8.8 km updip of Assynt well 204/18-1, but targeting essentially a similar amplitude-defined feature.

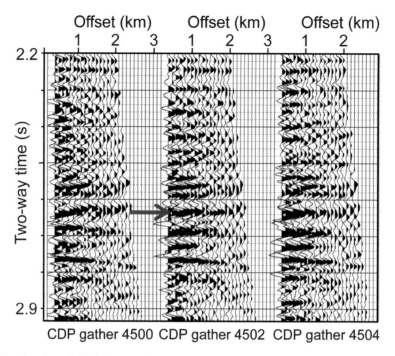

Fig. 25. CDP gathers for well 204/17-1: note decrease in amplitude with offset (the data are heavily muted beyond 2 km offset, and the target reservoir is indicated by the red arrow).

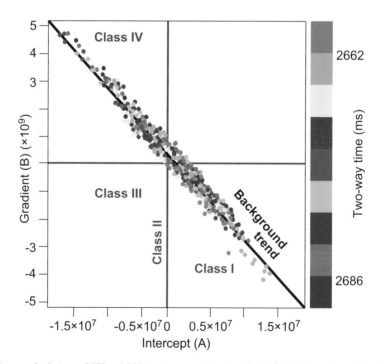

Fig. 26. AVO cross-plot between 2657 and 2691 ms two-way time in well 204/17-1, illustrating a Class I response associated with its non-hydrocarbon bearing reservoir.

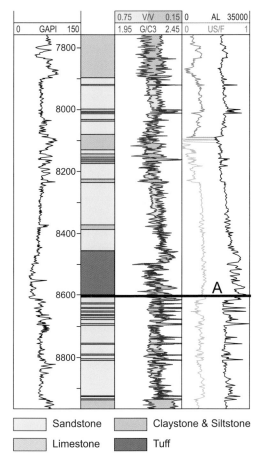

Fig. 27. Log response across the target interval in well 204/17-1. The target seismic amplitude anomaly can be seen post-drill to have been generated by the acoustic impedance contrast (at A) between the thick volcaniclastic tuffs and the underlying sandstones.

lowstand wedge provides an ubiquitous top seal. The location of Assynt shows it to be downslope of, or even within, the basinward equivalent of this package. The Paleocene T36 shale/tuff sequence would therefore be required to provide an ultimate top seal to the Assynt prospect.

Examination of the near and mid offset stacks at the Assynt 204/18-1 well (375–2241 m offset c. 2.7 ms) shows a high-amplitude response. Conversely, on the far offsets (2241–3174 m) the amplitudes are much weaker. Moreover, a review of the gathers shows that the high amplitudes switch off at 1775 m offset (Fig. 22), which is further evidence that the anomaly is non-hydrocarbon-bearing. More significantly, post-drill AVO analysis of the Assynt amplitude anomaly indicates that it has a predominantly Class 1 type AVO, which is generally not indicative of a hydrocarbon-bearing interval (Fig. 23).

Well 204/17-1 is located just 8.8 km to the SW and marginally updip of the Assynt well on a smaller amplitude anomaly known as the North Fleet Prospect (Fig. 24), and was also dry. Post-drill analysis of 204/18-1 shows that overlying the Assynt anomaly is a non-sealing sequence comprising predominantly thick sandstone and siltstone lithologies. Bearing this in mind, at the 204/17-1 well location the 3D seismic data do not categorically show any evidence that the lithology overlying or updip of North Fleet would be any different to that penetrated at the Assynt well.

Analysis of the offset stacks for the 204/17-1 well shows that the data are heavily muted to c. 2540 m offset, with the amplitudes observed at c. 2.66 ms decreasing with offset (Fig. 25). Amplitude extraction over the high-amplitude anomaly between 2660 and 2690 m clearly shows the key characteristics of a classic Class I type AVO anomaly (Fig. 26). Coincidentally, the results of the well confirm very similar lithologies immediately above the anomaly to those encountered at the 204/18-1 well (c. 160 m of sandstones overlying the anomalous zone). The pre-drill prognosis from the operator was that the 'high amplitudes' represented low impedances that were interpreted to be oil-filled siliciclastic sediments. The North Fleet Prospect was interpreted as a DHI, although only very partial conformance between amplitude and structure was observed. A well post-mortem by the operator suggested that failure was mainly attributed to the absence of a topseal. The seismic anomaly was actually induced by an interbedded succession of sandstones, shales and volcaniclastic tuffs at this level (Fig. 27).

In another example Flett Sub-basin well 205/10-5A was positioned on an amplitude anomaly, with the hope of finding hydrocarbons being largely based on the presence of oil shows in offset well 205/10-B2 only 7 km to the north that produced oil on test from 6 m of sands. Unfortunately, the 205/10a-5 anomaly turned out to be a 90-m-thick igneous intrusive (quartz porphyry), which coincidentally shows radial geometry on a time slice that should have implied a non-sedimentary feature.

The geological setting and AVO analysis for 204/17-1 and 204/18-1 showed that both failed wells were positioned on invalid traps with anomalous amplitude anomalies (not to be confused with valid 'AVO' anomalies) where the operators' work mistakenly indicated a strong Class III/IV AVO response. What is unusual about the operators' interpretation of these anomalies is that their amplitudes categorically decrease with offset and, in particular, unlike nearby Foinaven where amplitudes increase with offset. Moreover, the Foinaven amplitudes have good conformance to structure. Both 204/17-1 and 204/18-1 categorically conform to Class I AVO (non-hydrocarbon-bearing) interpretation. Not surprisingly, since 2007 seven Paleocene exploration wells have been positioned on robust traps with good conformance to structure and genuine 'AVO anomalies', and all have been successful in finding hydrocarbons.

Conclusions

Hydrocarbons have been encountered in almost every interval within the Paleocene succession west of Shetland, but there are still prospective areas that remain sparsely or under-explored that could potentially contain valid hydrocarbon traps. However, it is envisaged that many of these are quite subtle and would require advanced seismic and other techniques to suitably de-risk and validate their prospectivity. The exploration history to date has shown that appropriate design of parameters for seismic acquisition and imaging can significantly enhance the imaging of seismic attributes to provide a much clearer picture of exploration potential. The recent discoveries of Tormore, Glenlivet and Tornado have brought the region into the spotlight, prompting a reassessment of the hydrocarbon prospectivity of the area by many operators.

From a total of 79 Paleocene exploration wells there have been 56 failures. Eighty percent of the failures are attributed to unreliable traps. Not surprisingly, those wells located on more reliable, robust structures achieved a higher success rate of c. 66%. The sound mapping of a valid trap is therefore viewed as the key component to increasing exploration success. Many of the failed wells were not optimally positioned to test a valid stratigraphic trap. With this in mind, exploring for valid structures requires the best analysis possible to fully validate all of the ingredients that might constitute a successful prospect.

Evaluation of proven examples of successful Paleocene traps such as Foinaven, Laggan, Tormore, Glenlivet and Tornado, which have a strong stratigraphic component, can add to the understanding of why a large number of stratigraphic wells have failed. Furthermore, appropriate techniques such as AVO analysis are valuable tools for exploration. To maintain the recent success rate, and to improve the future exploration success rate, there is however a need to utilize the highest quality possible of long offset 3D seismic data together with all additional appropriate techniques to fully validate robust Paleocene traps pre-drill with a high degree of confidence.

Special thanks are extended to I. Andrews of the British Geological Survey for his contribution to the evaluation of the Paleocene West of Shetland play. I am also grateful to D. Cameron and J. Brzozowska for providing reviews of this paper. I acknowledge permission from DECC to publish the paper.

References

CARRUTH, A. G. 2003. The Foinaven Field, Blocks 204/19 and 204/24a, UK North Sea. In: GLUYAS, J. G. & HICHENS, H. M. (eds) *United Kingdom Oil and Gas Fields, Commemorative Millennium Volume*. Geological Society, London, Memoirs, **20**, 121–130.

CASTAGNA, J. P., SWAN, H. W. & FOSTER, D. J. 1998. Framework for AVO gradient and intercept interpretation. *Geophysics*, **63**, 948–956.

COOPER, M. M., EVANS, A. C., LYNCH, D. J., NEVILLE, G. & NEWEL, T. 1999. The Foinaven Field: managing reservoir development uncertainty prior to start-up. In: FLEET, A. J. & BOLDY, S. A. R. (eds) *Petroleum Geology of Northwest Europe: Proceedings of the 5th Conference*. Geological Society, London, 675–682.

DORÉ, A. G., LUNDIN, E. R., JENSEN, L. N., BIRKELAND, Ø., ELIASSEN, P. E. & FICHLER, C. 1999. Principal tectonic events in the evolution of northwest European Atlantic margin. In: FLEET, A. J. & BOLDY, S. A. R. (eds) *Petroleum Geology of Northwest Europe: Proceedings of the 5th Conference*. Geological Society, London, 41–61.

EBDON, C. C., GRANGER, P. J., JOHNSON, H. D. & EVANS, A. M. 1995. Early Tertiary evolution and sequence stratigraphy of the Faeroe-Shetland Basin: implications for hydrocarbon prospectivity. In: SCRUTTON, R. A., STOKER, M. S., SHIMMIELD, G. B. & TUDHOPE, A. W. (eds) *The Tectonics, Sedimentation and Palaeoceanography of the North Atlantic Region*. Geological Society, London, Special Publications, **90**, 51–69.

GORDON, A., YOUNIS, T. ET AL. 2010. Laggan; a mature understanding of an undeveloped discovery, more than 20 years old. In: VINING, B. A. & PICKERING, S. C. (eds) *Petroleum Geology: From Mature Basins to New Frontiers – Proceedings of the 7th Petroleum Geology Conference*. Geological Society, London, 279–297.

ILIFFE, J. E., ROBERTSON, A. G., WARD, G. H. F., WYNN, C., PEAD, S. D. M. & CAMERON, N. 1999. The importance of fluid pressures and migration to the hydrocarbon prospectivity of the Faeroe-Shetland White Zone. In: FLEET, A. J. & BOLDY, S. A. R. (eds) *Petroleum Geology of Northwest Europe: Proceedings of the 5th Conference*. Geological Society, London, 601–611.

LAMERS, E. & CARMICHAEL, S. M. M. 1999. The Palaeocene deepwater sandstone play West of Shetland. In: FLEET, A. J. & BOLDY, S. A. R. (eds) *Petroleum Geology of N W Europe: Proceedings of the 5th Conference*. Geological Society, London, 645–659.

LEACH, H. M., HERBERT, N., LOS, A. & SMITH, R. L. 1999. The Schiehallion development. In: FLEET, A. J. & BOLDY, S. A. R. (eds) *Petroleum Geology of Northwest Europe: Proceedings of the 5th Conference*. Geological Society, London, 683–692.

LINNARD, S. & NELSON, R. 2005. Effects of Tertiary Volcanism and later Events upon the Faroese Hydrocarbon system. *In*: ZISKA, H., VARMING, T. & BLOCH, D. (eds) *Faroe Islands Exploration Conference: Proceedings of the 1st Conference*. Annales Societatis Scientarum Faerensis Supplementum, **43**, 41–53.

LOIZOU, N. 2005. West of Shetland exploration unravelled – an indication of what the future may hold. *First Break*, **23**, 53–59.

LOIZOU, N., ANDREWS, I. J., STOKER, S. J. & CAMERON, D. 2006. West of Shetland revisited – the search for stratigraphic traps. *In*: ALLEN, M., GOFFEY, G., MORGAN, R. & WALKER, I. (eds) *The Deliberate Search for the Stratigraphic Trap: Where are We Now?* Geological Society, London, Special Publications, **254**, 225–245.

LOIZOU, N., LIU, E. & CHAPMAN, M. 2008. AVO analysis and spectral decomposition of seismic data from four wells west of Shetland, UK. *Petroleum Geoscience*, **14**, 355–368.

MARGESSON, R. W. & SONDERGELD, C. H. 1999. Anisotropy and amplitude versus offset: a case history from the West of Shetlands. *In*: FLEET, A. J. & BOLDY, S. A. R. (eds) *Petroleum Geology of Northwest Europe: Proceedings of the 5th Conference*. Geological Society, London, 634–643.

PARR, R. S., COWPER, D. & MITCHENER, B. C. 1999. The search for mountains of oil: BP/Shell's exploration activity in the Atlantic Margin, West of Shetland. *Society of Petroleum Engineers SPE* 56897, 1–4.

RODRIGUEZ, J. M., PICKERING, G. & KIRK, W. J. 2010. Prospectivity of the T38 sequence in the Northern Judd Basin. *In*: VINING, B. A. & PICKERING, S. C. (eds) *Petroleum geology: from mature basins to new frontiers – Proceedings of the 7th Petroleum Geology Conference*. Geological Society, London, 245–259.

RUTHERFORD, S. R. & WILLIAMS, R. H. 1989. Amplitude-versus-offset variations in gas sands. *Geophysics*, **54**, 680–688.

SMALLWOOD, J. R. & KIRK, W. J. 2005. Paleocene exploration in the Faroe–Shetland Channel; disappointments and discoveries. *In*: DORE, A. G. & VINING, B. A. (eds) *Petroleum Geology: North-West Europe and Global Perspectives; Proceedings of the 6th Conference*. Geological Society, London, 977–991.

STEPHENSEN, M. W., LARSEN, M., DAM, D. & HANSEN, T. 2013. The Glenlivet gas discovery an integrated exploration history. *First Break*, **31**, 51–64.

SULLIVAN, M., COOMBES, T., IMBERT, P. & AHAMDACH-DEMARS, C. 1999. Reservoir quality and petrophysical evaluation of Paleocene sandstones in the West of Shetland area. *In*: FLEET, A. J. & BOLDY, S. A. R. (eds) *Petroleum Geology of Northwest Europe: Proceedings of the 5th Conference*. Geological Society, London, 627–633.

WHITCOMBE, D. N., CONNOLLY, P. A., REAGAN, R. L. & REDSHAW, T. C. 2002. Extended elastic impedance for fluid and lithology prediction. *Geophysics*, **67**, 63–67.

Basement exploration, West of Shetlands: progress in opening a new play on the UKCS

ROBERT TRICE

Hurricane Energy plc, The Wharf, Abbey Mill Business Park, Lower Eashing, Godalming, Surrey GU7 2QN, UK (e-mail: robert.trice@hurricaneenergy.com)

Abstract: Commercial production of hydrocarbon from fractured crystalline basement is well documented, with petroleum basins across the globe hosting fractured basement fields. The UK is an anomaly within this global phenomenon as, despite numerous serendipitous discoveries of basement oil in the North Sea and the West of Shetlands, there is currently no commercial field on the United Kingdom Continental Shelf (UKCS) that is reliant on oil production from fractured basement. Recognizing that this situation presented an exploration niche, Hurricane Energy plc (Hurricane) was formed to focus on UKCS basement exploration and concentrated its efforts in acquiring exploration acreage in the West of Shetlands. In 2009 Hurricane drilled the first well designed to explore the basement play on the UKCS, leading to the Lancaster Discovery with a contingent resource range (1C–3C) of 62–456 million barrels of oil equivalent (MMboe). The Lancaster Discovery is presented to summarize the challenges and processes that have been applied in the exploration and evaluation of the West of Shetlands basement play. Conclusions from this work indicate that basement hydrocarbon resource potential is of such significance that it may represent a strategic resource for the UK, with over 1154 MMboe of 2C and mean unrisked prospective resources so far identified.

Gold Open Access: This article is published under the terms of the CC-BY 3.0 license.

The Lancaster Discovery is located in Quad 205 West of Shetlands (WoS), SE of the Foinaven and Schiehallion fields and NE of the Solan Field (Fig. 1). Lancaster is located on the Rona Ridge, a prominent NE–SW-trending basement high which acts as a structural feature separating the Faroe–Shetland Basin from the West Shetland and the East Solan basins (Figs 1 & 2). The Lancaster Discovery is the result of the first UK exploration well drilled specifically to evaluate basement as an exploration play. While Lancaster is considered to be the first UKCS basement exploration well, it is important to note that it is not the first basement discovery. Many UK wells have been drilled into the basement with a significant number of these being cored and tested, each adding to an important database of serendipitous UK basement discoveries. With the exploration success of Lancaster and the nearby Whirlwind discoveries (Fig. 1), the remaining challenge to proving a UK basement play is to demonstrate proven basement reserves. That stated, the data and story behind the Lancaster Discovery is encouraging, and Hurricane believes that realization of UK basement reserves is simply a function of further investment in appraisal drilling and planning for field development. This perspective is not considered to be optimistic when considered in context with the global experience of basement discoveries, many of which were made accidentally and yet have still led to world-class producing assets. A summary of the rationale, planning and operations that resulted in the Lancaster Discovery is provided as well as insight into some of the key challenges that have been overcome in evaluating one of the UK's few remaining unproven hydrocarbon plays.

The basement play

Naturally fractured crystalline basement reservoirs (basement reservoirs) are a global phenomenon, as depicted in Figure 3 (see also Aguillera 1995; Nelson 2001; Petford & McCaffrey 2003; Gutmanis *et al.* 2012). Despite the proven commercial success of basement reservoirs it has long been recognized that basement reservoirs are a globally underexplored play (Eggleston 1948; Hubbert & Willis 1955; Landnes 1959; Landnes *et al.* 1960; P'an 1982). Support for this perspective arises from the fact that basement reservoir discoveries have historically occurred more by chance rather than as a result of basement-focused exploration programmes. A successful reverse in this trend has however been seen in recent years resulting in numerous basement oil discoveries and increasing numbers of basement field developments. Examples can be found in Yemen and Vietnam, both of which have enjoyed material impacts to reserves as a result of basement field developments (Areshev *et al.* 1992;

From: CANNON, S. J. C. & ELLIS, D. (eds) 2014. *Hydrocarbon Exploration to Exploitation West of Shetlands.*
Geological Society, London, Special Publications, **397**, 81–105.
First published online February 28, 2014, http://dx.doi.org/10.1144/SP397.3
© 2014 The Authors. Publishing disclaimer: www.geolsoc.org.uk/pub_ethics

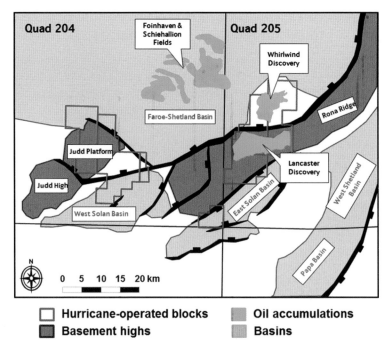

Fig. 1. Location of the Lancaster and Whirlwind basement discoveries, Quad 205 West of Shetlands.

San *et al.* 1997; Tran *et al.* 2006; De Urreiztieta *et al.* 2007; Gutmanis 2009; Cuong & Warren 2011; Gutmanis *et al.* 2012).

The petroleum system components of basement reservoirs are no different from those of conventional clastic reservoirs. The basement trap is typically a four-way dip closed structure, either a faulted block or buried hill (or combination of the two) or the trap can be formed as the flank of a tilted fault block. Source rock is generally found proximal to the trap juxtaposed as onlapping or capping successions. The reservoir typically comprises lithologies such as granite, granodiorite and gneiss with secondary lithologies including dolerite and basalt. The basement reservoir poro-perm system is provided by fractures ranging in scale from faults detectable from seismic data to those fractures that can be detected and quantified from the evaluation of core. Basement reservoirs tend to have complex geological histories with fracture properties resulting from a combination of tectonic, hydrothermal and epithermal processes.

Analysis of well data in the West of Shetlands led Hurricane to the conclusion that, of the wells with serendipitous basement discoveries, very few had penetrated significant intervals of basement. From the perspective of basement exploration a significant penetration is considered to be 100–300 m as analogue data indicates that any penetration less than this is unlikely to confirm the presence of permeability or mobile hydrocarbon (Aguillera 1996). At the time of study, the majority of basement penetrations in the WoS were typically less than 50 m TVT. One important exception to this trend is the Clair Field discovery well 206/7-1 which, although targeting a sedimentary succession, additionally penetrated a potential oil column of at least 200 m in the Lewisian Basement. An appraisal of fractured Devonian sandstone and the underlying fractured Lewisian was undertaken by a subsequent deviated well, 206/7a-2, which tested the basement through a 530-m-long horizontal section. The basement was characterized as faulted, fractured and weathered. After acidification, production rates from the basement were reported as 2110 barrels of oil per day (bopd) from an interval having five fracture zones.

As part of Hurricane's data review an analysis of the fracture trends recorded from the 206/7a-2 well and review of the well's position relative to mapped seismic faults was undertaken. A conclusion of this work was that 206/7a-2 had not been located or oriented optimally to evaluate the basement potential. This observation, coupled with the limited number and limited depth of basement penetrations in the WoS, provided support to Hurricane's supposition that the basement in the West of Shetlands was a material play requiring a bespoke exploration well.

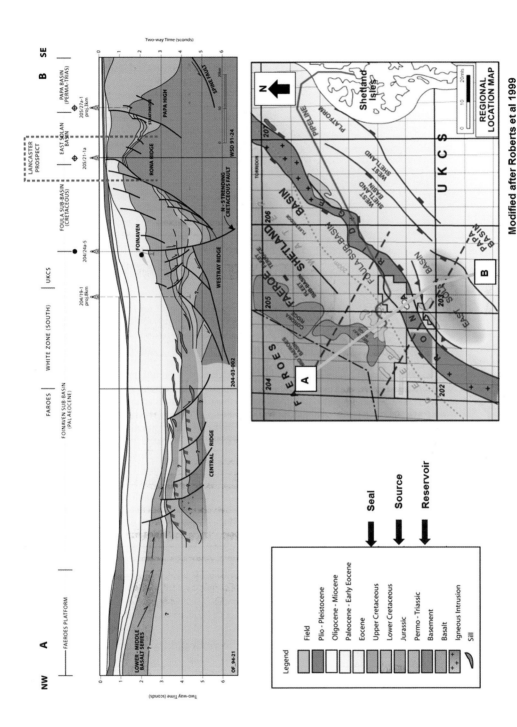

Fig. 2. Regional cross-section across the Rona Ridge portraying the structural disposition of the Lancaster Prospect and its relationship to the 'hydrocarbon kitchens' of the Foula Sub-Basin and East Solan basin.

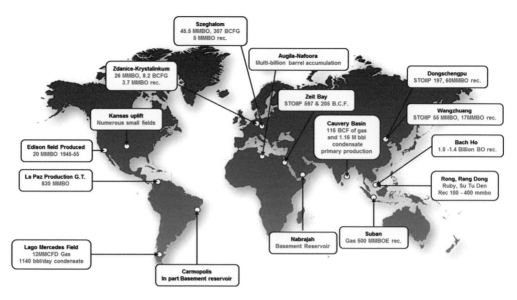

Fig. 3. Examples of global basement fields and basement reservoirs.

After completing a narrowing process that involved the evaluation of 2D seismic, offset well data and regional geological analysis, an application was made for a nine-block Frontier Licence in the UK Offshore 23rd Licencing Round. The licence application area included a large basement high, the Lancaster Prospect, that had been previously drilled in 1974 (by well 205/21-1a) to evaluate a Mesozoic clastic succession which onlapped the basement. The 205/21-1a well was a discovery, testing oil to surface and recovering core which supported oil-bearing clastics and oil associated with fractures in the underlying basement. The well was plugged and abandoned with the operator concluding that the clastics were of non-reservoir quality. Hurricane's preliminary interpretation of composite log and end-of-well reports indicated that not only was there potential for commercial volumes of oil to be present in the basement, but that the oil would most likely be light. A provisional assessment of the Lancaster Prospect resource potential was achieved through a thorough evaluation of producing basement field analogues and WoS well data. Global analogues provided insight into basement reservoir properties, exploration histories and appraisal/development strategies, all of which combined to provide a template of basement reservoir characteristics and a basement exploration strategy (see Wu *et al.* 2005 for exploration applications of analogues). Analogue information also provided ranges and averages applied in assessing basement resource potential. Core data from the WoS was used as a control on analogue porosity types and porosity ranges. The initial assessment of the Lancaster Prospect resource potential was a P50 of 191 million barrels (MMbbl) of oil recoverable which consisted of a conventional closure model and a combined conventional closure/flank upside model, the latter reflecting the stratigraphic trapping nature of buried hill traps and the potential for extensive hydrocarbon columns and flank accumulations reported in analogue datasets.

Despite Hurricane's perceived technical low risk of the Lancaster Prospect, the mechanics of making a successful licence application raised the first challenge in exploring the UK basement play: the challenge of funding. Funding was difficult as the concept of basement was untried in the UK, Hurricane as a company had no track record and the company had no licence. Hurricane was faced with the classic 'Catch 22' conundrum of no licence no funds, no funds no licence. Ultimately sufficient funds were raised to accommodate the proposed licence work programme and therefore secure the licence, but not without overcoming challenges raised by geological advisors to the finance community. Many advisors, professional geoscientists, doubted the validity of a UKCS basement play. Furthermore they considered the West of Shetlands to be a heavy oil province which would add significant risk to the basement play concept. Many potential investors heeded their geological advisors and decided that the Hurricane story was not for them; however, a core group of investors bought into the story and, thanks to their vision and support, Hurricane has now achieved two significant basement discoveries.

Having overcome the financial hurdle, the remaining pre-drill challenges distilled to three key points: (1) identifying a robust drilling target that would effectively evaluate the basement reservoir and its hydrocarbon bearing potential; (2) establishing the source of the 205/21-1a basement oil; and (3) establishing a safe and optimum method of drilling the basement. Before describing how these technical challenges were overcome it is prudent to review the petroleum system in which the Lancaster prospect resides.

Lancaster petroleum system

The petroleum system elements that are relevant to the WoS basement play are associated with the Rona Ridge, a major tectonic feature depicted in Figure 2. The Rona Ridge is composed of a NE–SW-trending basement structure stratigraphically classified as the Lewisian Basement (see inset, Fig. 2). The Lewisian Basement is a remnant of Pre-Cambrian crust. Recovered core and cuttings from offset wells indicate the Rona Ridge to be composed of gneiss and granite with subordinate basic rock (dolerite and diorite). The basic rock is either the result of segregation in the early melt or post-cooling intrusions. Local deformation to cataclasite and tachylite is also noted proximal to fault/shear zones.

The trap at Lancaster is a buried hill with a four-way dip closed crest, formed as a result of the rifting and opening of the NE Atlantic during Jurassic–Cretaceous time and subsequent uplifting during phases of Tertiary compression. Specifically, the Lancaster buried hill trap results from the Cretaceous marine drowning of the Rona Ridge that had previously existed as an uplifted region from at least Devonian–Carboniferous time through to Jurassic time. Top and lateral seal is provided by Upper Cretaceous mudstones. The source for Lancaster is provided by the Kimmeridge Clay formation which onlaps on the Rona Ridge. The Kimmeridge Clay Formation has been at sufficient depth for hydrocarbon to be generated on both sides of the Rona Ridge, specifically within the Foula Sub-Basin to the north, in the main Faroe Shetland Basin and also in the East Solan Basin to the south (Fig. 2).

The Lancaster reservoir is the Lewisian Basement and is considered a Type 1 fractured reservoir (Nelson 2001). Type 1 fractured reservoirs owe their productivity and storativity to a hydrodynamic fracture network which is a subset of the natural fracture network, consisting of fractures that are spatially connected and collectively capable of transmitting fluid. Fluid flow in the hydrodynamic network is controlled by: (a) fracture connectivity; (b) the relative magnitude of fluid pressure and lithostatic pressure; and (c) the magnitude and orientation of the mean stress across the fracture network (Jolly & Cosgrove 2003).

Although fractures have traditionally been considered by geoscientists and reservoir engineers as a series of smooth-sided parallel plates (Reiss 1980), the reality is far more complex as fluid flow within the hydrodynamic network is a function of channelling. Fracture fluid flow cannot therefore be assumed to be related to aperture at a specific location. Channelling within a fracture is controlled by a critical path of connected void space (Pyrak-Nolte et al. 1987; Tsang & Tsang 1989); consequently, fracture network flow properties are more like those of fluvial systems.

The Lewisian Basement comprises a number of specific fractures identified from subsurface and outcrop analogue data, each of which has the potential to contribute to the hydrodynamic fracture network: (a) faults; (b) regional joints; (c) shear fractures; and (d) sheet fractures. From the perspective of basement, exploration faults are partitioned as seismic- and subseismic-scale features. Seismic-scale faults represent specific mappable features that are considered as potential drilling targets and can be readily interpreted at the basement level from the analysis of 3D seismic data. Subseismic faults are detectable from image logs and core and consequently do not represent potential drilling targets. Faults are associated with a fault damage zone (fault zone) which includes a variety of fractures and a range of rock fabrics associated with fault rock. Specific rock fabrics of exploration interest include fault breccia, which can be associated with pervasive fracturing at the scale of core measurement. Faults within the basement that are most likely to act as super-conduits for fluid flow are those that have been heavily reactivated by several phases of brittle deformation (Holdsworth et al. 2007). Regional joints, detectable from image log data, are fractures having no detectable shear movement and that have a specific orientation that is persistent across a given structure or geological/geographical region. Regional joints are considered to include cross joints and regional longitudinal fractures as described by Cloos (1922). Regional joints are not treated as a specific drilling target but are given consideration in the planning of exploration and appraisal well paths. Shear fractures have a detectable shear displacement and are often associated with fault rock. The identification of shear fractures is important as an aid in defining fault zones and subseismic faults. Sheet fractures are a phenomenon associated with the unloading of basement during uplift. Sheet fractures do not form an explicit exploration target although it is recognized that, as in the case of regional joints, sheet fractures have the potential to improve the connectivity of the hydrodynamic fracture network and thereby increase the drainage potential of the Lewisian Basement.

The identification of the prospective fracture types present in the Lewisian basement is a key step in developing a conceptual exploration model; however, it is also important to evaluate the potential for epithermal and hydrothermal processes to have contributed to the properties of the hydrodynamic fracture network, as the analysis of global analogues indicate that such processes are important geological components in producing basement fields (Dimitriyevskiy et al. 1993; Petford & McCaffey 2003; Li et al. 2004; C&C Reservoirs 2005; Geoscience 2007). The Lewisian Basement has a long geological history of subaerial exposure and orogenic activity; there is therefore significant potential to enhance fracture network drainage through epithermal and hydrothermal processes. Minerals associated with epithermal processes have been recorded from basement cores recovered from the Rona Ridge and include feldspar and calcite dissolution (porosity enhancing), kaolinite and pervasive haematite cements (porosity reduction). Weathering of basement reservoirs can represent preferential exploration targets (Areshev et al. 1992), but while the process of subaerial exposure can result in enhanced porosity and permeability, it can also generate porosity and permeability inhibiting clays which reduce reservoir potential. Offset and analogue data indicated that at Lancaster such processes are expected to be concentrated within the longer and better-connected fault planes as well as any joint systems hydraulically connected to unconformity surfaces. Evidence of hydrothermal alteration of the Rona Ridge basement is explicitly noted in offset well 208/27-2 by the presence of minerals such as epidote, iron pyrite and chlorite. These minerals are commonly associated as vein fill from Rona Ridge basement core. Hydrothermal products including epidote, opaque magnetite or goethite and sericite/talc are also recorded from sidewall cores recovered from the Lancaster Discovery. It is also worth noting that hydrocarbon stained calcite vein fill is ubiquitous in Rona Ridge basement cores and is also described in basement cuttings and sidewall cores recovered from Lancaster. Dissolution of vein calcite is also recorded from Lancaster sidewall cores and from the offset well 204/23-1, which has whole core measurements describing 2–3 cm vugs in calcite vein fill with up to 20% porosity as inter-crystalline void space.

While it is considered important to understand the potential fracture types and processes that could enhance fracture channelling, it was also considered essential to evaluate offset data to evaluate where the best poro-perm systems are likely to be located within the Rona Ridge and, explicitly, within the Lancaster prospect. After comparing offset composite logs, basement cored intervals, basement well tests and seismic sections from the Rona Ridge it became evident that seismic-scale faults will likely have the greatest potential to exhibit enhanced permeability in the basement. Such an observation is consistent with analogue data which document the importance of faulting in basement rock, as it is the fault zones and fault zone intersections that are associated with the best-developed fracture networks (Younes et al. 1998; C&C Reservoirs 2005). Consequently, defining a robust drillable fault zone became the first step in planning a basement exploration well.

Planning the basement exploration well

Given that seismic-scale fault zones were considered to be the primary exploration target, the mapping of seismic-scale faults became an essential element in the exploration work flow. It was apparent from the most rudimentary seismic interpretation of the Lancaster prospect that the basement was highly faulted, so the challenge was to find a seismic fault that would deliver an exploration success. This first step in the process of identifying a drilling target was to develop a conceptual fault target, that is, what combined fault properties are considered optimum to make a discovery? The five key criteria for a drillable seismic fault at Lancaster were considered to be: (a) robust seismic feature; (b) fault is within proven oil-bearing rock volume; (c) fault forms part of a connected fault network; (d) age of the fault is consistent with a history of reactivation; and (e) fault preferentially oriented to the current-day stress field. A robust seismic feature was considered to be a fault trace associated with an obvious offset at the basement reflector and that was also clearly mappable as an extensive lateral feature. The interpreted fault could be identified by both manual and automatic methods (Slightam 2012). The Lancaster structure benefited from the earlier exploration well 205/21-1a drilled in 1974 (Fig. 4). The 205/21-1a well encountered hydrocarbon in the basement and overlying clastic succession. Oil bleeding from fractures was described in the end-of-well report and these data were used to define an oil-down-to contour (Fig. 4) and therefore an associated gross-rock-volume (GRV) in which to locate a suitable fault drilling target. Defining a well-connected fault was achieved by detailed seismic mapping of the fault network, integrating manual and coherency picked faults (Slightam 2012). This approach resulted in a map of the fault network which was then subjected to specialist software designed to provide a numeric connectivity analysis. The results of the connectivity analysis allowed for specific well-connected faults to be highlighted from which

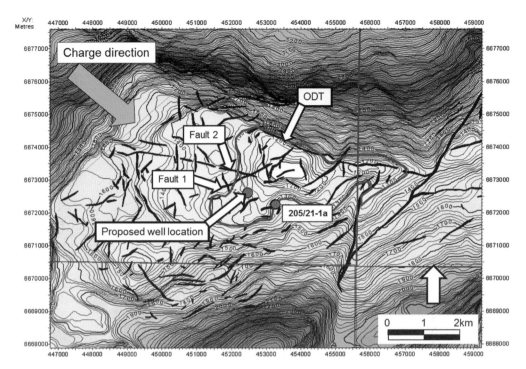

Fig. 4. Top basement surface of Lancaster Prospect portraying depth contours, faults considered as prospective drilling targets, the proposed Lancaster well (205/21a-4) and the two exploration targets Fault 1 and Fault 2. The discovery well 205/21-1a is shown and its related oil down to contour is marked as a red contour. Hydrocarbon charge direction from the Foula Sub-Basin is marked with a green arrow.

potential drilling targets could then be chosen. An important element in the connectivity analysis was high-grade long faults, as such faults had the greatest potential for extension into the basement and were therefore more likely to be subjected to hydrothermal processes. Fault age is an important consideration as older faults have the greatest potential to be reactivated, and reactivated basement faults are documented as being associated with enhanced permeability (Holdsworth et al. 2007). A further consideration was that older faults are more likely to have undergone more extensive periods of hydrothermal/epithermal enhancement given that the Lewisian Basement has a long history of orogenic activity and subaerial exposure.

It is well documented that there is a significant relationship between the maximum horizontal stress and fracture fluid flow (Heffer & Lean 1993; Ameen 2003; Rogers 2003; Henriksen & Brathen 2006). Stress impacts on fault fluid flow have also been documented (e.g. Barton et al. 1995; Canchani et al. 2003). However, it can be considered that in situ stress has a more limited effect upon the permeability of faults and fault zones in crystalline rocks as the rock mass within the fault zone can accommodate rock stresses in ways other than through the mechanical reduction of fracture aperture. An additional consideration is that channelling enhanced by dissolution/abrasion is less likely to respond to effective aperture reduction through stress-related processes than fracture apertures generated through purely mechanical processes. If seismic-scale faults are being pursued as the most productive targets, considerations of the current in situ stress field may therefore be of secondary importance to the structural history and the possibility of a history of fault reactivation. Predicting the stress magnitude and orientation of the basement's stress field at a specific proposed well location is a challenge, due to the paucity of bespoke stress measurements available for the Lancaster prospect. Recognizing the difficulty in assessing the current-day stress field, a variety of stress evaluation methods were employed and the results compared with the objective of establishing end-members for the most likely direction of maximum horizontal stress.

Evaluation of the seismic fault map using the five key criteria for a drillable fault target led to a specific well location that accommodated the drilling of two seismic-scale faults (Fig. 4). The faults were interpreted as being vertical based on inferences from outcrop analogues; this interpretation was subsequently supported by the Lancaster well result. Given the uncertainty of the stress conditions acting on the reservoir, and indeed the degree by which stress would affect fluid flow, it was decided that drilling across faults with different strike orientations (Fig. 4) would provide an improved chance of encountering permeable fault zones and thereby achieving oil to surface. This interpretation was further supported by coherency analysis and connectivity analysis that indicated the target faults to be well connected within the identified fault network. Furthermore, Fault 2 was interpreted as being connected directly to a kitchen area to the NW of Lancaster and was consequently considered to be a potential conduit for migrating hydrocarbons from the kitchen to the crest of the Lancaster structure (Fig. 4).

The various geological observations were combined and used to develop two conceptual models: one of the fracture network (Fig. 5) and one of the reservoir within structural closure (Fig. 6). The conceptual models of the fracture network utilize data from producing basement field analogues, seismic mapping of the Lancaster structure and outcrop observations from the Isle of Lewis (Slightam 2012) combined with the previously described observations of: (a) fracture type; (b) potential for hydrodynamic and epithermal enhancement; (c) structural disposition most likely to favour enhanced poro-perm systems. A further key assumption portrayed in Figure 5 is that maximum horizontal stress would be oriented NW–SE, thus preferentially enhancing the fluid flow potential of NW–SE-striking faults and similarly striking fractures within the fracture network. The conceptual fracture network model also indicates that orthogonal fracture sets and faults could contribute to background flow through providing additional connectivity. The 'background' fracture network is anticipated to support reasonable fluid flow through the provision of frequent and local high connectivity. Implicit in the Conceptual Fracture Network Model is the assumption that should the actual maximum horizontal stress differ from that modelled, then there would be corresponding effects on the conceptualized fracture behaviour to fluid flow such that preferential fluid flow would be noted within fractures and faults with strike directions aligned to within 45° of S_{Hmax} (Fig. 5).

The conceptual model of the 'basement reservoir within structural closure' was derived from observations gained from reviewing field analogue data and offset well data. It was used primarily to convey understanding of fluid distribution and the concept of a threefold facies scheme, which was

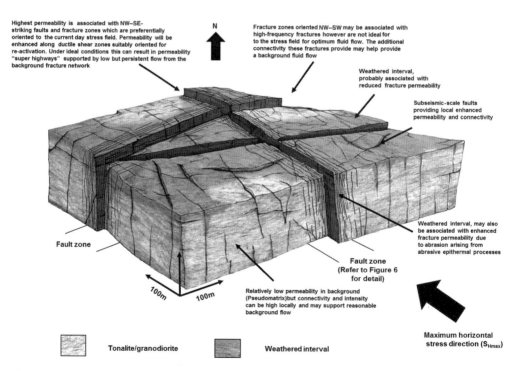

Fig. 5. Conceptual model of fracture network within Lancaster reservoir, with annotations reflecting aspects of the fracture network pertinent to exploration well planning.

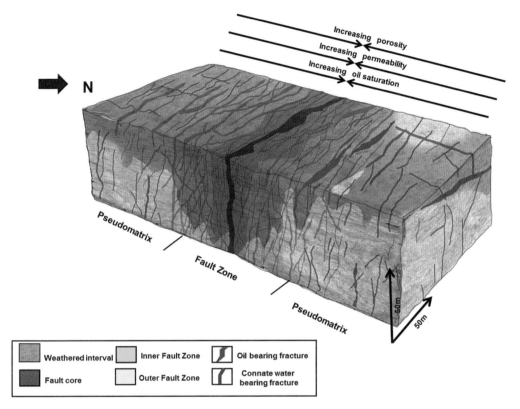

Fig. 6. Conceptual model of basement reservoir within structural closure. The formation is divided into three facies: (1) an Inner Fault Zone; (2) an Outer Fault Zone (both of which combine to make up a Fault Zone); and (3) Pseudomatrix. Fault Zones are associated with a specific GRV related to mapped seismic-scale faults and the Pseudomatrix represents the remaining GRV which consists of subseismic-scale faults and protolith.

used as a template for pre-drill volumetric analysis, and to act as a reference for well planning. The facies, which are based on the faults identified from seismic data, include Pseudomatrix, Inner Fault Zone and Outer Fault Zone (Fig. 6).

The conceptual model considers seismic-scale faults to be linear bodies and associated with volumes of faulted and fractured rock distributed within a fault zone. Fault zones are characterized as two facies: Inner Fault Zone and Outer Fault Zone (Fig. 6).

The Inner Fault Zone is effectively the volume of rock associated within the immediate vicinity of the fault plane and is associated with the greatest degree of fracturing. Such fracturing is considered to be at a range of scales with fracture lengths ranging from the core scale (pervasively fractured rock) up to decametre fracture lengths. Small-scale fractures are anticipated to have highly variable fracture strike azimuths whereas longer fractures are anticipated to have fracture strike azimuths parallel to the seismically mapped fault trace. The fracture network within the Inner Fault Zone was anticipated to be highly connected with relatively enhanced apertures. The Inner Fault Zone aperture enhancement and connectivity are physical properties related to preferential fluid flow associated with hydrothermal and epithermal processes. The Inner Fault Zone would also include volume(s) of fault core which was anticipated to be cataclasite and consequently a local seal or low permeability volume.

The Outer Fault Zone is the GRV volume located between the Inner Fault Zone and the Pseudomatrix. The Outer Fault Zone was assumed to be fractured with fracture lengths predominantly on the decametre scale oriented with their strike parallel to the seismic scale fault. Fracture frequency was anticipated to diminish as a function of distance away from the Inner Fault Zone.

The Pseudomatrix is the facies term applied to the remaining GRV and comprises fractures and subseismic-scale faults. Fracture orientations within the Pseudomatrix were anticipated to be dominated by regional fracture strike trends but also to be influenced by local seismic-scale and subseismic-scale faults (Fig. 5). The Pseudomatrix was considered to be associated with a wider range of fracture strike azimuths than Fault Zones resulting in a fracture network with local volumes of high connectivity, which could contribute to fluid flow (Fig. 5). However, it was anticipated that the Pseudomatrix would have a significantly less-well-connected hydrodynamic network than the Fault Zones and this would be reflected by significantly smaller effective apertures and lower fracture frequencies.

The nature of the fracture network, described by the conceptual model, would result in significant permeability anisotropy and heterogeneity which in turn would lead to specific volumes of the GRV having relatively poor fracture connectivity. Such regions could be bypassed during phases of hydrocarbon fill, resulting in volumes of perched water. This phenomenon would be exacerbated closer to any aquifer with the fractures within the Pseudomatrix retaining significant volumes of connate water. In contrast, the Inner Fault Zones would be associated with the highest oil saturations with residual water confined to the fracture population having the smallest aperture fractures and, of this water volume, a significant proportion would be considered immobile. Local alteration of the facies reservoir properties was anticipated to occur where the basement surface had been modified by subaerial exposure resulting in a weathered interval as depicted in Figures 5 and 6.

In summary it was anticipated that porosity permeability and oil saturation would increase significantly towards seismic-scale faults with the Inner Fault Zone representing the best-quality reservoir rock. This in turn would result in a variable 'oil down to' across the Lancaster prospect (Fig. 7) with an anticipated deeper oil-down-to associated with the structural flank proximal to the Foula Sub-Basin kitchen area (Fig. 9). Such a distribution of oil would in turn lead to the potential of oil being found outside the structural closure. Extensive hydrocarbon column heights are anticipated to be preferentially located proximal to the Faroe Shetland Basin (Foula Sub-Basin) along the NW flank of the Lancaster structure. Hurricane considered that a significant charge, or multiple charges, would be required to generate the necessary oil volumes for oil to accumulate outside of Lancaster's structural closure. A significant charge was considered unlikely to be provided solely by the East Solan Basin, which meant that the best

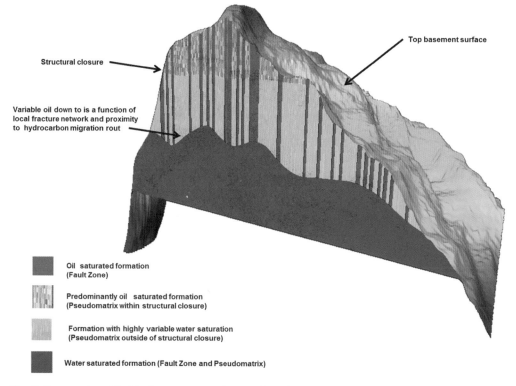

Fig. 7. Conceptual model of the Lancaster reservoir across the main crestal area depicting fluid distribution as a function of the reservoir facies Fault Zone and Pseudomatrix.

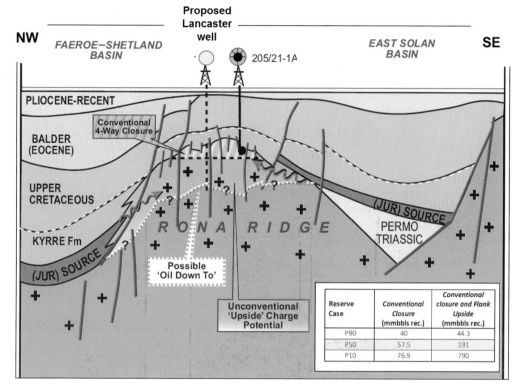

Fig. 8. Schematic NW–SE cross-section across the Lancaster Prospect portraying hydrocarbon associated with conventional four-way dip closure and oil trapped outside of conventional structural closure described as unconventional upside. Insert table is a summary of pre-drill prospective resource estimates for Lancaster Prospect.

chance for Lancaster to work as a prospect filled beyond structural spill would be for it to be charged from the Faroe Shetland Basin.

The second exploration challenge was therefore to establish the source of the 205/21-1a oil. Consequently a geochemical analysis of the oil samples recovered from 205/21-1a core was undertaken. This work was aided by an understanding of the regional basin development (Dean et al. 1999) and a burial history analysis of the Lancaster structure and the local kitchen area proximal to the Lancaster NW structural flank. The conclusions from the oil sample analysis were that the 205/21-1a oil was typical of oils sourced for the Faroe–Shetland Basin and distinct from oil sourced from the East Solan Basin. Hydrocarbon expulsion and timing was modelled using offset well data and pseudo-wells (PW1 and PW2, Fig. 9) located proximal to Lancaster in the Foula Sub-Basin of the Faroe–Shetland Basin (Farrimond 2007).

The results from this work were also correlated to burial history analysis (inset Fig. 9) and indicated that Lancaster had undergone at least two periods of charge with the latest charge occurring during a period of significant uplift post-Neogene time (Figs 9 & 10). An additional observation was that, prior to uplift, Lancaster had been buried to a sufficient depth that it crossed the 80 °C biodegradation threshold and had effectively become 'pasteurized,' killing any bacteria within the reservoir. Subsequent charge of the uplifted structure would therefore be unaffected by biodegradation and so the resultant accumulated oil would include a remnant of earlier biodegraded component, but be predominantly a light oil charge. The burial/expulsion history model and the predicted oil type fitted well with the oil density of 0.839 g mL^{-1} recorded from the testing of well 205/21-1a and published data describing regional fluid pulses (Parnell et al. 1999). These and other data were compared as a summary table (Fig. 10) to describe charge and tectonic history of the Lancaster prospect which indicated the favourable combination of coeval uplift and charge.

The combination of seismic mapping and geochemical analysis meant that a provisional well location and well path could be delineated. The exploration objective was to drill and test a 300 m

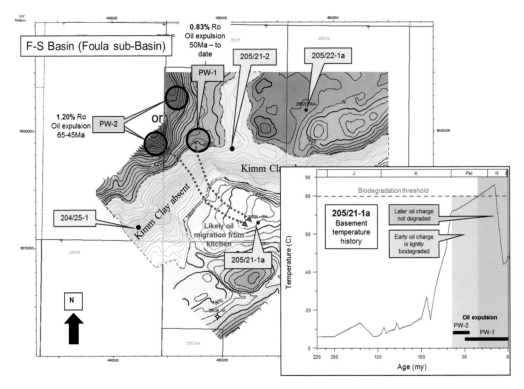

Fig. 9. Modelled hydrocarbon expulsion and timing from the Foula Sub-Basin to the Lancaster Prospect. Actual and modelled wells (PW-1 and PW-2) are marked. Inset graph depicts the temperature history, biodegradation threshold and timing of two phases of oil expulsion for the Lancaster prospect.

measured depth (MD) section which would cross two subvertical seismic-scale faults. The proposed deviated well path would penetrate the first fault (Fault 1) within the oil-down-to GRV and the second fault (Fault 2) would be penetrated at a depth outside of the structural closure, thus allowing for the testing for oil outside of the structural closure (Figs 4, 8 & 11).

The remaining technical challenge was to determine whether a 300 m basement section could be drilled effectively. Acceptable rate of penetration (ROP), vibration and mud loss management would be required in order to drill the well safely and cost-effectively and to provide quality logging while drilling/measurement while drilling (LWD/MWD) data. Offset and analogue data were used to evaluate bit technology and preferred bottom hole assemblages (SPD 2007, see also Nooraini et al. 2009). Techniques employed for controlling mud loss were reviewed from published examples from the Cuu Long Basin, Vietnam and correlated to drilling fluids performance. One of the challenges of drilling basement is to reduce drilling fluid ingress into open fractures. This is for two reasons: firstly, any solids associated with drilling fluid can reduce fracture permeability; and secondly, fluids can enter the fractured formation and require extensive 'clean up' during production testing. This not only adds time to an operation but provides ambiguity in distinguishing formation water from drilling fluids. While there are solutions to fluid loss, such as underbalanced drilling and managed pressure drilling, deployment of such engineering solutions is impractical when an operator is faced with a very limited rig market. Accepting that basement drilling could be a challenge and that losses were to be expected, a review of drilling fluids was undertaken and the optimum solution was determined to be use of a mixed metal oxide mud (MMO), which is a shear thinning fluid and as such has excellent solid suspension when being circulated by mud pumps. When an MMO penetrates a fracture, the fluid is out of the path of the circulating drilling fluid column and is consequently removed from shear processes. It subsequently solidifies into a highly viscous 'gel'. This solidifying effectively stops mud loss into the fractures. Post-drilling, the MMO can be removed from the fractures by deploying, through coiled tubing, weak acid which breaks down the MMO gel on contact.

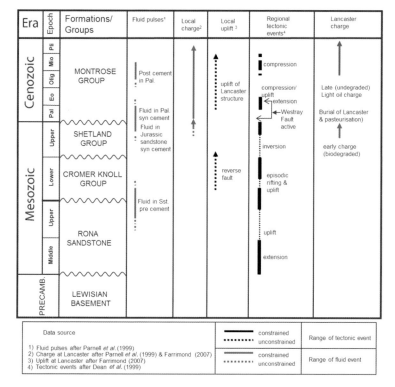

Fig. 10. Summary of local and regional tectonic events in relations to fluid pulses and local hydrocarbon charge to the Lancaster.

Part of Hurricane's operational planning was to have company staff present offshore. This approach arises from the author having the old-fashioned view that a geoscientist should take responsibility for his/her data, and that responsibility starts with data acquisition. While data acquisition includes wireline, testing and vertical seismic profile (VSP) witnessing, the broader context is to ensure that communication between the various disciplines for drilling, mud logging and LWD was optimized. Such an effort is considered paramount for fractured reservoir operations, as early integration of operational data is an essential step in decision making at the well site and in providing supporting information during post-operational data analysis.

With all technical and operational preparations in place, Hurricane was ready to drill what was, to our knowledge, the first well ever planned with the explicit intention of exploring the UK basement play.

Lancaster operations

The operational plan was to drill the basement section to the planned total depth (TD) equating to the base of the fault zone associated with Fault 2 and, if there was sufficiently encouraging signs of hydrocarbon presence, to extend the well to a depth recommended by the offshore team. The TD decision would be aided by high-resolution gas chromatography augmented by traditional mud logging techniques. On reaching TD, the well would be cleaned using acid deployed through coiled tubing and tested via a single open hole drillstem test (DST). During testing a production logging tool (PLT) would be run and down-hole fluid samples taken. Prior to testing, the reservoir section would be evaluated through an offset VSP and wireline logging, the latter to include image logs, conventional porosity and resistivity logs and wireline conveyed pressure testing. The latter technique was targeted at measuring pressure within individual fractures.

The operational plan started well with drilling through the basement proceeding better than expected. Clear signs of fractures were detected through drilling breaks and LWD and, despite these positive signs of fracture permeability, mud losses were kept at manageable levels. Additional support for formation permeability was provided by gas peaks, oil-stained cuttings and gas ratios typical of mobile oil. On reaching the planned TD

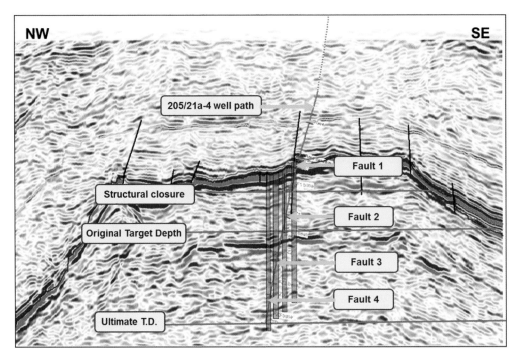

Fig. 11. Annotated seismic line oriented NW–SE through well 205/21a-4. Fault 1 and Fault 2 represent the original exploration drilling targets for well 205/21a-4. Fault 3 and Fault 4 are additional faults penetrated as a result of extending the well to a deeper TD.

the mud logging/gas chromatography data were supportive of hydrocarbon presence. The well was therefore extended by an additional 248 m, over which a further two faults were crossed (Fig. 11). During operations ROP and vibration while drilling were better than anticipated and, on inspection, the drill bit was seen to have very little wear. This was later determined to be a function of the Lewisian Basement being significantly less abrasive than the granites reported from the Cuu Long Basin, which Hurricane had referred to as one of the most recent analogues for drill bit selection.

At the final extended TD, everything looked extremely positive and wireline operations proceeded with the anticipation of detecting a highly fractured basement succession. The first wireline run included the FMI (Formation Microimager, electrical imaging log) and wellsite interpretation of this data supported the presence of numerous permeable fractures based on the integration of the FMI, gas chromatography, drilling breaks and LWD data. The presence of permeable fractures was later confirmed by running a wireline tester (Schlumberger Modular Dynamic Tester in dual packer mode) and drill stem testing.

Despite the very positive indicators of a highly permeable reservoir, the realization of a significant discovery was diluted by a testing programme fraught with operational difficulties. The first issue arose while running coiled tubing with the inability to deploy the tubing into the open hole due to a blockage within the test string. Despite significant effort, the blockage could not be overcome and testing proceeded without the benefit of cleaning the fractures of the drilling fluid.

After a short flow period to clean up the well, the test produced oil to surface (measured at 38° American Petroleum Institute); however, no stability in the produced fluid was achieved with oil rates significantly lower than water rates. In addition to oil and water production, the well flowed drilling fluid indicating that the well was cleaning up. It was also noted that the produced oil had mixed with the drilling fluid to create an emulsion, the flowing properties of which were unknown. Figure 12 presents representative operational photographs of how the MMO drilling fluid interacted with oil. Added to this unexpected turn of events, the weather was turning for the worst and it was clear that the well would need to be unlatched. In an attempt to reduce the water cut it was decided to cement the open hole to the base of Fault 2 (a depth consistent with very good indicators of a hydrocarbon column) and then unlatch the well

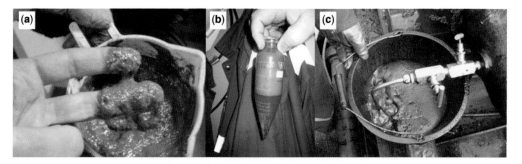

Fig. 12. (a) Characteristics of mixed metal oxide drilling mud during testing aerated foam; (b) fluid sample after gravity segregation demonstrating emulsion and separated water; and (c) foaming oil emulsion at the test separator.

with the intention of undertaking a second DST. The second DST would hopefully not be associated with a blockage, and coiled tubing could therefore be deployed to clean the fractures of drilling fluid. The second DST would be confined to the interval of the original drilling targets Fault 1 and Fault 2.

On laying out the DST string on deck and later inspection under laboratory conditions it was determined that the SenTree component of the DST string was packed with rubber and the ball valve would only partially open. DST 1 had therefore been undertaken through highly restricted tubing which compromised the ability to achieve an optimum flow. Analysis of the rubber and inspection of the blowout preventer (BOP) confirmed that the rubber was sourced from the BOP due to an earlier stripping operation.

DST 2 was also undertaken without the benefit of running coiled tubing. In this instance the coiled tubing was not run due to time constraints imposed by anticipated bad weather. The DST 2 results showed no material improvement despite a short period of 100% oil flow. Analysis of BS&W (bottom sediment and water) indicated that produced fluids were being impacted by drilling fluid and that the drilling fluid was reacting with the oil to create an emulsion, as in DST 1. The produced fluids from DST 2 can be summarized as 200 bopd, (15% of total flow) v. 1200 barrels of water per day (bwpd), 85% of total flow. Given the compromised nature of the DSTs, the decision was made to suspend the well with the intention of returning to Lancaster to undertake a side-track. In summary, the key conclusions from testing the Lancaster well were:

(1) a light 38° oil was produced;
(2) no H_2S or CO_2 were produced;
(3) no depletion was evident;
(4) the reservoir was highly permeable;
(5) flow rates were limited by a tubing restriction;
(6) the well was not cleaned up partly due to limited drawdown due to the rubber debris within the test string and partly due to the return permeability properties of the drilling fluid (which was a significant inhibitor to oil flow and likely to have been a major factor in the poor production performance of the test); and
(7) formation water was produced.

Given the 205/21a-4 results the Lancaster side-track (205/21a-4z) was planned with the primary objective of undertaking a test that would significantly reduce the ambiguity of reservoir produceability and fluid distribution. The side-track was drilled in the following year (2010) with the aim of evaluating both Fault 1 and Fault 2, which were penetrated in the original well. The side-track was designed to drill and test Fault 1 and then drill to the base of Fault 2 and then test Fault 1 and Fault 2 as a combined test (Fig. 13). A key part of the rationale behind this plan was to check for mobile water within the Pseudomatrix and each fault zone as the mud logging data acquired during the original well indicated that mobile water could be associated with Fault 1 but was less likely within Fault 2. During both DSTs production logging and fluid sampling was to be undertaken. Seabed pressure gauges were also deployed as it was anticipated that reservoir pressures would be affected by semi-diurnal and diurnal tides; with the provision of accurate seabed data, the reservoir pressures could be corrected for tidal effects. In addition to the testing programme, an extensive wireline programme was planned, expanding on that already procured in 205/21a-4 with the addition of sidewall cores, acoustic imaging and nuclear magnetic resonance.

The side-track logging and testing proceeded as planned with no material incident and procurement of excellent quality static and dynamic data. Wireline data, including wireline pressure

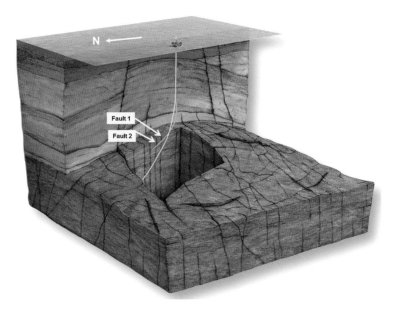

Fig. 13. Representation of the Lancaster crestal area portraying the original 205/21a-4 (yellow) and side-track, 205/21a4z, (green) wells and their relative position to Fault 1 and Fault 2.

measurements, supported the presence of numerous permeable fractures and the well test results over the comingled interval (geological equivalent to the original well DST 2) providing encouraging results with oil flow rates significantly in excess of that achieved in the original well and water rates reduced to 8% of flow being confined to a single fracture associated with Fault 2. A comparison of the DSTs over the Fault 1 and Fault 2 interval from the original and the side-track are provided in Figure 14. Analysis of the side-track well test data supports the assumption that the majority of the reservoir is fractured and productive. The well was also associated with significant skin attributed to cuttings ingress and the effects of bull-heading; despite this, a maximum flow rate of 2500 bopd was achieved (the original well achieved a maximum oil flow rate of 200 bopd). Post-well analysis indicates that should well damage and the associated skin be overcome, the reservoir should be able to achieve sustainable rates in excess of 10 000 bopd. Given that the skin appears to be related to drilling and well control methods, it is anticipated that skin can be materially reduced in future wells. Data to support this supposition is provided by a comparison of the flow rates achieved in the side-track DST 1 and DST 2. DST 1 was associated with seven distinct flowing intervals; however, when the equivalent interval is compared to DST 2, specific intervals flowing during DST 1 were not flowing during DST (Fig. 15). As there is no evidence of cross flow it has been concluded that the reduced flow potential between DST1 and DST 2 is a result of an operational phenomenon rather than a product of reservoir characteristics.

DST 2 is also associated with water production with water from a single fracture being produced on a 48/64" choke; however, this fracture does not represent an oil-down-to as dry oil was produced from this fracture on a 16/64" choke and from fractures located deeper in the well (Fig. 15). The origin of the water is yet to be confirmed and it is postulated that the water originates from a depth below the side-track TD, with the water mobility encouraged by preferential 'injection' of bull-headed fluid through the wide aperture fracture (aperture estimate accounting for hole deviation and fracture angle is 0.6 m) prior to testing. This thesis is supported by the change in water chemistry during DST 2, which showed a gradation in salinity from that specific to the drilling fluid to that of formation water recorded during testing of the original well.

The key outcome from evaluating the original and side-track data is that the reservoir is highly permeable and far more fractured than predicted in the conceptual model, particularly within the reservoir volume assigned as Pseudomatrix. Permeable and productive fractures have been established by the integration of image logs, PLT and wireline tester data and are found to comprise fractures associated with seismic-scale fault zones and also

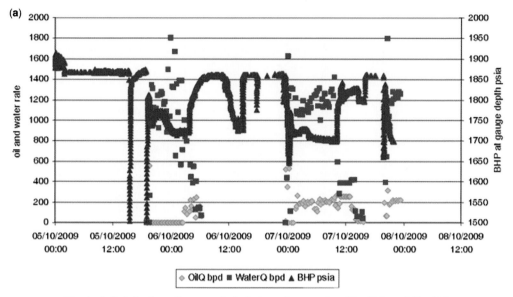

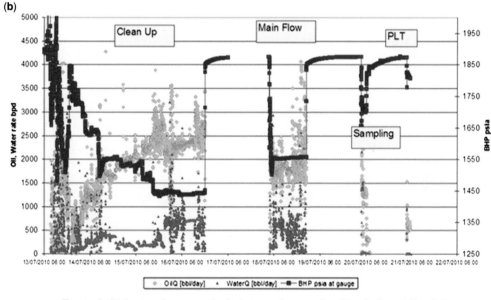

Fig. 14. Comparison of (**a**) original and (**b**) side-track drill stem test measurements over the interval of Fault 1 and Fault 2. Y-axis: oil water rate (bpd), scale 0–2000 for original and 0–5000 (bpd) for side-track. The marked improvement in the side-track DST is attributed to a test string devoid of blockage and use of brine as a drilling fluid.

regional joints sets that are distributed throughout the penetrated intervals. Figure 16 provides examples of representative image log responses for the Lancaster reservoir. Each interval portrayed in Figure 16 is associated with fluid flow as determined from the PLT (refer to Fig. 14). Further details on Lancaster fracture characteristics can be found in Slightam (2012).

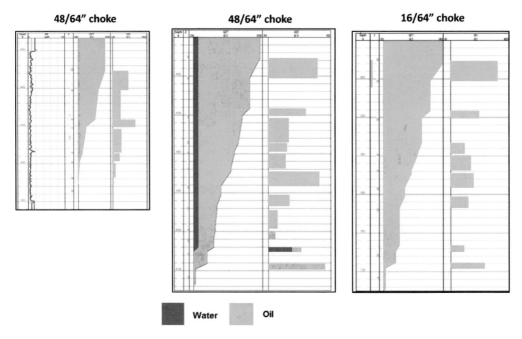

Fig. 15. Comparison of cumulative and zonal flow from PLTs recorded in the side-track (205/21a4z) over Fault 1 and Fault 1 comingled with Fault 2. DST 1 represents flow from top basement to the base of Fault Zone 1 and DST 2 represents flow from top basement to the base of Fault Zone 2. Refer to Figure 13 for relative position of Fault 1 and Fault 2 to the side-track well path.

Further to analysis of both the original and side-track well data, a surprising and key observation is that the Fault Zone/Pseudomatrix transition cannot be defined by fracture frequency and that no distinction between an Inner and Outer Fault Zone can be made. This conclusion reflects the very high intensity of fractures throughout the logged intervals rather than poor-quality or ambiguous data. This observation is further supported by post-side-track fieldwork undertaken on the Isle of Lewis. As a

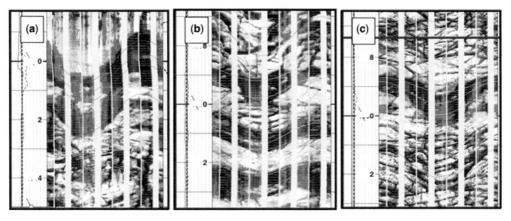

Fig. 16. Representative image logs responses of the Lancaster reservoir. Images are dynamically enhanced over the presented image interval and presented depth increments are 2 ft. Dark colouration is electrically conductive formation; light colour is electrically resistive formation. (**a**) Relatively wide aperture fracture associated with oil and water production; (**b**) formation within a Fault Zone close to the Fault Zone/Pseudomatrix boundary; and (**c**) an example of a relatively wide aperture fracture and an image log fabric interpreted as representing fault rock. Image (b) and (c) are associated with oil inflow identified from the PLT.

result of these well- and field-based observations, a new conceptual model was generated in which the GRV volume is split into Pseudomatrix and Fault Zone (Fig. 17). This bimodal facies split correlates well with wireline mudlogging and fracture aperture data and replaces the tripartite split of the original predrill conceptual model as depicted in Figure 6. The bimodal split is also refined by the presence of a pervasive NE–SW regional joint set which is present within Pseudomatrix and Fault Zone intervals. A further refinement of the original conceptual model is that the effects of weathering have been removed from the conceptual model as neither the side-track nor the original well show any significant rock mass weathering effects despite acquisition of sidewall cores, natural gamma and chemical logging data. It is accepted that weathering of the rock mass may be present elsewhere on the Lancaster structure, however it is also acknowledged that granites having a low biotite mica content are more resistant to weathering and may form hills or mountains with abundant rock outcrop and thin, sandy, easily eroded soils. Other refinements to the conceptual model include lithology, with the original conceptual model based only on tonalite (a quartz-rich granite) being refined to include dolerite. The dolerite is distributed with a well-defined North–South strike orientation. Dating of a dolerite sample from the side-track well provides a minimum age of 2.3 billion years, indicating that the dolerite is interpreted as being part of the initial melt rather than a post-cooling intrusion. Image log data indicate dolerite intervals to be highly fractured and the PLT data indicate that fractured dolerite contributes to hydrocarbon flow.

Establishing the Lancaster Discovery resource potential

Fracture reservoirs are a challenge to evaluate from the perspective of establishing hydrocarbon resource volumes. Currently there is no

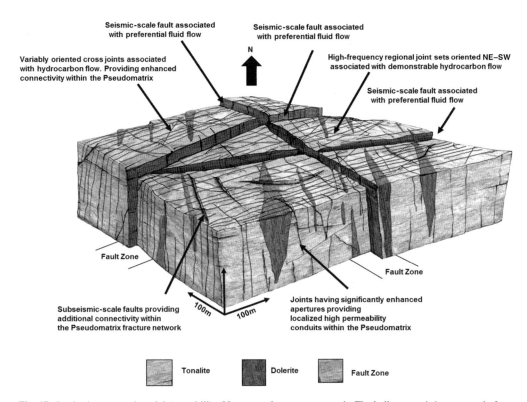

Fig. 17. Revised conceptual model (post drill) of Lancaster basement reservoir. The bulk reservoir is composed of tonalite which comprises both a foliated and un-foliated rock fabric. Dolerite is subordinate to tonalite and is present as lens-shaped bodies, interpreted to be associated with the original melt rather than as post cooling intrusions. Productive fractures are present within the Pseudomatrix and Fault Zones. Fault Zone reservoir quality is preferentially enhanced by fractures associated with fault rock and fractures having enhanced aperture/connectivity.

acknowledged industry standard for evaluating fractured reservoirs; those companies with fractured reservoir assets develop their own methodology for assessing resource volumes. Type 1 fractured reservoirs bring their own specific challenges and Hurricane has worked closely with third-party specialists to develop a robust methodology for establishing fractured basement resource volumes. The details of this approach remain proprietary; however, the method applied provides a mechanism for evaluating and risking hydrocarbon within structural closure and hydrocarbon outside of structural closure. This bilateral split provides a very practical method of accommodating Hurricane's exploration and appraisal strategy, as well as focusing on the specific resource volume that could be fast tracked to field development. The basic steps to estimating resource volumes are summarized in Figure 18. Figure 18a depicts the faults across the Lancaster structure and is derived from manual fault picks and coherency analysis (Slightam 2012).

Faults are treated as vertical features based on the practicality of modelling software and the interpretation of seismic fault attitude inferred from the evaluation of image log, VSP and outcrop data and the interpretation of intra-basement reflectors. Any disparity between modelled fault geometries and reality is considered to have an insignificant error on calculated in-place volumes. While it is accepted that fault and fracture geometry plays a significant role in the connectivity of the fracture network and consequently the ultimate recoverable volume, the geometry of the fracture network has not explicitly been applied in establishing an effective recovery factor at the time of writing.

A subset of the faults is extracted and transferred to a static 3D model (Fig. 18b). This subset is used to model the two facies component of Fault Zone and Pseudomatrix, each of which has distinct input parameters. This subset is also used to define a dynamic sector model, which is used to predict well rates as well as act as a corroboration of the static model. The modelled GRV is confined by a top surface which is provided by the top basement reflector (Fig. 18c) and a bottom surface represented by an ODT. The ODT is variable to allow for the modelling of oil within structural closure and without structural closure. A depiction of a single realization of a model based on a deep oil contact is seen by reference to Figure 19.

Input parameters for the facies are chosen to reflect the conceptual model and are constrained by static and dynamic data recorded from the sidetrack and original wells. Net to gross is assumed to be 100%. This reflects the conceptual model of a Type 1 fractured reservoir and is consistent with PLT and image log data which indicate that permeable fractures are pervasively distributed throughout the drilled sections and, by inference, throughout the Lancaster structure. Effective fracture porosity (porosity) is considered to be entirely associated with a hydrodynamic fracture network with no material contribution from the matrix. It was therefore considered that bulk porosity derived from a combination of neutron/density and nuclear magnetic resonance would provide end-member values from which to estimate porosity ranges. The established porosity ranges are consistent with reported analogue data (C&C Reservoirs 2005). Based on the established wireline-derived porosity ranges the average porosity for the gross structure is calculated to be 4.7%. This average porosity value is consistent with porosity estimates derived from history matching of the Lancaster well data and estimates of average fracture porosity derived from image log fracture apertures calibrated to PLT data.

For the purposes of volumetric modelling, bulk porosity measurements are used to populate the facies input parameters. Water saturation (S_W) cannot be readily measured in a Type 1 fractured reservoir as the relevant wireline tools run to infer S_W will be affected by near wellbore invasion of saline drilling fluids (Tuan et al. 1995; Ngoc et al. 2007). This fact, coupled with disparities in the measured volume for formation between resistivity tools and porosity tools, further compounds the problem, also exacerbated by the heterogeneous and commonly anisotropic poro-perm systems present in Type 1 fractured reservoirs. While it can be inferred that the irreducible water saturation within a fracture measured at the core plug scale is of order 0–20% (Aguilera 1999; Ngoc et al. 2007), the effective porosity system at Lancaster is not, and cannot be, represented by core or core-plug-scale measurements as the flowing fractures as detected by image log and PLT comparison cross-cut the borehole as discrete planes. Such fractures are commonly associated with apertures of sufficient magnitude to cause discernible responses on conventional and high-resolution wireline logs; with apertures estimated to be commonly in excess of 2 cm, such fractures are anticipated to be associated with zero irreducible water (Aguilera 1999). Despite the inability to measure S_W it is used in the volumetric analysis; however, it is applied primarily to provide an estimate of perched water distribution. Perched water has not been confirmed with the available well data; however, perched water is predicted to be present by the conceptual model and therefore requires inclusion. The thickness of the Fault Zone is determined from a combination of well and outcrop data while the upper and lower boundaries of the Fault Zone within the model are terminated by either basement top

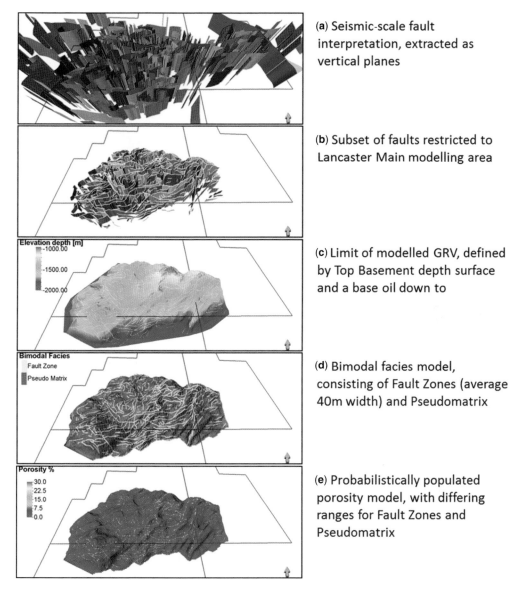

Fig. 18. The basic five steps used in evaluating the resource potential of the Lancaster Prospect. (**a**) Seismic-scale faults mapped across the Lancaster structure. Faults planes are interpreted to be vertical. (**b**) A subset of the fault volume extracted for modelling purposes. (**c**) The modelled GRV. The upper surface is defined by the top basement reflector and the lower surface is defined by oil-down-to. (**d**) Portrays the GRV split into the two reservoir facies Pseudomatrix and Fault Zones. (**e**) Reservoir properties are probabilistically populated within the model and, in this example, porosity is presented within Fault Zones and Pseudomatrix.

surface or the modelled ODT. Tables 1 and 2 provide summaries of the input parameters applied in the evaluation of the Lancaster GRV within and outside of structural closure. For GRV outside of structural closure, the input parameters are varied to reflect geological uncertainty and the concept that oil distribution will have increased irregularity as a function of the connected fracture pathways, relative proximity to any aquifer and the relative position on the structure with flank locations anticipated to be associated with more extensive oil columns (Table 2).

Once constructed the model was populated with input parameter ranges and various volumetric

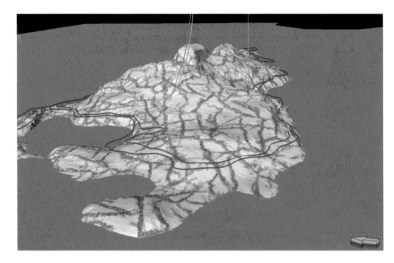

Fig. 19. Example 3D model realization of the Lancaster basement structure portraying modelled faults, wells and top basement surface in context to a shallow, intermediate and deep oil down to (ODT).

modelling estimates were run using a combination of Monte Carlo and static-model-based realizations. Third-party volumetric evaluation, as part of a formal competent persons report, estimate that Lancaster has a contingent (1C-3C) resource range of 62–456 MMboe. The next step towards further de-risking of the Lancaster resource potential is to undertake an appraisal drilling program.

Appraisal wells will be placed and designed so that they can, in the case of success, be readily converted to producing wells. Appraisal well drilling will be targeted to reduce uncertainty on the 1C-2C volumes by further evaluating the ODT and to drill to a sufficient depth that will allow for any aquifer to be evaluated. In addition, appraisal drilling will investigate the potential of achieving improved productivity through a horizontal well.

The first horizontal well will be planned to evaluate the 1C volume, specifically the productivity, fracture network characteristics and potential for perched water within structural closure. If both of these wells prove to be successful it is anticipated that Lancaster will be progressed to a first phase of field development. This first phase will concentrate on accessing the estimated 60 million barrels of recoverable oil within structural closure and in the recognition that excessive 'high or aggressive' flow rates will shorten the productive life for the reservoir and result in a lower ultimate recovery. An additional consideration for the successful management of the Lancaster Discovery is that reservoir performance will depend on the drive mechanism. It is therefore critical to gain an understanding of the hydraulic fracture network and

Table 1. *Input parameters applied to resource evaluation within structural closure*

	Low	Best	High
Contact range (m)	1333	1340	1380
Proportion PM (%)		64	
Proportion FZ (%)		36	
Porosity PM (%)	1	2	4
Porosity FZ (%)	3	7	10
S_W PM (%)	5	10	20
S_W FZ (%)	2	5	10
B_0 (rb/stb)	1.2	1.2	1.2

B_0, oil formation volume factor; PM, Pseudomatrix; FZ, Fault Zone; rb, barrel(s) of oil at reservoir conditions; stb, stock tank barrel.

Table 2. *Input parameters applied to resource evaluation within GRV outside of structural closure*

	Low	Best	High
Contact Range Discovered (m)	1475	1597	1781
Contact Range Undiscovered (m)	1781		2000
Proportion PM (%)		64	
Proportion FZ (%)		36	
Porosity PM (%)	0.5	1	3
Porosity FZ (%)	2	5	8
S_W PM (%)	30	50	70
S_W FZ (%)	2	5	10
B_0 (rb/stb)		1.2	

PM, Pseudomatrix; FZ, Fault Zone.

the drive mechanism ahead of a field development plan. This will require:

(a) the ability to map drillable targets from the available seismic data;
(b) acquisition of sufficient static and dynamic data to effectively quantify the properties of the fracture network; and
(c) acquisition of sufficient static and dynamic data to quantify the properties of an aquifer or gas cap.

Conclusions

The exploration concept that fractured Lewisian basement has the potential to be a productive and material play has been evaluated by two wells on the Lancaster prospect. The exploration program successfully tested the concept that basement would be associated with preferentially permeable seismic-scale fault zones and that oil columns would extend outside of structural closure. The well results also confirmed geochemical modelling that indicated light oil would be present within the basement reservoir. The strategy of spending time in acquiring high-quality and wide-ranging data and the investment in ensuring that company staff have a visible offshore presence has paid off, resulting in a robust estimate of Lancaster's resource potential. Furthermore, the testing of a pre-drill conceptual model has resulted in a refined geological model from which future operations can be planned with increased confidence. Correlation of well data with seismic data indicates that seismically mapped faults can be identified from wireline data and that such faults are associated with preferential reservoir properties. Comparison of subsurface data with outcrop analogues has proven invaluable in developing conceptual and numerical reservoir models. The observation that the basement play may represent a material resource for the UK is given further encouragement by Hurricane's second basement discovery, Whirlwind, which was drilled back to back with the Lancaster sidetrack in 2010. The Whirlwind discovery located some 12km north of Lancaster has very similar reservoir properties to Lancaster and is associated with either volatile light oil or a gas condensate. Contingent resource volumes for Whirlwind are of a similar magnitude to Lancaster ranging from 98 to 373 million stock tank barrels (MMstb) of oil and 236–1017 billion standard cubic feet (Bscf) of gas for the volatile oil case and 91–301 MMstb oil and 437–1308 Bscf of gas for the gas condensate case. The current evaluation of two basement discoveries, Lancaster and Whirlwind, and two basement prospects, Lincoln and Typhoon, indicates that basement has the potential to deliver on Hurricane's acreage 710 MMboe 2C contingent resource estimate and 444 MMboe of un-risked mean prospective resource. In considering such estimates it is important to recognize that the two prospects are associated with oil on structure as confirmed by previous operators' exploration activities.

In evaluating the basement, Hurricane has attempted new and experimental techniques and has been rewarded by some valuable lessons and insight into a reservoir that has the potential to be a strategic resource for the UK. Lancaster would not have been possible without the philosophy promoted by DTI/DECC to the PESGB in January 2002 in which the clear message was that 'we want to get licenses in the hands of those who are hungry and innovative'. It was this message that encouraged Hurricane to explore the basement of the Atlantic Margin and it is continued application of this philosophy that will be required to take the UK Basement Play to the next stage and progress Lancaster to reserves and field development.

I would like to thank the reviewers of this manuscript for their comments and suggestions which have helped improve this paper and bring clarity to key issues.

References

AGUILLERA, R. 1995. *Naturally Fractured Reservoirs*. 2nd edn. Pennwell Publishing Company, Tulsa, OK.

AGUILLERA, R. 1996. *Hydrocarbon Production from Naturally Fractured Granite*. Technical Note 2. April 18, 1996. http://www.servipetrol.com/oldtech.htm

AGUILERA, R. 1999. Recovery factors and reserves in naturally fractured reservoirs. *Journal of Canadian Petroleum Technology*, **38**, 15–18.

AMEEN, M. S. (ed.) 2003. *Fracture and In-situ Stress Characterization of Hydrocarbon Reservoirs*. Geological Society, London, Special Publications, **209**.

ARESHEV, E. G., DONG, T. L., SAN, N. T. & SCHNIP, O. A. 1992. Reservoirs in fractured basement on the continental shelf of Southern Vietnam. *Journal of Petroleum Geology*, **15**, 451–464.

BARTON, C. A., ZOBACK, M. D. & MOOS, D. 1995. Fluid flow along potentially active faults in crystalline rock. *Geology*, **23**, 683–686.

CANCHANI, S. K., ZOBACK, M. D. & BARTON, C. 2003. A case study of hydrocarbon transport along active faults and production-related stress changes in the Monterey Formation, California. *In*: AMEEN, M. S. (ed.) *Fracture and In-situ Stress Characterization of Hydrocarbon Reservoirs*. Geological Society, London, Special Publications, **209**, 17–26.

CLOOS, H. 1922. Uber Ausbau und Anwendung der granittek-tonischen methode. *Preussischen Geologischen Landesan-stalt*, **89**, 1–18.

C&C RESERVOIRS 2005. *Chapter five, Type 1 Fractured Oil reservoirs in Fractured, Tight and Unconventional Reservoirs*. Published by C&C Reservoirs Inc. http://ww3.ccreservoirs.com/unconventional_reservoirs_outline.html

CUONG, T. X. & WARREN, J. K. 2011. Bach Ho Field, a fractured granitic basement reservoir, Cuu Long Basin, Offshore SE Vietnam: A "Buried Hill" Play. *Journal of Petroleum Geology*, **32**, 129–156.

DE URREIZTIETA, M. ET AL. 2007. Formation testing strategy based on quick-look data evaluation from a fractured basement reservoir: A case study from Kahirir Field Yemen. *International Petroleum Conference IPTC 11518*, 4–6 December 2007, Dubai, U.A.E.

DEAN, K., MCLACHLAN, K. & CHAMBERS, A. 1999. Hydrocarbon migration history, West of Shetland: integrated fluid inclusion and fission track studies. *In*: FLEET, A. J. & BOLDY, S. A. R. (eds) *Petroleum Geology of Northwest Europe*. Proceedings of the 5th Conference, Geological Society, London, 533–544.

DIMITRIYEVSKIY, A. N., KIREYEV, F. A., BOCHKO, R. A. & FEDEROVA, T. A. 1993. Hydrothermal origin of oil and gas reservoirs in basement rock of the South Vietnam continental shelf. *International Geology Review*, **35**, 621–630.

EGGLESTON, W. S. 1948. Summary of oil production from fractured rocks reservoirs in California. *AAPG Bulletin*, **32**, 1352–1355.

FARRIMOND, P. 2007. *Petroleum systems of southern Quadrant 205, West of Shetlands, UK*. Consultancy report commissioned by Hurricane Exploration plc.

GEOSCIENCE 2007. *Analogue studies in support of basement fracture and flow characterization in Block 205, West of Shetlands*. Consultancy report commissioned by Gutmanis, J. C. (author) at Hurricane Exploration plc.

GUTMANIS, J. C. 2009. *Basement reservoirs-a review of their geological and production characteristics*. *In*: International Petroleum Technology Conference, IPTC 13156, 7–9 December 2009, Doha, Qatar.

GUTMANIS, J., BACHELOR, J., COTTON, L. & BAKER, J. 2012. Hydrocarbon production from fractured basement formations. Available at http://www.geoscience.co.uk/assets/file/Reservoirs%20in%20Fractured%20Basement%20Ver%2010_JCG.pdf

HEFFER, K. J. & LEAN, J. C. 1993. Earth stress orientation – a control on, and guide to, flooding directionality in a majority of reservoirs. *In*: LINVILLE, B. (ed.) *Reservoir Characterization III*. PennWell Books, Tulsa.

HENRIKSEN, H. & BRATHEN, A. 2006. Effects of fracture lineaments and in-situ rock stresses on groundwater flow in hard rock: a case study from Sunnfjord, Western Norway. *Hydrogeology*, **14**, 444–461.

HOLDSWORTH, B., MCCAFFREY, K., JONES, R. & IMBER, J. 2007. *Outcrop Data as an Analogue to HEX Lewisian Basement Acreage: Characterisation of Lewisian Fracturing in Scotland & the Møre-Trøndelag Fault Zone in Norway*. Consultancy report commissioned by Hurricane Exploration plc.

HUBBERT, M. K. & WILLIS, D. G. 1955. Important fractured reservoirs in the United States. Proceedings of 4th World Petroleum Conference, Rome, Section A-1, 57–81.

JOLLY, R. J. H. & COSGROVE, J. W. 2003. Geological evidence of patterns of fluid flow through fracture networks: examination using random realizations and connectivity analysis. *In*: AMEEN, M. S. (ed.) *Fracture and In-situ Stress Characterization of Hydrocarbon Reservoirs*. Geological Society, London, Special Publication, **209**, 177–187.

LANDNES, K. K. 1959. *Petroleum Geology*. 2nd edn. John Wiley and Sons, Inc, New York.

LANDNES, K. K., AMORUSO, J. J., CHARLESWORTH, L. J., HEANY, F. & LESPERANCEP, J. 1960. Petroleum resources in basement rocks. *AAPG Bulletin*, **44**, 1682–1691.

LI, B., GUTTORMSEN, J., HOI, T. V. & DUC, N. V. 2004. *Characterizing Permeability for the Fractured Basement Reservoirs*. SPE 88478, SPD Ltd, Aberbeen.

NELSON, R. A. 2001. *Geologic Analysis of Naturally Fractured Reservoirs*. Gulf Professional Publishing USA.

NGOC, S. L., JAMIOLAHMADY, M., QUESTIAUX, J. M. & SOHRABI, M. 2007. SPE 107141. An integrated geology and reservoir engineering approach for modelling and history matching of a Vietnamese fractured granite basement reservoir. *In*: EUROPEC/EAGE Conference and Exhibition, London, UK, 11–14 June 2007.

NOORAINI, M. B., TAN, J. H. P., HASMATALI, D., MING, L. K. & HEISIG, G. 2009. *Promising results from first deployments of rotary steerable technology in Vietnam basement granite*. Conference paper from Asia Pacific Oil and Gas Conference & Exhibition, 4–6 August, Jakarta, Indonesia. Published by Society of Petroleum Engineers, SPE-122690-MS.

P'AN, C. H. 1982. Petroleum in basement rocks. *AAPG Bulletin*, **66**, 1597–1643.

PARNELL, J., CAREY, P. F., GREEN, P. & DUNCAN, W. 1999. Hydrocarbon migration history, West of Shetland: integrated fluid inclusion and fission track studies. *In*: FLEET, A. J. & BOLDY, S. A. R. (eds) *Petroleum Geology of Northwest Europe*. Proceedings of the 5th Conference, Geological Society, London, 613–625.

PETFORD, N. & MCCAFFREY, K. (eds) 2003. *Hydrocarbons in Crystalline Rocks*. Geological Society, London, Special Publications, **214**.

PYRAK-NOLTE, L. J., MYER, L. R., COOK, N. G. W. & WITHERSPOON, P. A. 1987. *Hydraulic and mechanical properties of natural fractures in low permeability rock*. Proceedings International Society for Rock Mechanics, 6th International Congress on Rock Mechanics. A.A. Balkema, Rotterdam, Montreal, 225–231.

REISS, L. H. 1980. *The Reservoir Engineering Aspects of Fractured Formations*. Gulf Publishing, Houston, Texas.

ROGERS, S. 2003. Critical stress-related permeability in fractured rock. *In*: AMEEN, M. S. (ed.) *Fracture and In situ Stress Characterization of Hydrocarbon Reservoirs*. Geological Society, London, Special Publications, **209**, 7–16.

SAN, N. T., GIAO, N., DONG, T. L. & SON, H. P. 1997. *Pre-Tertiary basement: the new objective for oil and gas exploration in the continental shelf of South Vietnam*. Proceedings of the Petroleum System of SE Asia and Australasia Conference, May 1997.

SLIGHTAM, C. 2012. Characterizing seismic-scale faults pre- and post-drilling; Lewisian Basement, West of Shetlands, UK. *In*: SPENCE, G. H., REDFERN, J., AGUILERA, R., BEVAN, T. G., COSGROVE, J. W., COUPLES, G. D. & DANIEL, J. M. (eds) *Advances in the Study of Fractured Reservoirs*. Geological Society, Special Publications, **374**. First published online September 11, 2012, http://dx.doi.org/10.1144/SP374.6

SPD 2007. *Basement Drilling Optimization Review West of Shetland*. Consultancy Report Commissioned by Hurricane Exploration Plc.

TRAN, D. L., TRAN, H. V. ET AL. 2006. *Basement Fractured Reservoir of Bach Ho oil field – a Case Study. The Basement Reservoir*. HCM Science and Technics Publishing House, Ho Chi Mihn.

TUAN, P. A., MARTYNTSIV, O. F. & DONG, T. L. 1995. *Heterogeneity of fractured granite: effects on petrophysical properties of water saturation of preserved cores*. Society of Core Analysts Conference Paper Number 9525. http://www.scaweb.org/assets/papers/1995_papers/1-SCA1995-25.pdf.

TSANG, Y. W. & TSANG, C. F. 1989. Flow channelling in a single fracture as a two dimensional strongly heterogeneous permeable medium. *Water Resources Research* **25**, 2076–2080.

WU, S., SUN, S., TRICE, R. & GUZMAN, J. 2005. *Application of geological analogs in the exploration workflow: drawing on decades of knowledge to identify the critical elements of a play or trap*. Presented at the AAPPG Calgary Meeting, June 19–22, 2005. http://www.searchanddiscovery.com/abstracts/html/2005/annual/abstracts/wu.htm

YOUNES, A. I., ENGELDER, T. & BOSWORTH, W. 1998. Fracture distribution in faulted basement blocks: Gulf of Suez, Egypt. *In*: COWARD, M. P., DALTABAN, T. S. & JOHNSON, H. (eds) *Structural Geology in Reservoir Characterization*. Geological Society, Special Publications, **127**, 167–190.

Laggan Field 3D geological modelling: case study

L. H. KHEIDRI[1]*, F. CAILLY[3], K. JONES[1], A. DE SENNEVILLE[2] & R. GRAY[1]

[1]Total E&P UK Limited, Crawpeel Road, Altens, Aberdeen AB12 3FG, UK

[2]Total E&P S.A., 2 place de la Coupole, La Defense, Paris 92078, France

[3]Total E&P Avenue Raymond Poincaré B.P. 761 Pointe Noire, République du Congo

*Corresponding author (e-mail: hamza.kheidri@total.com)

Abstract: Drilling of Laggan development wells commenced in summer 2013 and finalization of well locations required updated reservoir models, fully incorporating the 2009 baseline seismic. Consequently, geomodelling workflows have been developed based on the 3D seismic, petrophysical and production test data. The reservoir model building methodology for Laggan consists of the following steps: (1) construction of a classical structural framework using seismic interpreted fault sticks, top reservoir depth surface and the overall reservoir thickness map estimated from seismic data; (2) building of zones model integrating well data, seismic and sedimentological concepts; and (3) generation of a 3D facies and petrophysical model constrained by seismic and well data, including well test results to calibrate the permeability.

A reasonable match of the well test data was achieved without any major adjustments to the reservoir model. This methodology has led to the construction of a robust reservoir description, capturing the geological knowledge of the turbidite system and constrained by well and seismic data.

The new Laggan model is now considered 'fit for purpose' for well locations update, well type and well number confirmation and development of the Reservoir Management Plan.

Laggan is a gas condensate field located on the UK continental shelf (UKCS) around 120 km WNW of the Shetland Islands in a water depth of 600 m (Fig. 1). It appears as a very strong anomaly in Top Reservoir (T35.2 ICHRON sequence) amplitude map and minimum of integrated Amplitude Versus Offset (AVO) (Pseudo Poisson's ratio) of Laggan Sand (Fig. 2a, b). The trap is combined structural and stratigraphic and the reservoir comprises sand-rich stacked and laterally extensive Paleocene turbidite lobes (T35.2 ICHRON sequence) (Fig. 3a) with good reservoir properties (Fig. 3b, c), separated by hemipelagic shales (Gordon et al. 2010).

The Laggan discovery well 206/1-2 was drilled in 1986 by Shell and encountered 34 m of net gas sand in the T35 'Laggan' ABC sands (Fig. 3a). Subsequently, well 206/1-3 was drilled by Total E & P UK in 1996, encountering 25 m of sand with 10 m of gas section. A dual leg appraisal programme from one surface location was then completed in 2004 with each branch, namely 206/1a-4a and 206/1a-4aZ, encountering a 20 m and 46 m column of gas, respectively (Gordon et al. 2010).

The Laggan reservoir has a gross interval thickness of 23–63 m. It consists of three main sand units named from top to bottom as A, B and C sands which are separated by shale intervals of a few metres thickness (Fig. 3a). The reservoir quality is very good with an average porosity of 25%. Porosity is preserved by chlorite coating but can decrease locally due to carbonate cementation (Fig. 3a). A few calcite nodules can be identified in the core or from the neutron-density curves and are interpreted as being scattered or of small extent.

A new 3D seismic coverage over Laggan field was acquired by Petroleum Geo-Services (PGS) in 2009. The survey was acquired prior to the commencement of development drilling and the installation of the subsea facilities. This survey was specifically designed to optimize reservoir characterization: it was processed late in 2009 just prior to project sanction, with sufficient time to allow full 3D model updates and review of well placement prior to drilling commencement in summer 2012. The expectation was also that this survey would form the baseline survey for future repeat seismic surveys (4D).

Different datasets were generated from the 2009 3D seismic data and were extensively used in the Laggan field study. The first dataset is Post Stack Time Migration (PSTM) with whitening which was used for the structural interpretation, principally top reservoir and faults picking (Fig. 4). Following a deterministic inversion study, the second dataset including reflectivity from acoustic impedance, acoustic impedance and relative Poisson ratio was used to pick intra-reservoir horizons, populate the porosity in 3D and generate net gas sands maps,

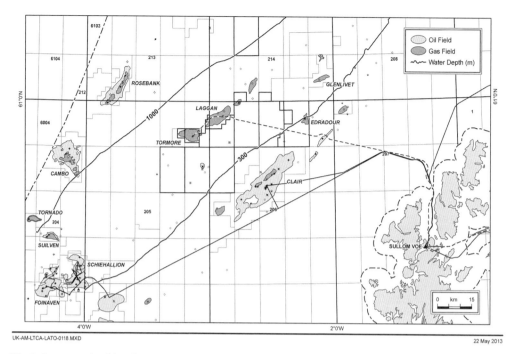

Fig. 1. Laggan regional location map.

respectively (Fig. 5) as explained in section 'Base reservoir construction'.

Subsequently, it was decided to rebuild new static models incorporating: (1) the new structural interpretation based on the 2009 3D seismic data, integrating structural concepts to help the well test match and to define downside case compartmentalization; (2) an internal study which helped to define a 3D sedimentological model used as deterministic input for the static model; and (3) a revised petrophysical dataset (permeability and water saturation model).

Structural interpretation

The structural pattern defined in the Laggan area appears to be controlled by the deep structure of the basement. Several structural elements were defined using coherency and dip maps (Fig. 6) as follows.

(1) *Seismic faults*: the main structural feature, which have observable throw on the seismic sections and also appear very clearly on the coherency maps.

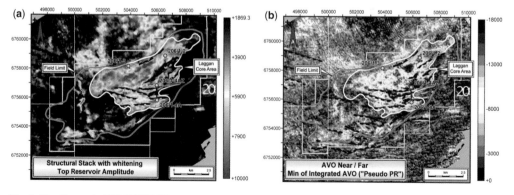

Fig. 2. Top Reservoir (T35.2 ICHRON sequence) (**a**) amplitude map and (**b**) minimum of integrated AVO (Pseudo-PR) of Laggan Sand.

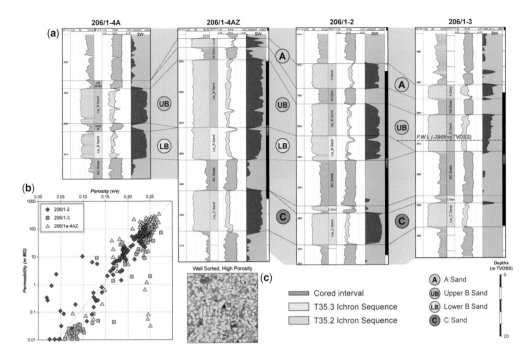

Fig. 3. (**a**) Laggan field well correlation showing sand-rich, stacked, laterally extensive Paleocene turbidity lobes (T35.2 ICHRON sequence), separated by hemipelagic shale. (**b**) Thin section of the T35.2 ICHRON sequence sands. (**c**) Core porosity v. permeability plots in Laggan field.

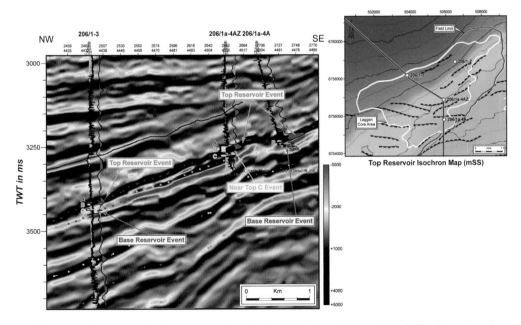

Fig. 4. Cross-section of PSTM with whitening used for the structural interpretation, principally Top Reservoir and fault picking.

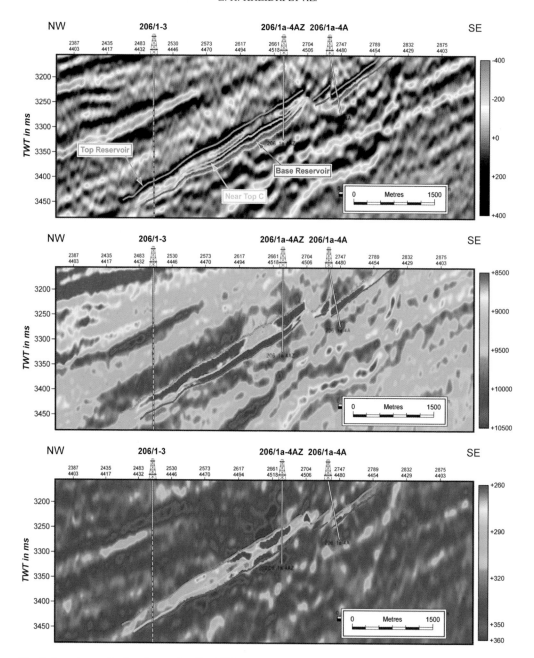

Fig. 5. Inversion results from Laggan baseline 2009 used to pick intra-reservoir horizons, populate the porosity in 3D and generate net gas sands maps. From top to bottom: reflectivity, acoustic impedance and relative minimum Poisson ratio.

(2) *Lineaments*: the second-order structural features which were identified on the attribute maps such as dip and curvature maps at the top of the reservoir; they cannot however be picked out on the seismic sections.

(3) *Fault extensions*: the third-order structural features which constitute the subseismic tips of the faults already interpreted from the seismic. These were mapped conceptually and based on the coherency maps whenever possible.

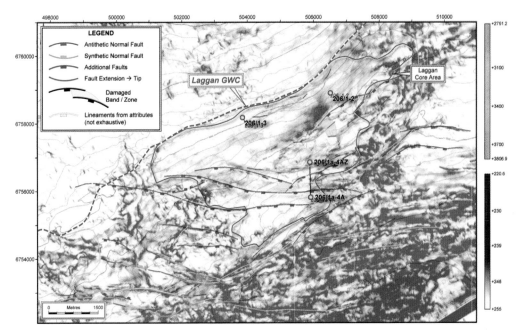

Fig. 6. Laggan structural pattern including faults picked on the seismic and lineaments and extensions identified using attribute maps (coherency and dip).

(4) *Damage zones*: the least supportable structural elements (from seismic) as they are purely conceptual, considered as relay zones between the seismic faults.

Laggan structural model building

Fault modelling

After the quality control and the data preparation the fault sticks were simply transformed to 3D faults as 'Listric Pillar'. The lineaments were built using the lines imported from the seismic. Lineaments in close proximity to a fault were given the same dip as the fault, otherwise they were considered as vertical. The extensions were built at the ends of the faults (as they are considered as the fault tips) using the lines imported from the seismic data, and were given the same dip as the associated faults. Both structural features were incorporated in the structural model without any throw.

No faults connections were constructed as most of the picked faults are naturally disconnected. The only connections integrated in the fault model derive from use of the lineaments or extensions. Few segments (potential compartments) can be defined in Laggan, even after incorporating all of the structural features.

Top reservoir construction

The only horizon modelled and incorporated in the grid directly from the seismic data is the top of the reservoir. This is the only horizon that can be interpreted across the complete area of extent of both fields. Locally, two other horizons (base reservoir and an intra-reservoir horizon) can be interpreted. In Laggan the intra-reservoir layer corresponds to a near Top C horizon.

The time to depth conversion and the well ties were completed in Sismage (Total proprietary software for seismic interpretation). The horizon was then exported as an Irap ASCII grid to Petrel (Schlumberger-owned Windows PC software for geology modelling).

The classic process 'Make Horizon' in Petrel was used to model the top of the reservoir which was adjusted to the well tops with an influence radius of 300 m. The three types of structural elements defined on the seismic data (faults, extensions and lineaments) were incorporated in the process. The faults were considered to be active in Petrel (considered as having a throw) and the fault throw was assigned and constrained by the fault polygons with Z (depth) value defined on the seismic. This constraint ensures that the throw within the model will be as close as possible to the seismic. The extensions and the lineaments were

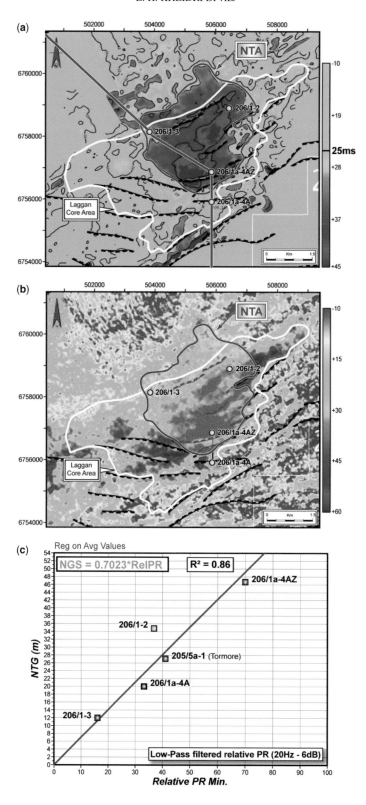

considered to be inactive; they will not display any throw but they will exist physically in the model as flow barriers, and could be used to define compartments or act as baffles when performing dynamic simulations.

Base reservoir construction

The main challenge with Laggan and Tormore modelling is the interpretation of the base of the reservoir; in most parts of both fields the gross reservoir thickness is below the tuning thickness, 20–25 ms equivalent. Locally in the central part of each field, within the area called the non-tuned area, the base reservoir can be picked on the seismic (Fig. 7a). In the tuned area, the true base reservoir cannot be picked on seismic and therefore needs to be back-calculated. To overcome this problem of tuning and build a realistic base reservoir horizon to be used in the entire model, a workflow was set up to combine different datasets. In the non-tuned area, as the gross reservoir thickness is above the tuning thickness the base reservoir was interpreted directly from the seismic. From this interpretation a gross reservoir thickness (in time) was calculated and then converted to depth using an interval velocity (Fig. 7a). In the tuned area, the gross reservoir thickness was back calculated from net gas sand (NGS) thickness maps (Fig. 7b). NGS thickness maps are generated using a relationship between inverted pseudo-Poisson ratio and well NGS thickness determined from a seismic inversion study (Fig. 7c). The back calculation of the gross reservoir thickness from NGS was completed in two steps: (1) the NGS is divided by the averaged Net To Gross (NTG) of the reservoir sands (0.95 in Laggan and 0.92 in Tormore) to obtain the total thickness of the net sand (Fig. 8), which represents the net pay and the cemented levels (calcite cementation); and (2) the AB and BC shales thicknesses are added to the net sand thickness obtained from the first step following the same sand/shale configuration seen at the wells. The AB and BC shales are considered to be hemipelagic and widespread across both fields, thickening down-dip. The AB and BC shales isochores were built accordingly using well data (Fig. 8).

The isochore obtained after the two above steps corresponds to the gross reservoir thickness in the tuned area. This is subsequently merged with the isochores of the gross reservoir thickness within the non-tuned area (picked on the seismic). The merge results in a single-field-wide map of the gross reservoir thickness (Fig. 9). Some editing was performed at the limit between the two isochores to obtain a smooth transition. Other editing was also performed within the tuned area to remove all the back-calculated thicknesses that were above the tuning thickness (considered as artefacts).

The final step is to incorporate the resulting isochores in the model to build the base map of the reservoir. This step is also performed in two stages: (1) using the isochores of the gross reservoir thickness to generate the base of the reservoir; and (2) the base is exported as a data point set and re-used in the 'Make Horizon' process with the top reservoir (both with the same constraint) to create a top and a base of the geological grid. This second step allows us to build more zones within the grid later.

Sedimentological model

In the Laggan field, the definition of the intra-reservoir sand-body geometry and its incorporation within the geological model was always considered as challenging because: (1) all the wells drilled in Laggan are concentrated in the central part of the field targeting the high amplitudes; (2) the thickness of all the sand bodies is below the seismic resolution, so the seismic data will likely reflect a cumulative response of all these sand bodies; and (3) no dynamic flow data exist that can be used to define the size of the sand bodies.

To overcome these challenges a sedimentological study was launched in the summer of 2010. The main objective of this study was to help in building a 3D sedimentological model to be used as the basis for the construction of the intra-reservoir zones. Different sedimentary bodies were identified based on the core description and the inter-well correlation and their lateral extension was constrained by the seismic data (amplitude and relative Poisson ratio maps). These results are summarized in Figure 10).

(1) Western channelized system: located to the west of the Laggan field and not penetrated by any well. This channel feature is thought to be older than the Laggan lobes, and its limits and geometry were picked directly on the amplitude map at the top of the reservoir. This channelized system includes (a) the main channel (which represents the central part of the system and appears in seismic as very bright amplitude) and (b) levees and

Fig. 7. Total reservoir thickness estimation methodology. (**a**) Gross reservoir thickness calculated directly from the seismic; (**b**) gross reservoir thickness back-calculated from net gas sand thickness (NGS) maps; and (**c**) relationship between inverted pseudo-Poisson ratio and net gas sand calculated at well.

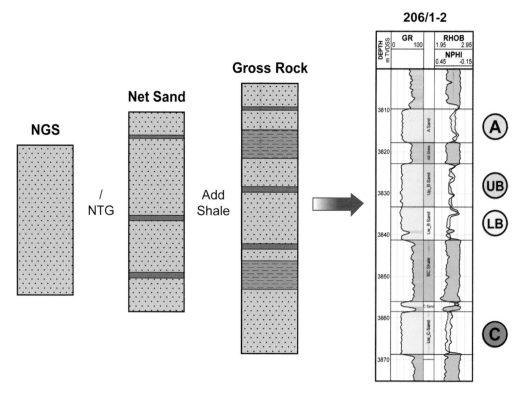

Fig. 8. Back-calculation of the gross reservoir thickness from NGS maps.

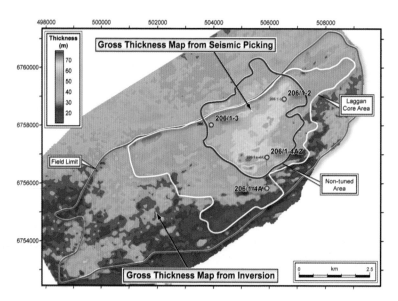

Fig. 9. Gross reservoir thickness map including the isochore back-calculated from NGS in the tuned area and the isochore of the gross reservoir picked on the seismic inside the non-tuned area.

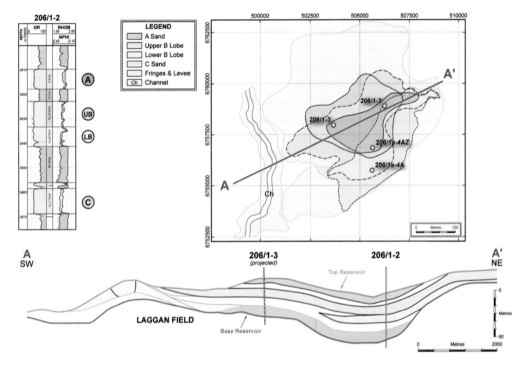

Fig. 10. Laggan field sedimentological model. The sedimentary bodies were defined on the core and their lateral extension was constrained by the seismic data (amplitude and relative Poisson ratio maps).

fringes (which form the edges of the channel and are characterized on seismic by a dimming of the amplitude).

(2) C sand is considered as a ponded lobe which fills a palaeo-topography created by the levee and the fringe of the Western channelized system. On the seismic an intra-reservoir horizon was interpreted (near Top C). Subsequently the limit of the C sands (onlaps) was identified on the seismic allowing the delineation of its spatial extension and a thickness map of the 'C sand' in time was produced. Combining the limit of the sand, the thickness of the C sand at the well and using the time isochores as trend, an isopach of the C sand was generated; this will reproduce its 3D geometry.

(3) Lower B Sand is considered as prograding lobe constituted by two facies: central lobe and lobe fringes. From well data/correlation, 'Lower B Sand' lobe does not exist in the well 206/1-3, indicating that this lobe has an elongated shape (NE–SW) and developed in the south-eastern part of the Laggan field, on the flanks of C Sand lobe. The lobe fringe pinches out towards the SE on a topography created by the levee of the Western channelized system.

(4) Upper B Sand corresponds to the maximum of progradation of the Laggan lobe complex. It is the thickest and the most extensive sand body, was penetrated by all the Laggan wells and is considered to develop across the whole field apart from the Western area. As for the Lower B Sand, Upper B Sand comprises two facies (central lobe and lobe fringes).

(5) The A Sand corresponds to a back-stepping phase of the Laggan lobe complex and develops in the central part of the field. Its limits are highly uncertain and were defined conceptually around the three wells which have encountered this lobe.

Thickness allocation

The 3D sedimentological model defined previously was used as a basis for the building of the intra-reservoir zones in Laggan. As a result of the sedimentological study and the four well penetrations, most of the zones will have a deterministic input (thickness maps). These zones are as follows:

(a) C sand: the thickness map is generated using the thickness values from the wells combined with the time thickness map of the C sand (used as trend) and the C lobe limits derived from the seismic data.

(b) A sand: the thickness map is generated using the thickness values from the wells combined with the limits built conceptually around the three wells which have encountered the 'A sand'.

(c) AB and BC Shale: these two shale layers were interpreted as hemi-pelagic shales extending across the entire field. The two thickness maps used as deterministic inputs were generated using the well thickness values with a trend of gentle thickening northwards towards the centre of the basin.

(d) Intermediate B shale: this is the most uncertain deterministic input as it was encountered in just one well, 206/1a-4A. The thickness map of this layer is generated using its thickness in well 206/1a-4A combined with the limits from the conceptual model. The Intermediate B shale was therefore modelled only in the southern part of the field and disappears northwards; in the northern part of the field this will be the equivalent of the BC shale.

(e) B sand: the remainder of the sand thickness is systematically allocated to the B sand and split into Upper and Lower B sand according to the proportion of the zone thickness seen at the wells. The limits from the conceptual model were used to constrain the spatial distribution of the lobes.

Outside the Core Area of the Laggan field the total thickness of the sand was vertically allocated to a single sedimentary body (western periphery).

Reservoir characterization

Before developing the facies and petrophysical modelling, it is worth describing the reservoir characteristics and the main heterogeneities present in the reservoir that are considered as having an impact on the Gas Initial In Place (GIIP) and on the fluid flow dynamics.

On the basis of the logs response, the core and the thin section descriptions, the Laggan field comprises a very simple reservoir characterized by fine sands, very well sorted and coated by chlorite. The reservoir is characterized by high preserved porosities of an average of 24–25% due to the chlorite coating which has prevented quartz dissolution and precipitation, common at such burial depths (Sullivan *et al.* 1999). As the sands are very well sorted the average permeability (70–80 mD) is also high.

The main reservoir heterogeneities are: (1) calcite nodules (or thin layers), easily identifiable on the neutron-density logs and more so on the core, which appear in two different geometries

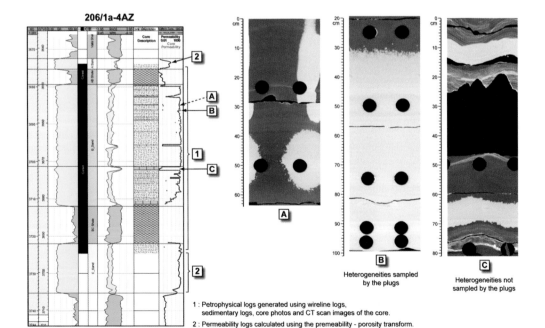

Fig. 11. Calcite nodules or thin layers identifiable on the neutron-density logs and on the core.

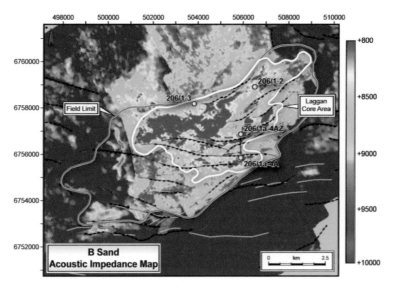

Fig. 12. Acoustic impedance map from the seismic capturing the spatial repartition of the reservoir degradation seen at well 206/1-2.

(layers of 30–50 cm thickness and as nodules of tens of centimetres of diameter; Fig. 11); and (2) localized porosity degradation, which can be seen in well 206/1-2 and is interpreted as a consequence of fluid movement (de-watering) in the early stages of the compaction. This reservoir-quality degradation appears to be captured by the seismic data (amplitude and acoustic impedance maps) as a shadow area around well 206/1-2 (Fig. 12).

As all the sand packages seen in all the wells in Laggan are characterized by fine sand, very well sorted, chlorite coated and with very good porosities and permeabilities, only one rock type was considered necessary. The calcite nodules and layers could be considered as another rock type; however this approach has not been take for the two following reason: (1) the size of these nodules is not big enough to be captured within the grid chosen for both fields; and (2) the calcite nodules do not have any significant effect on the fluid flow.

Facies modelling

As the main outcomes of the sedimentological study were already captured during the thickness allocation process, a simple facies modelling was implemented. The main objective of this very simple process is to differentiate between the central part of the lobes and their fringes. The limits defined in the sedimentological study were used directly in the model using the geometrical modelling process.

Petrophysical modelling

The petrophysical modelling was completed in two main stages: (1) core data quality control and correction of reservoir conditions and validation; and (2) 3D modelling of the different parameters such as net-to-gross, porosity, permeability and water saturation, used for the volume calculation and as the reservoir simulation input.

Data quality control and correction

The data quality control is based on the core photos and core description. Any ambiguous core plugs, such as plugs taken at the edges of the calcite nodules and on fractures, were removed from the Conventional Cora Analysis (CCAL) and Special Cora Analysis (SCAL) data sets. The main corrections applied are described in the following.

Depth corrections. Depth corrections were derived from the SCAL data and were applied on both porosity and permeability core measurements. The overburden correction relationships from ambient to reservoir conditions (c. 6200 psi) for porosity and permeability are:

$$\phi_{OVB} = 0.9753 \times \phi_{AMB} - 0.0004 \quad (1)$$

$$K_{OVB} = 0.7302 \times K_{air}^{1.0388} \quad (2)$$

where ϕ_{OVB} is overburden-corrected porosity, ϕ_{AMB} is ambient porosity; K_{OVB} is overburden-corrected permeability; and K_{air} is permeability to air.

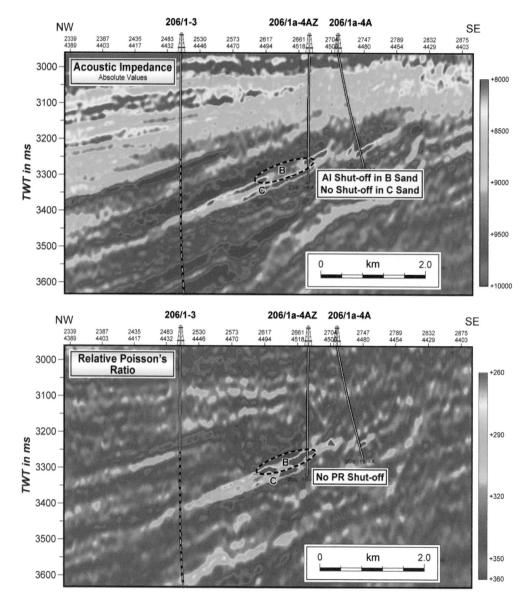

Fig. 13. Acoustic impedance v. relative minimum Poisson ratio section across the low-porosity area in Laggan field.

Klinkenberg correction. As the core permeability measurements were performed with gas, a Klinkenberg correction was applied:

$$K_L = 0.4844 \times K_{air}^{1.1235} \quad (3)$$

where K_L is liquid permeability and K_{air} is permeability to air.

Connate water saturation correction. This is applied to the permeability for comparison with well test permeability:

$$K_{L\text{-SWC}} = \left(1 - \left(3.5835 \times Swi^{2.5749}\right)\right) \times K_L \quad (4)$$

where $K_{L\text{-SWC}}$ is permeability corrected for initial water saturation.

After being quality controlled and corrected to reservoir conditions, the core measurements are extensively used to: (1) calibrate the calculated porosity logs; and (2) build log petrophysics in

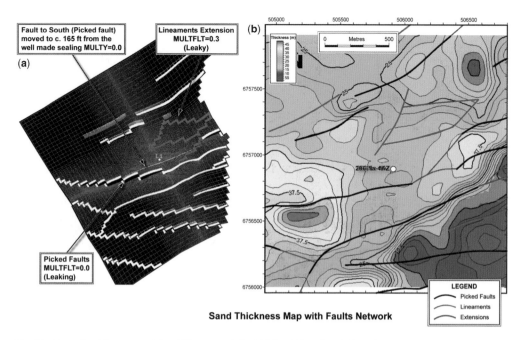

Fig. 14. Example of Laggan model muilding around the 206/1a-4AZ well. (**a**) Cartesian local grid refinement built around the tested well (206/1a-4AZ). (**b**) Thickness map and structural features from the static model (faults, lineaments and extensions).

the cored intervals and generate permeability logs using the equation derived from the porosity–permeability plot in the non-cored intervals.

Permeability and petrophysical logs. The permeability logs were generated using the equation derived from the core data corrected to the reservoir conditions:

$$K_{\text{Log}} = 10^{(-2.803 + 19.541 \times \phi)} \quad (5)$$

where K_{Log} is log permeability.

The wireline log resolutions are not sufficient to capture the true thickness of calcite beds and nodules and are prone to 'shoulder effects' at the edges of the sand bodies and calcite nodules. To overcome these difficulties, modelled petrophysical logs were generated based on the wireline logs, sedimentary logs, the core photos and CT scan images of the core. The resulting petrophysical porosity log was used to derive the permeability log to minimize artefacts in the modelled properties.

A new set of logs is generated using a combination of the petrophysical logs (in the cored interval) and the calculated logs using the permeability–porosity transform-based equation (in the non-cored interval; Fig. 11). These are used as input for the permeability validation and 3D permeability modelling.

3D petrophysical modelling

The 3D petrophysical modelling was performed using sequential Gaussian simulation algorithm. The first property to be modelled was the total porosity without cut-off. This was then modelled using the distribution from the well values as constraints, and a porosity map from the seismic as a trend (B sand only). This map is considered to capture the spatial distribution of the reservoir degradation seen at well 206/1-2 (Fig. 12). The amplitude maps as well as the acoustic impedance (AI) map show a shut-off in the NE part of the field (NE–SW trend). This dimming is interpreted to be related to the reservoir degradation seen at well 206/1-2. This shut-off is very clear on seismic cross-section, and affects only the B sand (Fig. 13).

The other properties were co-located and co-kriged with the total porosity as a secondary property (used as main drift). The different properties modelled in 3D are as follows:

(a) *Net-to-gross (NTG)*: this is modelled using SGS with the distribution from the well values as a constraint, and co-located and co-kriged with total porosity. Even although no relationship was identified between porosity and NTG, the use of this technique allows us to have a similar spatial distribution of the low porosity and the low NTG cells which correspond to

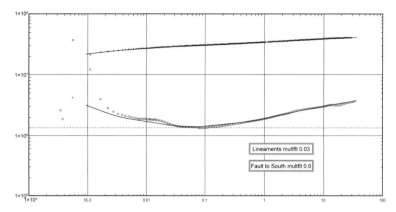

Fig. 15. Well test match of the 206/1a-4AZ well DST.

the calcite nodules. A very short range variogram is used to capture the nugget effect of the calcite nodules, which is randomly distributed.
(b) *Total net porosity*: 9% porosity and 0.3 Vsh cut-off were used to generate the net porosity property. This is modelled using SGS with the distribution from the well values as a constraint and co-located and co-kriged with total porosity.
(c) *Net permeability*: the match to the well test permeability is only obtained by using permeability without the calcite nodules, a net permeability was used. This latter is modelled using SGS with the distribution from the well values as a constraint and co-located and co-kriged with total porosity (a very good relationship is shown by the porosity–permeability plots).

Water saturation modelling

A saturation height model was used based on the J-function method. Input for the model is core-measured capillary pressure (Pc) curves and the measured porosity and permeability for each sample. To convert to reservoir saturations, the interfacial tension for Laggan reservoir fluids was determined at reservoir conditions and *in situ* porosity–permeability relationships were utilized. The J-function saturation height model used is:

$$S_W = A \times J^B \qquad (6)$$

where $J = $ (height above free water × Delta Rho fluids × Square root Permeability/Porosity)/ interfacial tension with a free water level identified from pressure data at a depth of -3909 TVDSS.

A and B, functions of the core plug porosity, are determined by varying the values until a minimum error between the water saturation S_W from the model and the actual measured water saturation was achieved using Excel solver, defined:

$$A = a_0 \times \text{Porosity}^{a1} = 0.41 \times \text{Porosity}^{-0.09}$$

$$B = b_0 \times \text{Porosity}^{b1} = -0.549 \times \text{Porosity}^{0.403}$$

Well test matching

To accurately model the transient pressure effects a Cartesian local grid refinement was built around the tested well (206/1a-4AZ) (Fig. 14a). The thickness map and structural features from the static model (faults, lineaments and extensions) (Fig. 14b) together with all static array data (porosity, NTG and permeability) were incorporated in this grid.

Picked faults on seismic data were considered sealing with zero transmissibility. Lineaments and fault extensions were allowed to leak with (0.3) transmissibility, and together their location and the global permeability became the tuning parameters to obtain an acceptable DST match.

A reasonable match of the well test data was achieved with no major adjustments of the reservoir model (Fig. 15). The only modification performed was the position of the fault to the south of the tested well (Fig. 14a). This was later moved to the north by 100 m, a modification which can be justified by the uncertainty on the fault picking. The ease of matching the DST data gave confidence in the static model.

Conclusion

Realistic structural and sedimentological conceptual models were defined and used as the basis for the static model building. Simple facies and petrophysical modelling processes were run based on the same data used for the building of the sanction model, but with new corrections applied (from laboratory to reservoir conditions).

A good match of the well-test of 206/1a-4AZ was achieved with minimal adjustment of the updated model parameters, giving increased confidence in the quality of the new model. The new model of Laggan is now considered 'fit for purpose' in terms of: (1) well locations update, well type and well number confirmation; and (2) development of the reservoir management plan.

The authors acknowledge the many geoscientists from Total whose contributions over almost two decades have led to the development of Laggan-Tormore fields. The authors would also like to thank Laggan-Tormore and Edradour co-venturer Dong Energy for permission to publish this paper.

References

GORDON, A., YOUNIS, T. ET AL. 2010. Laggan; a mature understanding of an undeveloped discovery, more than 20 years old. *In*: VINNIG, B. A. & PICKERING, S. C. (eds) *Petroleum Geology: From Mature Basins to New Frontiers: Proceedings of the 7th Petroleum.* Geological Society, London, Petroleum Geology Conference Series **7**, 279–297.

SULLIVAN, M., COOMBES, T., IMBERT, P. & AHAMDACH-DEMARS, C. 1999 Reservoir quality and petrophysical evaluation of Paleocene sandstones in the West of Shetland area. *In*: FLEET, A. J. & BOLDY, S. A. R. (eds) *Petroleum Geology of Northwest Europe: Proceeding of the 5th Conference.* Geological Society, London, Petroleum Geology Conference Series **5**, 627–633.

Laggan-Tormore: reservoir to sales production forecasting and optimization using an integrated modelling approach

K. JONES[1]*, D. MACKINNON[2], A. DE SENNEVILLE[2] & K. WATT[1]

[1]*Total E&P UK Limited, Crawpeel Road, Altens, Aberdeen, AB12 3FG, UK*

[2]*Total E&P S.A., 2 place de la Coupole, La Defense, Paris 92078, France*

*Corresponding author (e-mail: kevin.jones@total.com)

Abstract: Laggan and Tormore are gas-condensate fields in the West of Shetland region of the UK, to be developed as a 143 km subsea tie-back (twin 18″ flowlines) to a new onshore gas plant (Shetland Gas Plant, SGP). Condensate will be sent to the neighbouring Sullom Voe Terminal for stabilization and a new 30″ gas export pipeline will provide a route from the SGP to Total's St Fergus terminal on the UK mainland. The development is planned as a hub to enable future development of similar reservoirs in the region.

Production forecasting and field management are complex due to the evolving nature of the reservoir compositions over time, the back-pressure and flow assurance impacts of the long flowlines and final sales products being delivered at two remote terminals. This will be further complicated with the introduction of any additional fields, requiring an accurate understanding of field interactions, pipeline ullage, plateau extension and management of production allocation. To address these issues, Total E&P UK (TEPUK) have developed an integrated asset model (IAM) using state-of-the-art techniques to model the system from reservoir to final product sales. The IAM will be used to communicate a common geosciences and development understanding for forecasting and future development studies.

The Laggan and Tormore gas condensate fields comprise stacked mass-flow sheet sand reservoirs of Palaeocene age, located *c.* 125 km to the west of the Shetland Islands in *c.* 600 m water depth (Gordon *et al.* 2010). Development of the two fields was sanctioned by Total E&P UK (TEPUK) Ltd and partner DONG Energy (UK) Ltd in 2010.

The development will use subsea facilities connected to a new onshore processing facility, the Shetland Gas Plant (SGP) (Fig. 1). Two 18″ production flowlines will tie-back multiphase production from template-manifold structures at Laggan and Tormore to the SGP. An 8″ Monoethylene Glycol (MEG) pipeline will deliver MEG from the SGP for continuous hydrate inhibition with a 2″ service pipeline piggy-backed to the MEG pipeline. The subsea facilities will be controlled via a hydraulic/electric/communications umbilical which will also deliver chemicals. The SGP is a new TEPUK-operated facility being constructed adjacent to the existing Sullom Voe oil Terminal (SVT). Separated, dehydrated gas will be exported from the SGP through a new 30″ Shetland Island Regional Gas Export (SIRGE) pipeline, which will connect to the TEPUK-operated Frigg UK Association (FUKA) pipeline in the North Sea. The FUKA line transports gas to the St Fergus gas terminal in NE Scotland, which extracts natural gas liquids (NGLs) and provides sales gas to the UK grid. Condensate separated at the SGP will be routed to the neighbouring SVT for stabilization and export as part of the terminal's crude blend.

The SGP will have an initial capacity of 14.16 MS m^3 d^{-1} of gas production and associated hydrocarbon condensate. In the future, as the minimum flow rate at which the 18″ flowlines can operate at is reached, the SGP arrival pressure will be reduced to extend production further. In late life, production will be reduced to one flowline.

Original approach to reservoir modelling

For general sensitivity studies such as modelling reservoir behaviour, determining well planning and assessing well numbers, location and type, a simple Eclipse-based model was developed by the Reservoir Engineer (RE). The model comprised:

(a) coupled Laggan-Tormore Eclipse models;
(b) black oil pressure volume temperature (PVT) representation of the reservoir fluids;
(c) pipeline network (using pressure loss tables generated from a separate pipeline modelling study); and
(d) programmed control of pipeline turndown limits and switches of SGP operating pressure modes.

As a day-to-day modelling tool for reservoir engineers, the above representation was sufficient.

From: CANNON, S. J. C. & ELLIS, D. (eds) 2014. *Hydrocarbon Exploration to Exploitation West of Shetlands.* Geological Society, London, Special Publications, **397**, 123–129.
First published online February 20, 2014, http://dx.doi.org/10.1144/SP397.4
© The Geological Society of London 2014. Publishing disclaimer: www.geolsoc.org.uk/pub_ethics

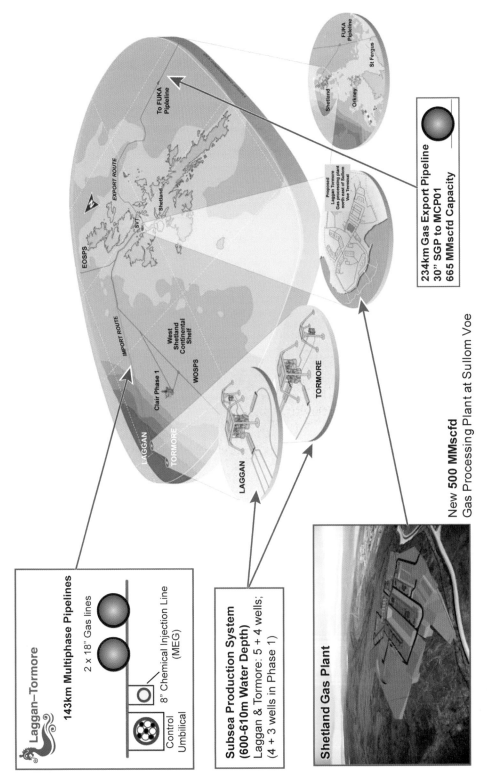

Fig. 1. Laggan-Tormore development showing field location, pipeline routes and new gas processing plant at the Shetland Islands.

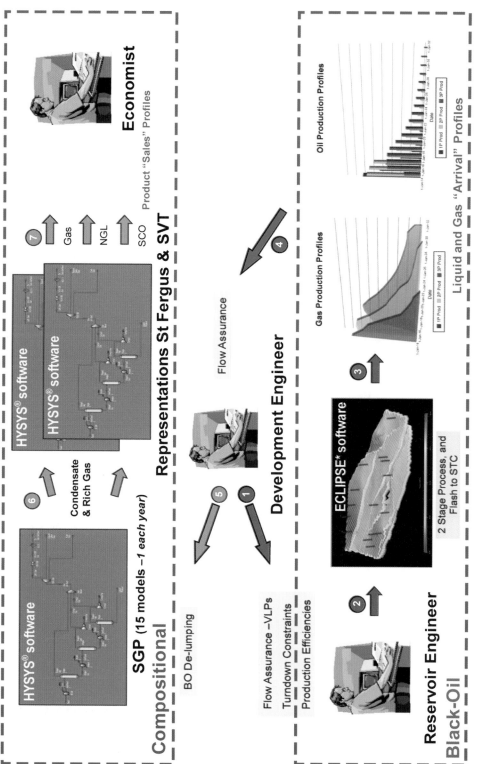

Fig. 2. Sequential modelling approach from initial development assumptions, to reservoir modelling, to process simulation and eventual sales product forecasting.

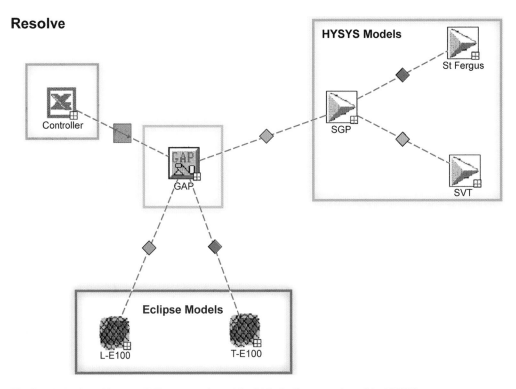

Fig. 3. IAM basic architecture: Eclipse reservoir models, GAP pipeline network models, HYSYS process representations and external spreadsheets linked together and controlled using Resolve.

Transfer of gas and condensate profiles to the Development Engineers (DE) for more accurate product stream modelling and economic analysis was however complicated, as Eclipse-based black oil output comprised condensate and gas profiles at arbitrary reference conditions. The steps involved in the procedure used include (Fig. 2):

(1) DE confirms turndown constraints on pipeline/plant and production efficiencies to the RE;
(2) RE imposes constraints in Eclipse/network model;
(3) RE generates gas and condensate profiles based on black oil representation of PVT;
(4) RE gives profiles to the DE for black oil de-lumping;
(5) based on reservoir pressure depletion, DE determines composition of reservoir fluid for process plant simulation, honouring the Eclipse gas/liquid profiles;
(6) DE conducts detailed process modelling to extract the sales product streams (gas, natural gas liquids and stabilized crude oil); and
(7) product streams passed to the economist.

The time-consuming sequential nature of this process meant that detailed compositional profiles were not easily generated. The approach was therefore not easily suited to running multiple scenarios and would be even further complicated to study in the event of additional field tie-ins to the system.

West of Shetland integrated asset model (WoS IAM)

To address these issues, TEPUK have developed a WoS IAM using integration and optimization software to model the entire system from the Laggan and Tormore reservoirs to the final product streams at the downstream terminals.

This approach allows the software to complete the data transfer between individual simulations, a task which was previously done manually. In addition, the integration software can be used to automatically schedule changes in operating mode or start-up of wells, triggered by events. Multiple scenarios can be evaluated.

The WoS IAM is now used as a tool for:

(a) assessing reserves and production profiles;

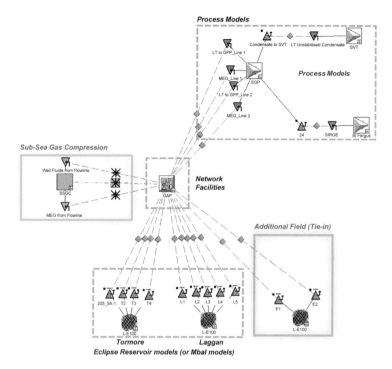

Fig. 4. Examples of WoS IAM studies: subsea gas compression and additional field tie-in linked into the GAP pipeline network.

(b) completing development project assessment and screening;
(c) determining the likely interactions between existing and additional field tie-ins to the pipeline system; and
(d) production optimization (such as optimal well routing per flowline).

WoS IAM software and workflow

The base software chosen for development of the WoS IAM was the Integrated Production Modelling (IPM) software suite from Petroleum Experts Ltd (PETEX), comprising: MBal (material balance reservoir modelling) which allows integration of

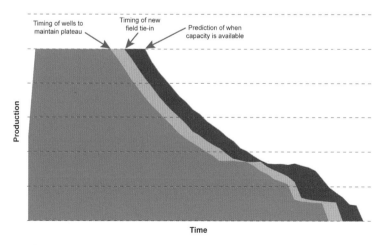

Fig. 5. Gas rate production plateau extended with additional wells and an additional field tie-in.

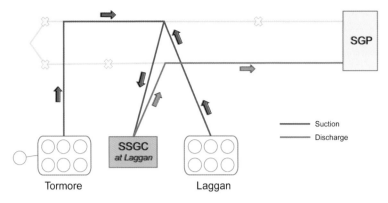

Fig. 6. Subsea gas compression located at Laggan: schematic of pipeline routing changes.

simple reservoir models for development studies; Prosper, well performance modelling; GAP, flowline network modelling and optimization software; and Resolve, which acts as a system integration and optimization application, coordinating the data flow between modules and controlling the management constraints concerning wells, pipeline turndowns, SGP operating mode, event scheduling and well/pipeline routing optimization.

The following software is also used in the WoS IAM alongside the PETEX tools: Eclipse, black oil reservoir models supplied by Schlumberger; HYSYS, process simulation modelling (SGP and representations of downstream terminals) supplied by AspenTech; and Microsoft Excel spreadsheets for external calculations.

The basic architecture of the WoS IAM is depicted in Figure 3. Resolve transfers the Eclipse black oil model output into a full fluid composition in the GAP model, which subsequently determines the achievable production rate from the wells to the SGP. Resolve transfers the flow rate and composition of the production arriving at the SGP into the HYSYS model of the onshore processing facilities, which in turn calculates the amount of export gas and condensate that are transferred to the SVT and St Fergus terminals. This procedure is repeated at each time step to build a life-of-field production forecast.

A significant benefit of the IAM approach is the automated handling of fluid PVT behaviour. The condensate and raw gas input streams from Eclipse are automatically 'de-lumped' into 30+ pseudo components based on automated matching of the gas-oil-ratio from Eclipse, while respecting the mass balance.

The Resolve software is used to optimize production on each time step to a defined target,

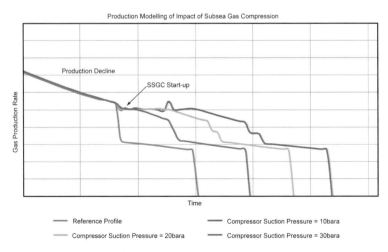

Fig. 7. Estimate of production extension with subsea gas compression: impact of different compressor suction pressure.

change plant operating pressure or flowline routing conditional on production rates or events and optimize the routing of wells to individual flowlines at specified times or events.

Typical WoS IAM case studies

A main advantage of the IAM approach over coupled Eclipse simulation modelling is the ease with which complex issues such as additional field tie-ins and facilities reconfigurations can be thoroughly evaluated. The impact on field recoveries, well start-up and project timings and production plateau extensions are easy to observe, allowing informed understanding and decision making. Figure 4 depicts the architecture for studying additional field tie-ins and subsea gas compression. Two examples are discussed in the following sections.

Case study 1: additional field tie-in

Once a potential tie-in point is identified for a new field (at an existing manifold, or at some appropriate entry point along the production flowline route) the reservoir model for the new field (either MBal based or Eclipse) is connected within the Resolve model. The GAP model network is updated to include any new flowline(s) or facilities. Constraints on field production rates or timing are then scheduled with Resolve. The model can then be run to determine full production profiles for the new field and the existing fields in the system. Full screening of field interactions, start-up date sensitivities and impact of the additional field can be evaluated, leading to informed project planning.

An example of an additional field tie-in is shown in Figure 5. Additional wells were added to the existing Laggan-Tormore fields (Phase II drilling campaign), extending the gas plateau duration. The additional field start-up was timed to coincide with end of the post phase II plateau, thereby extending the plateau even further. The prediction of when plant capacity becomes available is determined as a result.

Case study 2: subsea gas compression

Laggan-Tormore will not have any imposed pressure support, for example gas or water injection, and consequently the driving force for the flow from reservoir to SGP is solely dependent on the stored (compression) energy within the reservoir gas and any connected aquifer, together with rock compaction effects during depletion. One approach to increase field recovery factors and allow the wells to flow at lower reservoir pressure would be to install subsea gas compression (SSGC) during late field life. The suction side of the gas compressors would reduce the wellhead back-pressures and allow the wells to flow beyond their normal expected flow life. The energy to transport the fluids to shore is now supplied by the gas compressors.

The benefit of subsea gas compression has been reviewed with the WoS IAM. Typical variants for the investigation included the physical position of the compressors (at Laggan or Tormore), compressor suction pressure and timing of installation. Figure 6 shows a pipeline routing schematic with SSGC installed at Laggan. Hydrocarbons from Tormore and Laggan flow to the Laggan header where they are routed to the suction side of the compressors. The discharged hydrocarbons flow down a single flowline to the SGP.

The impact on the combined Laggan-Tormore late life gas production profile is shown in Figure 7.

Conclusions

Production modelling and forecasting for a long subsea tie-back of gas-condensate reservoirs is complicated, particularly due to the changing fluid composition over time as the reservoirs deplete. Many technical discipline activities are required including reservoir modelling, well performance, flow assurance and process modelling. The WoS IAM integrates the individual models from each of these technical disciplines to provide a combined model. Implementation of system design and operating constraints has then allowed life-of-field production forecasts to be generated. The tool is now used for this purpose and as a development option screening tool able to assess optimal well routings, changes to operating modes and the introduction of new fields.

The authors acknowledge the many geoscientists from Total whose contributions over almost two decades have led to the development of Laggan-Tormore fields. The authors would also like to thank Laggan-Tormore and Edradour co-venturer Dong Energy for permission to publish this paper.

Reference

GORDON, A., YOUNIS, T. ET AL. 2010. Laggan; A mature understanding of an undeveloped discovery, more than 20 years old. In: Proceedings of the 7th Petroleum Geology Conference, London, 279–297.

The discovery and appraisal of Glenlivet: a West of Shetlands success story

CATHERINE HORSEMAN[1], ALWYN ROSS[1] & STEVE CANNON[1,2]*

[1]*DONG E&P (UK) Limited, 33, Grosvenor Place, London SW1X 7HY*

[2]*Present address: Steve Cannon Geoscience Limited, 63 Box Lane, Wrexham, LL12 8BY*

*Corresponding author (e-mail: steve@stevecannongeoscience.co.uk)

Abstract: The Glenlivet gas field was discovered in 2009 with well 214/30a-2, drilled to test a seismically defined stratigraphic and faulted closure at intra-Tertiary level. A sequence of gas-bearing Vaila Formation sands was penetrated comprising high-density turbidite flows with excellent reservoir properties. The discovery was immediately appraised by two sidetracks that tested the updip and downdip extent of the accumulation: the downdip location penetrated the water leg of the field. A comprehensive programme of data acquisition was carried out on all three penetrations, leading to a series of post-well studies designed to fully appraise the field. Based on post-drilling seismic interpretation and extensive analysis of the well data, a revised depositional model was formulated that allowed the construction of a robust 3D static model used in development planning.

Licence P1195 in the West of Shetlands (WoS) sector of the UKCS was awarded in 2004 with a one year extension granted in 2008 to facilitate the drilling of a well. A proven gas accumulation, Laxford, was discovered in 1984 by well 214/30-1 drilled by British Gas Corporation, but was considered too small for development at the time. Danish Oil and Natural Gas Consortium (DONG) is the operator of Glenlivet with 80% equity in the field; partners at the time of discovery were First Oil and Faroe Petroleum each with 10% equity. Figure 1 is a regional location map showing the licensed areas and major discoveries in the Flett Sub-basin.

Glenlivet was discovered in 2009 with well 214/30a-2 and immediately appraised by two sidetracks, 2Z and 2Y. Gas was encountered in the Late Paleocene (T34.2) Vaila Sandstone Member of the Faroe Group (Lamers & Carmichael 1999); the sands are interpreted as proximal to medial deposits of high-density turbidite flows with excellent reservoir properties. The gross reservoir interval is 64.9 m thick; net thickness is 63.0 m, and average effective porosity is 29.0% in the discovery well. The upper part of the reservoir was cored, from which much of the conceptual geological understanding is derived. Detailed geological and petrophysical analysis indicates that the reservoir is fairly homogenous. The well was not tested; however, multi-Darcy permeability and high gas saturation reduces the uncertainty in producibility from the reservoir.

Seismic interpretation and structural configuration

Three 3D seismic datasets have been used in the interpretation; these are products that are derived from the original 3D survey acquired by Shell in 1996 and include the PGS Mega-Merge dataset. The data offer continuous 3D coverage over the Glenlivet and Laxford fields within the P1195 licence area (Fig. 2).

From a regional structural perspective, Glenlivet lies in the Flett Sub-basin of the of the Faroes–Shetland Basin. A deep Cretaceous and Paleocene trough formed as a result of lithospheric stretching along the proto-North Atlantic prior to opening of the ocean to the north of the Faroes (Dean *et al.* 1999). Glenlivet is located in the shelf–slope transition zone and there is no structural closure; it has an updip stratigraphic pinch-out (which may have a subtle structural component) onlapping the base Tertiary unconformity. Trapping is provided by a combination of faulting and sand pinch-out. The sands pinch-out laterally along-strike to the SW; the NW field limit is however defined by a later channel system that has cut down and eroded Vaila sands. This channel is presumed to have back-filled with shale providing a lateral reservoir seal. The Glenlivet field lies on a broad fault terrace; a series of relay faults connect these terraces down into the next intra-slope low in the basin where the Laxford Field sits (Fig. 3).

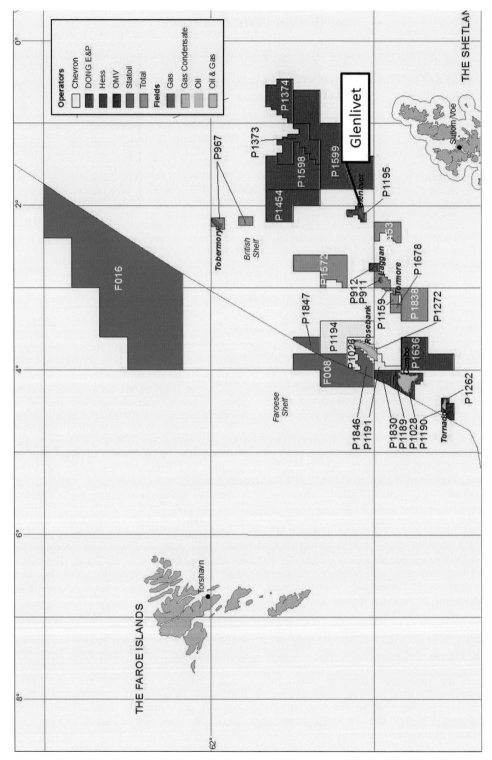

Fig. 1. West of Shetlands regional location map showing the Glenlivet Field and the major hydrocarbon discoveries in the Flett Sub-basin.

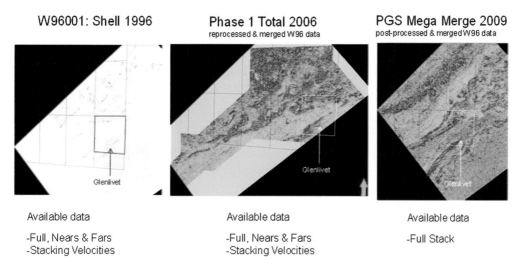

Fig. 2. The 3D seismic datasets available over the P1195 licence area.

Formation of the Glenlivet gas field has been intimately linked to several episodes of faulting. Growth faults developed during the Late Paleocene created and controlled laterally restricted local lows in the Glenlivet area along the shelf–slope transition zone. Sediment derived by slope failure on adjacent delta fronts (located to the SE) was transported into the Glenlivet area by gravity flows; deceleration of sediment flow caused the fall-out of coarse sediment from turbulent suspension resulting in a succession of amalgamated high-density sandstones of exceptional reservoir quality. As a result, sediment ponded on the down-thrown side of the growth faults, pinching out updip to onlap the Cretaceous subcrop. Once the available accommodation space was full, the turbidite flows 'spilled

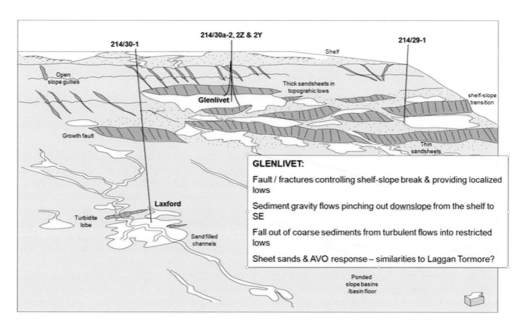

Fig. 3. Depositional setting of the Glenlivet Field.

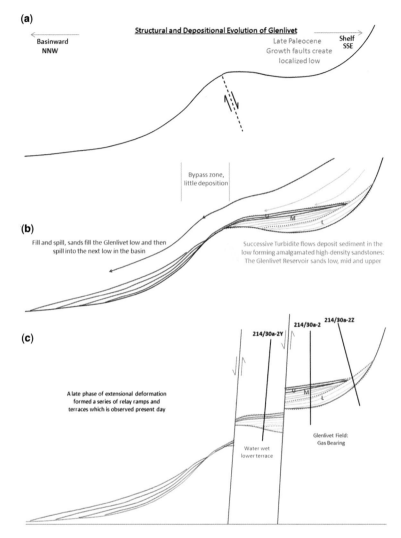

Fig. 4. Structural and depositional evolution of Glenlivet. (**a**) Late Paleocene growth creates local low; (**b**) successive turbidite flows fill low and then spill into the next basin low; (**c**) later extension forms relay ramps and terraces.

and filled' into the next low. A later phase of extensional deformation formed a series of relay ramps and terraces (Glenlivet is contained within one such terrace) (Fig. 4).

The field is well resolved on seismic data; the top and base reservoir are mapped confidently, supported by an excellent synthetic-to-seismic tie at the discovery well location (Fig. 5). The discovery well was drilled in the thickest part of the field (the 'sweet spot') to the NE; the field is thinner to the SW. Within this sweet spot a second seismic doublet is observed that is restricted to this thicker field area (Fig. 6). Integrating the conceptual depositional model together with the synthetic tie and observed seismic data enabled the internal reservoir architecture to be further refined on seismic data, providing control on the distribution of this unit in 3D.

Overall reservoir thickness is a critical parameter in the evaluation of Glenlivet. There is an anomalously bright amplitude response on seismic data with a strong peak amplitude anomaly at top reservoir (negative acoustic impedance or −ve AI) and strong trough at base reservoir (+ve AI). As the stratigraphic wedge pinches out updip the apparent reservoir thickness in time is no longer proportional to the true thickness; this is a consequence of the constructive/destructive interference effects at top and base reservoir reflections, referred to

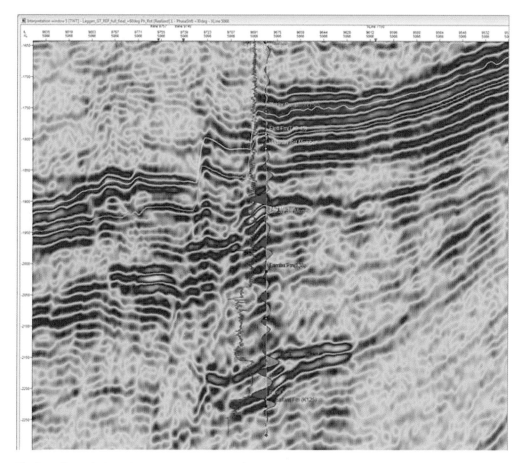

Fig. 5. Well-to-seismic tie for the discovery well 214/30a-2, providing excellent calibration of the two data types.

as seismic tuning. A net-pay analysis study has been used to better constrain the gas-bearing sand distribution and thickness in the field. The analysis is based on the assumption that, for the band-limited case, the average band-limited impedance over the apparent pay thickness is proportional to 'seismic net-to-gross' for a fixed apparent thickness (Connolly 1999). The results have been used to define the net rock volume as part of the initial estimates of gas in place.

Pre-drill, the modelled amplitude variation with offset (AVO) response for Glenlivet was a weak Class III or Class IV anomaly (Rutherford & Williams 1989). Subsequent AVO modelling with the 214/30a-2 well results indicates a slight decrease in amplitude with offset suggesting a Class III anomaly; this is consistent with the observed Class III response on seismic data (Fig. 7). This contrasts with Laxford which has a Class II AVO response.

The Glenlivet discovery and two sidetrack wells were drilled based on surfaces depth-converted using a standard approach in exploration consisting of: constant-velocity water column; velocity gradient $(V_0 + k)$ approach from seabed to top Balder; constant interval velocity (V_{INT}) from top Balder to top reservoir; and constant V_{INT} top–base reservoir.

This is a common and perfectly adequate depth-conversion approach for predicting stratigraphic tops in the exploration environment. For Glenlivet, the resulting actual v. prognosis gives a good prediction of top reservoir but a poorly constrained base reservoir in the discovery well due to the velocity selected in the reservoir being too fast. The downdip water-wet sidetrack prognosis was very poorly constrained, probably a combination of poor imaging and a more complex velocity structure associated with the well trajectory penetrating several fault terraces, offset one from the other.

The final depth conversion model for the Greater Glenlivet area was produced following tests on a number of model building strategies.

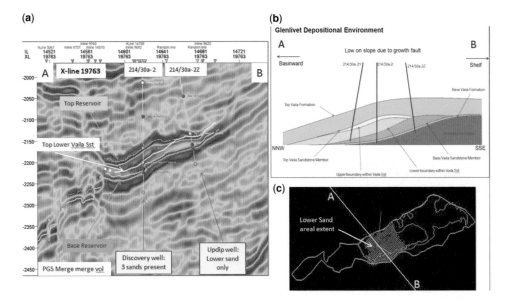

Fig. 6. (**a**) Seismic dip-line through Glenlivet discovery well showing the internal reservoir architecture of the lower sand unit; (**b**) Glenlivet deposition model based on chemical stratigraphy; and (**c**) map showing the location of the display line through the thickest part of the field and how the lower sand unit is restricted to this 'sweet spot'.

These tested various combinations of layer-based V_{INT} maps derived from stacking velocities with differing calibration methodologies, single-layer constant interval velocities and a model built using the $V_0 k$ approach (Al-Chalabi 1979). The final depth-conversion velocity model was built using the following methodology.

- *Layer 1 Sea-level–seabed*: water column uses constant interval velocity (1466.7 m s^{-1})

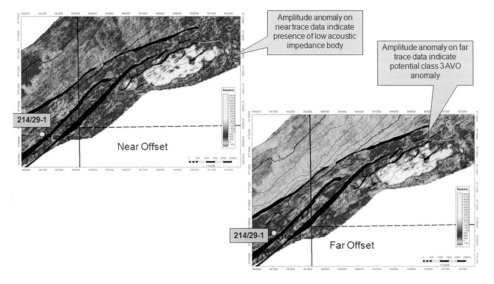

Fig. 7. AVO modelling pre-drill on Glenlivet indicates a bright amplitude response on both the near and far amplitude stacks, consistent with a Class III AVO anomaly.

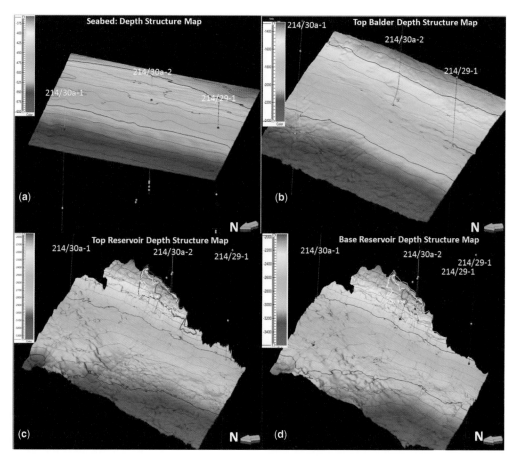

Fig. 8. Depth structure maps (exaggerated vertically ×10) for the Greater Glenlivet area: (a) seabed; (b) top Balder; (c) top reservoir; and (d) base reservoir.

- *Layer 2 Seabed–top-Balder*: V_{INT} velocity maps derived from stacking velocities are calibrated and corrected to well velocities by flexing the V_{INT} velocity map
- *Layer 3 Top-Balder–top-reservoir*: V_{INT} velocity maps derived from stacking velocities are calibrated and corrected to well velocities by flexing the V_{INT} velocity map
- *Layer 4 Top–base reservoir*: constant V_{INT} (2630 m s^{-1})
- *Layer 5 Base-reservoir–base model (3200 ms)*: constant V_{INT}.

The depth structure maps for the Greater Glenlivet area including Laxford are displayed in Figure 8. They have been vertically exaggerated (×10) to convey the distinction between the Glenlivet terrace at reservoir level and as the surfaces 'step' into the next basin low. Laxford appears as an isolated promontory building into the basin, a consequence of faults 'popping' the structure up. Figure 8a–d show depth structure maps for the seabed, top Balder and top and base reservoir horizons across the Greater Glenlivet area.

Seismic interpretation recognized 34 faults, five of which are considered to be 'major', extending the length of the field and exhibiting significant structural offset. These faults were depth converted and used in the subsequent static model build. Five primary surfaces were also provided from the seismic interpretation and used to construct the static model; two defined the 'bounding' surfaces of the model, within which the top and base reservoir surfaces are located. Within the reservoir interval itself, an internal reflector that only exists over the central area of the field was modelled.

Depositional model

The Glenlivet reservoir comprises turbidite flows that can be split into three zones based on

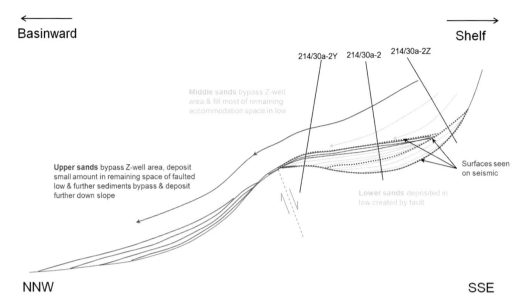

Fig. 9. Conceptual zone-fill model.

the seismic interpretation and a series of detail studies based on core, cuttings and log data (Fig. 9). While all three zones are recognized in the main well and the downdip sidetrack, only the lower zone is encountered in the updip penetration. Core sedimentology was instrumental in understanding the overall depositional setting, while image log analysis and chemical stratigraphy carried out on the main discovery well were used to calibrate the sidetracks.

Lower zone

The lower zone is not laterally extensive across the entire field area; it can be mapped seismically and covers c. 2.1 km^2 of the whole field area (10 km^2).

Seismic data flattened on this pick suggests deposition of this sand unit occurred in a restricted low within the main field low on the slope. The areal extent of the lower sand zone represents the 'sweet spot' in the field, where the sands are of the greatest thickness. The petrophysical properties of the lower zone degrade significantly between the updip sidetrack and the main discovery well. Two separate areas within the low are proposed, resulting in fall-out of the higher-quality reservoir sands from the turbidite flow into the updip area, which leads to lower-quality reservoir material being deposited further down the slope from the first influx of sands forming the lower zone. These 'mini-basins' can be identified on net-pay maps as areas of thickened gas-bearing reservoir. The

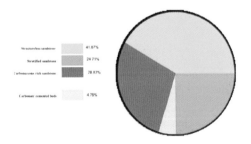

Proportion of image lithofacies in Upper Unit A – 214/30a-2

UPPER UNIT A:
Increased stratified sandstone facies

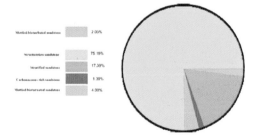

Proportion of image lithofacies in Lower Unit B – 214/30a-2

LOWER UNIT B:
Dominance of structureless sandstone facies

Fig. 10. Proportions of image lithofacies across the boundary between the lower and middle zones.

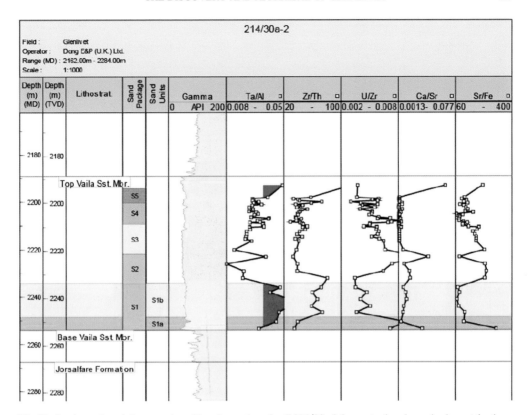

Fig. 11. Sandstone-based chemostratigraphic unit zonation of well 214/30a-2 demonstrating change in elemental ratios across the boundary of the lower zone (S1) and the middle zone (S2 and S3).

lower zone sand unit is characterized by high-quality reservoir sands in the updip well with an average porosity of 29% and average permeability of >5000 mD. Moving downslope, these values reduce to 25% and 85 mD, respectively, in the discovery well and 24% and 132 mD in the downdip sidetrack.

Middle zone

The middle zone is observed in all three penetrations of the Glenlivet Field and is postulated as being the most laterally continuous unit in the field. The reservoir properties of this unit are excellent with average porosity and permeability values

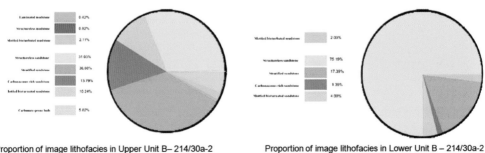

Fig. 12. Proportions of image lithofacies across boundary between the middle and upper zones.

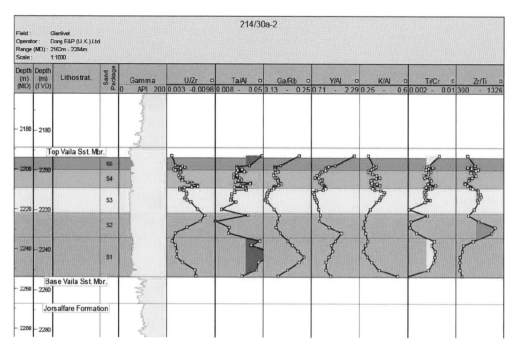

Fig. 13. Sandstone-based chemostratigraphic unit zonation of well 214/30a-2 demonstrating change in elemental ratios across the boundary of the middle zone (S2 and S3) and the upper zone (S4 and S5).

in the discovery well of 31% and >6000 mD respectively, and 29% and >7000 mD in the downdip sidetrack. The sands within this zone are characterized by very coarse-grained, featureless, massive units over 1 m thick, thought to be deposited from large, dense turbidity flows from which the

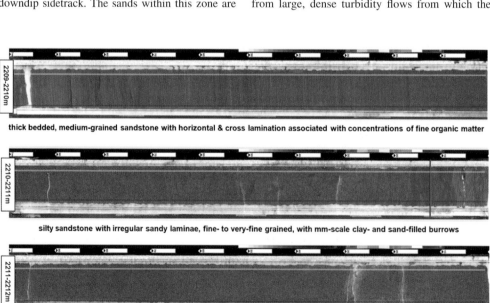

Fig. 14. Upper boundary between middle and upper zones observed in core from 214/30a-2 (depths in m MDBRT).

Lower Zone Sands Schematic

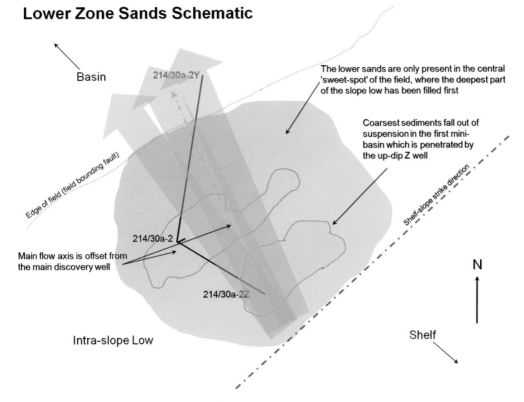

Fig. 15. Proposed deposition of lower zone sands.

largest particles have fallen out of suspension in the low on the slope as the flow progresses under gravity to the basin floor.

Upper zone

The upper zone consists of sediments filling the remaining accommodation space in the slope low and is far more variable in its composition. This is suggestive of multiple flow types – waxing and waning – with texturally diverse sands, resulting in a more variable range of reservoir quality properties. In the upper zone, porosity and permeability values average 31% and >6000 mD in the discovery well with 20% and 220 mD in the downdip sidetrack. Sand-prone layers with much lower values are observed but only in thicknesses of <1 m; they are therefore not considered to be laterally extensive or in danger of causing significant baffles to the flow of gas given the overall high reservoir quality.

Zone boundaries

Two main boundaries (middle–upper and lower–middle) have been highlighted by the main post-well studies within the Vaila Formation in addition to the top and base of the sandstone. Chemostratigraphy has demonstrated a change in sandstone character at these two boundaries; borehole image log (Fullbore Formation Micro-imager, FMI) analysis has indicated a change in depositional character at these two boundaries and petrophysical analysis has shown a change in porosity, permeability and saturation properties. These boundaries have been used to define the large-scale reservoir architecture prior to property modelling. This allows representative reservoir property distribution to be constrained by well and seismic data.

Lower–middle boundary

The boundary separating the lower and middle zones is identified on the FMI log as a possible flooding surface at 2232.8 m measured depth below rotary table (MDBRT) in the discovery well. The boundary separates the FMI upper unit A (Glenlivet lower zone) from the FMI lower unit B (Glenlivet middle zone). Figure 10 describes how the proportions of structureless sands and carbonaceous sands change across the boundary.

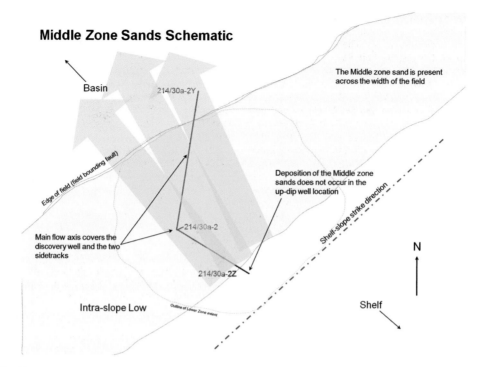

Fig. 16. Proposed deposition of middle zone sands.

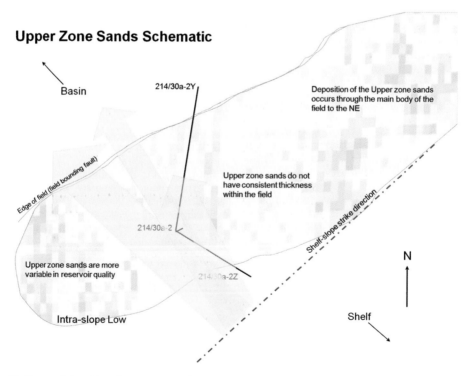

Fig. 17. Proposed deposition of upper zone sands.

The lower–middle boundary is identified in the chemical stratigraphy analysis by a significant change in elemental ratios. The boundary separates the chemostratigraphic unit S1 (lower zone) from S2 and S3 (middle zone) at 2234 m MDBRT in 214/30a-2, 3101 m MDBRT in 214/30a-2Y and 2380 m MDBRT in 214/30a-2Z. Unit S1 is characterized by high Ta/Al and Sr/Fe values and unit S2 by high Sr/Fe values and low Ta/Al values (Fig. 11). Unit S1 is thickest in updip Z well at 31 m true vertical depth sub sea (TVDSS), thinning down to 8 m TVDSS in downdip Y well.

Middle–upper boundary

The boundary separating the middle and upper zones is identified on the FMI log as a possible flooding surface at 2210.7 m MDBRT in the discovery well. The boundary separates the FMI upper unit B (Glenlivet upper zone) from FMI lower unit B (Glenlivet middle zone) and is expressed in the change in proportions of the key reservoir facies (Fig. 12).

The middle–upper boundary is identified in the chemical stratigraphy analysis by another significant change in elemental ratios. The boundary separates the chemostratigraphic unit S2 and S3 (middle zone) from unit S4 and S5 (upper zone) at 2209 m MDBRT in 214/30a-2 and 3051 m MDBRT in 214/30a-2Y. Unit S3 is characterized by high K/Al values and low Ga/Rb values, while Unit S4 is characterized by increasing Ga/Rb values and decreasing K/Al and U/Zr values (Fig. 13).

The cored interval in the discovery 214/30a-2 well also covers this boundary (Fig. 14); a clear change in facies can be seen in the digital images taken over this section. At the boundary, evidence of bioturbated mudstone and bioturbated sandstone is observed and cuttings in the downdip Y-well show a marked increase in sandstone proportions from 20% above the boundary in the upper zone to 85% beneath the boundary in the middle zone.

Conceptual model

The lower zone is only present in the updip sidetrack (214/30a-2Z) and is not recognized at the base of the field in the main discovery well and the downdip sidetrack. The middle and upper zones are observed in both the discovery well and the downdip sidetrack, but absent from the updip sidetrack well location. It is thought that the three zones represent three distinct turbidite flow episodes which vary compositionally.

A proximal-distal basinwards sedimentation trend is observed in the lower and middle zones; the upper zone may represent back-stepping within the system or a change to more lateral-distal deposits due to a shift in sedimentation direction, for example. The highly variable nature of the upper zone sand unit has also been interpreted to represent more channelized deposits. The updip well location is bypassed by the sediments forming the middle and upper zones. The axis of the turbidite flows must also be considered when examining the well data; the three separate flows appear to honour a westwards shift (Figs 15–17). The three sand zones are bound above and below by shale which forms the seal, in addition to structural pinch-out updip.

Conclusions

Exploration in the WoS is technically and commercially challenging; to make a viable discovery and to appraise the limits of the accumulation with one well (plus sidetracks) has been an unqualified success. The limits of the field are however defined seismically, not by the well alone. The capability of the well to set the limits depends on the field being characterized by the seismic anomaly and it is the integration of the two that define the field.

The application of a number of high-resolution analytical methods such as chemostratigraphy and log image analysis have enabled the well data to be integrated with the seismic interpretation and thus calibrate the internal architecture of the reservoir observed in seismic. From this robust interpretation, a static 3D model has been constructed with properly constrained property distributions leading to better estimates of initial hydrocarbons in place. Data acquisition at this early stage in a hydrocarbon discovery leads to quicker development decision-making and potential commercialization.

Factors contributing to this success were relatively benign weather conditions, a well-managed drilling operation and a management team prepared to make the most of the opportunity presented, supported by active partners. The future of the field is still dependent on third-party commercial arrangements, but the technical challenges have all been addressed by the well and the studies presented here and by the associated reservoir and development engineering projects designed to estimate deliverability and transportation of hydrocarbons.

The authors would like to thanks DONG E&P (UK) Limited and its' partners in the Glenlivet discovery for permission to publish this paper. We would also like to acknowledge that the exploration planning and successful discovery of the field was a result of the excellent work by our colleagues in Denmark led by S. Blegvad Anderson.

References

AL-CHALABI, M. 1979. Velocity determination from seismic reflection data. *In*: FITCH, A. A. (ed.) *Developments in Geophysical Exploration Methods*. Applied Science Publishers, Barking, 1–68.

CONNOLLY, P. 1999. Elastic impedance. *The Leading Edge*, **18**, 438–452.

DEAN, K., MCLACHLAN, K. & CHAMBERS, A. 1999. Rifting and development of the Faeroe-Shetland Basin. *In*: FLEET, A. J. & BOLDY, S. A. R. (eds) *Petroleum Geology of North West Europe. Proceedings of the 5th Conference*. Geological Society, London, 533–544.

LAMERS, E. & CARMICHAEL, S. M. M. 1999. The Paleocene deepwater sandstone play, West of Shetland. *In*: FLEET, A. J. & BOLDY, S. A. R. (eds) *Petroleum Geology of North West Europe. Proceedings of the 5th Conference*. Geological Society, London, 645–659.

RUTHERFORD, S. R. & WILLIAMS, R. H. 1989. Amplitude-versus offset variations in gas sands. *Geophysics*, **54**, 680–688.

Exploration and appraisal of a 120 km² four-way dip closure: what could possibly go wrong?

K. D. FIELDING*, D. BURNETT, N. J. CRABTREE, H. LADEGAARD & L. C. LAWTON

Hess Ltd, Adelphi Building, 1-11 John Adam Street, London WC2N 6AG, UK

*Corresponding author (e-mail: kevin.fielding@hess.com)

Abstract: Large four-way dip closures were obvious on multi-client 3D seismic data acquired during 1996–1998 over the unlicenced 'White Zone' between the Faroe and Shetland Islands. Cambo, located at the junction between the Westray and Corona ridges, is the largest of these features and covers an area of between 50 and 150 km². As of October 2011, Hess and its co-venturers had drilled four wells on the Cambo structure. This paper represents a brief summary of the results of these wells as an Exploration Case study. Key conclusions from this case study include: (1) it is the subsurface setting and not the drilling environment that controls uncertainty level in recoverable oil and requirement for appraisal; drilling in an environment where wells are expensive does not imply you will need fewer of them prior to making a development decision; (2) a volumetrically small oil field requires at least as much delineation as a large one of equal complexity and lateral extent; indeed, more data may be required in order to mitigate risk of commercial failure; and (3) crestal locations for wildcat exploration wells are often not optimal; this is especially the case if reservoir presence or column retention is the key risk of the play type being tested. Efforts at Cambo now focus on constraining the remaining subsurface uncertainties to an acceptable level and moving the project forward towards development sanction. Despite the knowledge acquired from the extensive exploration and appraisal campaign completed to date by the Cambo co-venturers, at the time of writing the timing of development and preferred development solution at Cambo remains unclear.

Multi-client 3D seismic data were acquired over the unlicenced 'White Zone' between the Faroe and Shetland Islands during 1996–1998. Large four-way dip closures were immediately identified on these data (e.g. Smallwood *et al.* 2004; Smallwood & Kirk 2005). The largest of these four-way dip-closed features is located at the junction between the Westray and Corona ridges, straddling blocks 204/4, 5, 9 and 10, and covers an area of between 50 and 150 km². The closure and its surrounding structurally high areas have been named 'Cambo' by Hess and its co-venturers (Fig. 1).

When the territorial dispute between the UK and the Faroes was resolved in 1999, Hess and its co-venturers decided to concentrate their efforts on structural traps rather than amplitude-driven stratigraphic plays, analogous to the Schiehallion and Foinaven fields. The Cambo acreage was acquired in the 19th and 22nd Licence Round with a commitment that included the drilling of two wells. The Licence Group consisted of Hess (32.5% equity and Operator), Chevron (32.5%), DONG (20%) and OMV (15%). The acreage covers some of the deepest parts of the Faroe–Shetland Channel, with water depth dominantly between 1000 and 1100 m. The area is renowned for operationally challenging weather and sea conditions.

Please see other papers in this volume for details of West of Shetland stratigraphy and structural setting (e.g. Ellis & Stoker in press). The Paleocene 'T scheme' created by BP staff during their evaluation of the Schiehallion and Foinaven area (Ebdon *et al.* 1995) is employed in this paper. Sandstones encountered within the Paleocene T40 Flett Formation sediments are referred to as the Colsay Member, and sandstones encountered within the T45 Flett formation are referred to as the Hildasay Member. Sandstones encountered within the Paleocene–earliest Eocene T50 Balder Formation sediments are referred to as the Cambo Member (Fig. 2). The licence history, and Hess's understanding of the evolution of petroleum plays, prospectivity and drilling on the Cambo High, are presented in Figure 3.

For the five years immediately prior to the resolution of the 'White Zone' boundary dispute, the industry in general (and BP in particular) had concentrated on stratigraphic traps within Paleocene T30 deep-marine basin floor and slope-fan turbidite sands. This resulted in the discovery and delineation of the major Schiehallion and Foinaven oilfields, as well as smaller discoveries such as Loyal, Suilven and Alligin. Total recoverable oil volumes were believed to exceed 1 billion barrels, and there was much speculation that the play would extend over the unlicenced White Zone with similar resulting resource gains. Perhaps unsurprisingly, this play became the focus of most industry interest when

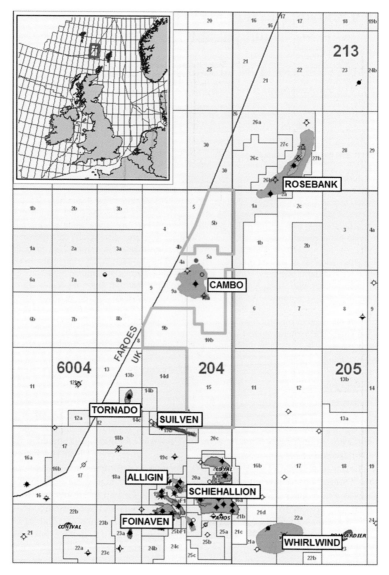

Fig. 1. Regional location map of the Cambo Area, showing Hess Operated Licences (in yellow), discoveries and developed fields. Newly acquired acreage from the 26th Round is outlined in green.

the first processed products from speculative multi-client 3D seismic survey shoots undertaken in 1995–1997 became available. Indeed, when geophysical interpretations were undertaken of the new seismic product, it became clear that seismic amplitude responses could be mapped that were, at least in some ways, consistent with the presence of large stratigraphic traps within the T30s sequences. However, trap definition for this play was recognized as difficult; seismic imaging techniques, including direct hydrocarbon indicator (DHI) analysis, are often ambiguous (Lamers & Carmichael 1999). It was also clear that other plays were present on the new data. Most obvious of these was the presence of four-way dip-closed structures on long-lived highs which appeared to be bald of T30s reservoir, but with potential in both shallower (T40–T50) and deeper (Mesozoic and older) reservoirs.

As part of the 19th Licence Round of offshore licensing in 2001, a number of licences were awarded covering blocks containing these long-lived structural highs. Cambo, on the P1028 Licence, was the largest of these closures. As operator for

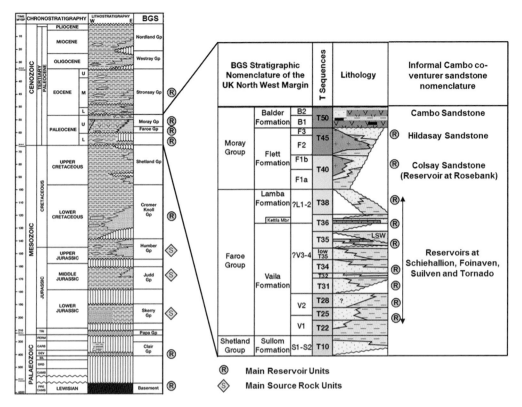

Fig. 2. Stratigraphy of the Cambo area.

the P1028 co-venturers in 2002, the first operational season after licensing, Hess spudded the 204/10-1 well and commenced evaluation of these dip closures. The well encountered hydrocarbons within the T45 sequence Hildasay Member sandstones, opening up a new play type for the area. This was followed up by a second exploration well on the Cambo structure (204/10-2) in 2003. Based on the results of these wells, the P1189 Licence covering the northern third of the closure was applied for by the Licence Group and awarded in the 22nd Licence Round. A further play was opened up in 2004 in the form of slightly older T40 Colsay Member sandstones at the Chevron-operated Rosebank discovery, some 40 km to the north.

After a period of five years of study of the results of the two exploration wells, Hess and co-venturers returned to drilling on Cambo in 2009 with the 204/10a-3 appraisal well, followed closely in 2011 by the 204/10a-4 delineation well and 4Z horizontal sidetrack. 2011 also saw a return of seismic acquisition, with the commencement of shooting of a state-of-the-art regional multi-client towed streamer 3D survey which was completed during the summer of 2012.

As with the advent of the first 3D seismic data over the area in the late 1990s, it is anticipated that the new Faroe-Shetland Basin (FSB) 2011–2012 3D Geostreamer seismic data, acquired in Q3 2012, will provide a step change in seismic image. Proprietary processing of these data is expected in-house by June 2013. This improvement in seismic image will, in turn, lead to a second wave of exploration and exploitation activity in the area, due both to an enhancement of our understanding of the existing play suite (progressing some existing discoveries towards project sanction) and potentially also providing evidence for new plays that can be matured for drilling later in the decade.

Pre-drill evaluation of the Cambo prospect

Unlike in many marine and land environments prospective for oil and gas, the first drilling in the vicinity of Cambo was targeted with the aid of good-quality 3D seismic data. Despite this high-quality imagery, the first well on Cambo, 204/10-1, was considered a rank wildcat with a 15% overall chance of success.

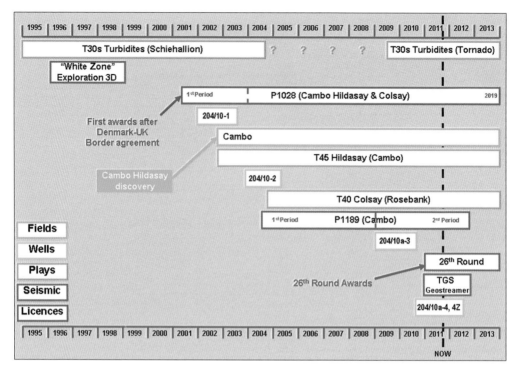

Fig. 3. A brief timeline of exploration and appraisal on the Cambo High.

The prospect was viewed as low risk from a structural perspective. For all the viable depth conversion methods employed, it forms a large-scale four-way dip-closed structure at all levels from the deepest coherent reflectivity to within a few hundred metres of sea bed. It was modelled as surrounded by mature Kimmeridge Clay Formation source rock. Under all circumstances envisaged, the structural high acts as a major focus for migrating hydrocarbons at a number of levels.

Figure 4 is a regional seismic line taken from the White Zone 3D seismic datasets which ties the closest pre-drill discovery well (Suilven) to the prospective Cambo High.

Prior to drilling, the critical risk of failure for the Cambo Prospect was viewed as reservoir presence. The Cambo High has been structurally prominent throughout the period that is imaged on seismic data; models predicting thick reservoir units over the structure were therefore challenging to create and difficult to sustain. Weak indications of brightened seismic amplitudes over the structure could be interpreted as thin hydrocarbon-bearing reservoir intervals or, conversely, could be attributed to volcanic and volcaniclastic lithologies associated with the North Atlantic Igneous Province (NAIP) or with Balder-age ash deposits from final rift break-up volcanism. Three thin reflective intervals within the post-NAIP section were interpreted as offering the best opportunity for containing reservoir units. These reflective intervals were tied to seismically defined sequences of believed Miocene, Eocene and Paleocene age (named respectively as the 'Rothbury', 'Upper Cambo' and 'Lower Cambo' targets). The remaining section above the NAIP lavas was thought highly likely to be dominated by shale-prone sediments.

In contrast to the well-imaged shallow stratigraphy, thick sandstone units with reservoir potential could be inferred within the poorly imaged pre-NAIP section. By analogy to other structural highs in the basin, Palaeozoic reservoirs such as the Devonian–Carboniferous at Clair (Clifford et al. 2005; Nichols 2005) and Mesozoic reservoirs such as those at Solan and Strathmore (Verstralen et al. 1995; Herries et al. 1999) were considered viable exploration targets. However, the presence of these units could not be confirmed through seismic character or direct mapping due to seismic imaging issues associated with the thin NAIP lavas that cover the structure. As a result, the presence of these postulated deeper reservoirs remained conjectural and so are considered very high risk.

As well as the reservoir risks, there were additional pre-drill concerns over seal integrity at

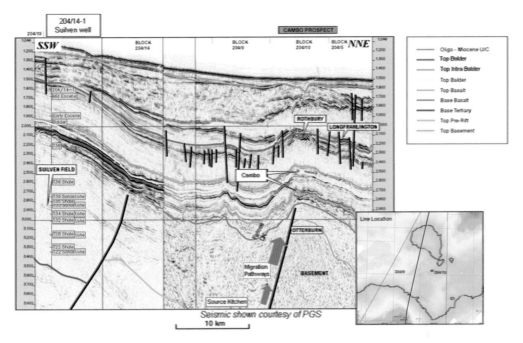

Fig. 4. A regional seismic line from the nearest pre-drill well control to the Cambo structure. Green arrows indicate the pre-drill belief that the Cambo High has acted as a focus for hydrocarbon migration from surrounding mature source rocks.

Cambo. Depth conversion of the seismic data suggested only 1000–1100 m of post-reservoir Eocene–Holocene age shale cover below the sea bed. The presence of seismic disruption above the crest of the Cambo closure was attributed to hydrocarbon leakage. The presence of pock marks or 'scoops' directly above the Cambo structure was also interpreted by some to represent fluidization of near-surface sediment by escaping hydrocarbons, and thus a risk that the trap would be found breached. An alternative explanation involving direct deep-water seabed erosion by turbulent deep-water currents associated with inversion of the high was also postulated (Smallwood 2004).

Finally, the team recognized significant potential issues on hydrocarbon-quality pre-drilling based upon the cold deep-water environment and limited sedimentary cover. Cold reservoirs and heavy, biodegraded oil were both expected and it was recognized that this could inhibit commercial viability of any accumulation found.

Figure 5 is a 3D seismic in-line from the White Zone data directly through the proposed location for the first planned well on the Cambo High. Potential hydrocarbon zones were recognized both directly above the NAIP volcanics (the Upper and Lower Cambo sands) and in the interval below them (the Otterburn). A shallower reflective interval of likely Miocene age was also identified as potential reservoir (named Rothbury) although this was seen to be as much a hazard zone to be overcome while drilling as it was a potentially commercially viable target.

In summary, prior to drilling the status of our understanding of the hydrocarbon potential of the Cambo Structure was based dominantly upon interpretation of 3D seismic data and regional exploration and appraisal well results. It could be summarized:

- 3D seismic confirmed the presence of a large four-way dip closure;
- models were invoked placing reservoirs within the pre-NAIP section and within the overlying Paleocene–Eocene sediments;
- seals were envisaged within Cretaceous shales, the volcanics and in post-Paleocene deep-marine shale cover; and
- potential volumes in place and recovery mechanisms were essentially unconstrained; this was a rank wildcat.

Well 204/10-1

The first well to be drilled on the Cambo High, 204/10-1, was drilled in late 2002. The well was drilled

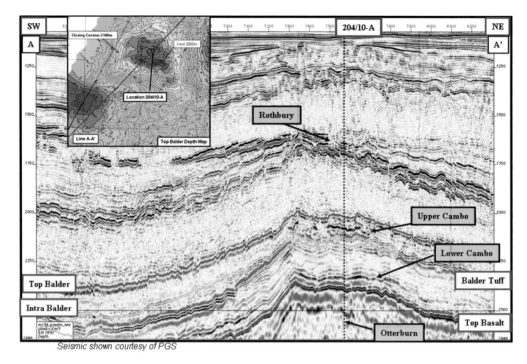

Fig. 5. A pre-drill seismic line from the White Zone 3D dataset indicating the main intervals believed to be prospective at the Cambo location.

successfully to a terminal depth (TD) in Precambrian basement rocks. Drilling, plugging and abandoning of the well took 38 days, including 6 days disruption associated with weather and sea state. The well encountered a number of thin reservoir intervals while drilling to TD. The Rothbury secondary target identified as a potential sandstone unit from seismic data was present but water-bearing. The Upper Cambo sandstone primary target for the well was present with a strong gas peak while drilling, but was shown to be water-bearing by wireline logs. The Lower Cambo sandstone target was also present, and was found to be oil- and gas-bearing but was thin. It was renamed Hildasay Member sandstones based upon T45 Paleocene–Eocene Flett Formation age confirmed by biostratigraphic data. The pre-NAIP volcanics Otterburn secondary target was completely absent.

Figure 6 illustrates the pre-drill and post-drill stratigraphy at 204/10-1. The major divergence from prognosis was the lack of any pre-volcanic clastic interval, with well TD in Precambrian granodiorite directly beneath the NAIP volcanics.

After wireline logging and incorporation of data from formation pressure tests, fluid sampling and sidewall core data, petrophysical evaluation of the Hildasay sandstone unit estimated an aggregate of 8.9 m net oil sand and 3.4 m net gas sand in three separate sand units (Fig. 7). Hydrocarbon shows were also encountered within and beneath the NAIP volcanics. While clearly technically a discovery, the presence of mixed oil and gas phase, complexity of the reservoir stratigraphy and thinness of the reservoir units were all viewed as disappointing by the co-venturers.

After drilling the 204/10-1 discovery well, the status of our understanding of the hydrocarbon potential of the Cambo Structure could be summarized as follows.

- Well 204/10-1 confirmed the presence of hydrocarbon-bearing reservoir sands and so was a technical 'success'.
- Seal was proven in both the NAIP volcanics and in post-Paleocene cover. However, efficacy of this seal in terms of the presence of a commercially viable column remained in doubt due to crestal location (limited column updip).
- Reservoir was proven within the Hildasay (Paleocene T45 age), however the viability of these sandstones as a commercial reservoir remained in doubt due to their thin and isolated nature.

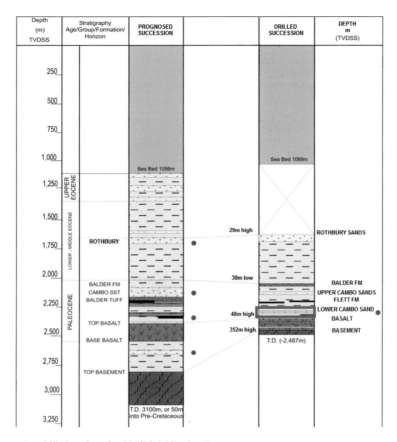

Fig. 6. Prognosed v. drilled sections for 204/10-1 (Cambo-1).

- Potential volumes in place were poorly constrained due to crestal location and lack of penetration of an oil–water contact.
- Commercial recovery appeared to be unlikely due to thinness and isolation of the reservoirs encountered, combined with the deep water and harsh environmental setting of the discovery.
- The well confirmed the pre-drill risk assessment for biodegraded oil, encountering moderately to heavily biodegraded crude with low gravity (in the range 22–25° API) and relatively high viscosity.

Although clearly our understanding of the petroleum prospectivity of the Cambo High had improved through the drilling of 204/10-1, one important lesson learned was that the crestal location for the discovery well was not optimal in this instance. The presence of a commercially viable hydrocarbon column (i.e. an effective seal), reservoir thick enough to be commercially viable and sufficient volumes for commercial development all remained unknowns due to the crestal location of the well.

Well 204/10-2

The second commitment well on the P1028 Cambo Licence, 204/10-2, was drilled primarily for the pre-volcanics pinch-out play. It was named Lindisfarne and was expressly drilled to assess reservoirs not encountered within the 204/10-1 discovery well, rather than to act as an appraisal for reservoirs encountered as oil-bearing in the first well.

Oil shows in the volcanic rocks and at top basement in 204/10-1 suggested the possibility of pre-volcanic reservoirs pinching out onto the Cambo Precambrian plug. With approaching a billion years of section missing at the structural crest, there were plenty of options for reservoir age for this prospect although Lower Paleocene sandstones (as seen at Schiehallion) probably represented the best opportunity for commercial reservoir.

As with the first well, reservoir presence was seen as the major pre-drill risk. Reprocessing of the seismic data and detailed mapping of reflectors and interval seismic character suggested that a

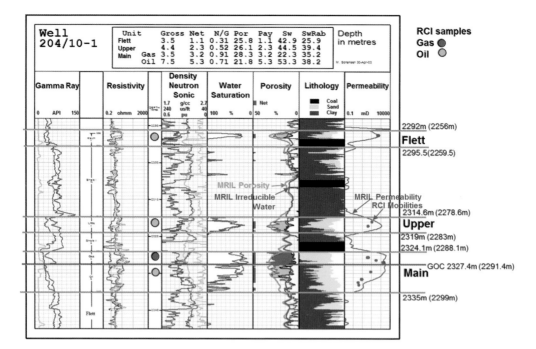

Fig. 7. Petrophysical evaluation of the oil- and gas-bearing Hildasay sandstones in 204/10-1 (Cambo-1).

thick sedimentary package existed on the flanks of the Cambo High, but still within overall closure of the structure. Figure 8 is a schematic section through Cambo illustrating our understanding of the potential for deeper reservoir targets on the closure. Figure 9 is a seismic line showing coherent reflectivity used as evidence to support this pre-drill interpretation.

In the event, 204/10-2 encountered a 1000 m Cretaceous shale section below the volcanic rocks which contained no viable reservoir. The well was still in Cretaceous shale at TD. Despite not being the main target for the 204/10-2 well, the Hildasay sand was again present and hydrocarbon-bearing, with a total of 11.1 m net oil sand and 3.4 m net gas sand again in three separate sand units. Unfortunately, due to a pronounced lateral velocity change, the Hildasay appraisal data were of limited use as the well was at exactly the same structural elevation as the discovery well. Additionally, an attempt to recover full bore core from the interval failed due to the loose, unconsolidated nature of the Hildasay sandstone reservoir encountered. Drilling, plugging and abandoning of the well took 45 days, with virtually no disruption to operations associated with bad weather and heightened sea state.

Figure 10 illustrates the pre-drill and post-drill stratigraphy at 204/10-2. The Miocene Rothbury secondary target was again present but water-bearing, Upper Cambo sandstone target was present but water-bearing and the Hildasay sandstones were again thin and oil- and gas-bearing. The pre-NAIP volcanic rocks Lindisfarne primary target was absent; over 1000 m of continuous Cretaceous shale was drilled before TD was called. The major divergences from prognosis were the lack of any pre-volcanic reservoir interval and the unexpectedly fast velocity of the shale-prone Cretaceous section encountered.

The drilling of the shallow 204/10-2 section confirmed that there is a significant depth conversion issue. Despite being imaged at significantly less two-way time (TWT) on 3D time-seismic data than the equivalent reflectors in 204/10-1, the Hildasay sandstone units in 204/10-2 were essentially at the same level in depth, indeed top main Hildasay sandstone was actually just a few metres deeper in the 204/10-2 well. This lateral velocity variability between the two wells was ascribed to a number of potential causes including an imaging artefact associated with shallow gas anomalies, lateral geological variability within the overburden stratigraphy and near-seabed low-velocity sediments present over parts of the Cambo structure (but absent at the 204/10-2 well location). Regardless of the cause, the variability in velocity in the overburden will continue to prove a key risk to forward evaluation in this low-relief closure environment.

The primary target section for 204/10-2 directly beneath the NAIP basalts was found to be absent.

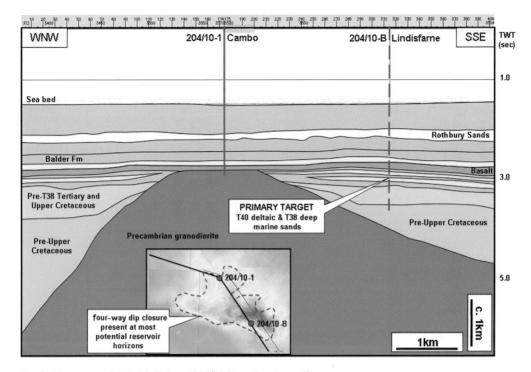

Fig. 8. Play concept for the Lindisfarne (204/10-B) exploration well.

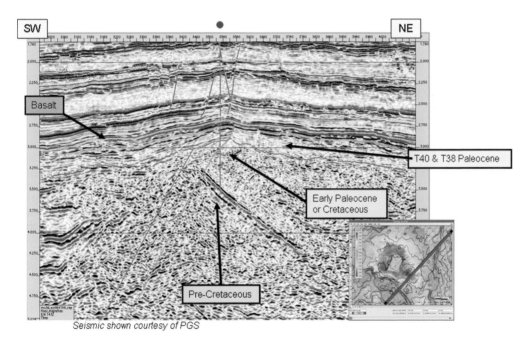

Fig. 9. Seismic line from the 3D through the proposed location for the Lindisfarne well illustrating the interpreted presence of a sedimentary section below the NAIP volcanic rocks based upon the presence of coherent inclined reflectors.

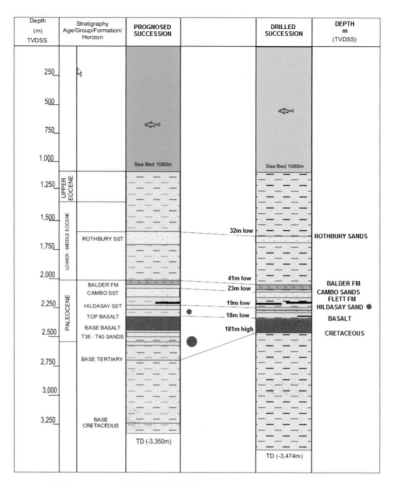

Fig. 10. Prognosed v. drilled sections for 204/10-2 (Lindisfarne).

Strong evidence for Cretaceous sediments within 15 m of the base NAIP volcanic rocks pick was obtained from wellsite biostratigraphy and confirmed by later laboratory studies. The Lindisfarne prospect failed not just because of a lack of sandstones within the Paleocene T30s, but because the whole pre-NAIP Paleocene sedimentary system is missing at this location. The lack of a significant pre-NAIP section implies that a larger area than just the pre-Cambrian granodioritic plug was structurally high during Paleocene time.

Figure 11 illustrates that, during post-well evaluation within Hess, it became apparent that the 204/10-2 well is unlikely to have reached the reflector identified pre-drill as a potential base of Cretaceous. This is because of a significantly higher velocity encountered within the indurated shales and carbonate stringers drilled within the 1000 m Cretaceous section. As a result of the TD in Cretaceous shale, the stratigraphy of the pre-Cretaceous in and around the Cambo High remains essentially unknown.

Figure 12 shows the petrophysical evaluation of the oil- and gas-bearing Hildasay sandstone unit encountered in 204/10-2. An aggregate of 11.1 m net oil sand and 3.4 m net gas sand were confirmed. As in 204/10-1, this net pay occurred in three separate sand units (Fig. 7). Pressure data and fluid recovery indicated the presence of stacked pay with vertical isolation and different pressures and fluid contacts indicated in the sand units. Again, as was found in 204/10-1, a small gas column was confirmed in the lower sand unit, underlying oil in the upper sand bodies.

After drilling the 204/10-2 well, the status of our understanding of the hydrocarbon potential of the Cambo Structure could be summarized as follows.

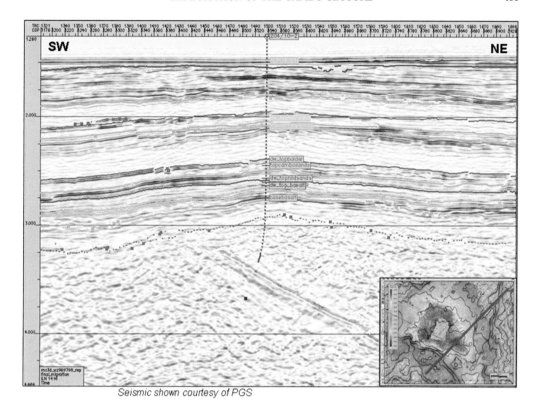

Fig. 11. 204/10-2 Lindisfarne post-drill NE–SW seismic line in TWT from the reprocessed 3D (IL1418), illustrating remaining deep Mesozoic potential at the Cambo High.

- 204/10-2 primary target pre-NAIP sediments contained no reservoir and so this was an exploration dry hole.
- An additional seal was proven in thick Cretaceous shales.
- There was only limited appraisal value from the well at Hildasay levels due to the combination of almost identical structural elevation (so that proven column height was not increased), similar thin hydrocarbon-bearing sandstones encountered with no oil–water contact (OWC) in a similarly crestal location and a failure to recover full bore core.

However, our understanding of the petroleum prospectivity of the Cambo High had improved as a result of drilling 204/10-2. On the downside we had learned that depth-conversion sensitivity was going to play an important role at Cambo. On a more positive note however, the main Hildasay sandstone unit had been shown to be continuous over a number of kilometres and we had proven the potential for stacked pay with different OWCs.

Well 204/10a-3

In 2009, a downdip appraisal well was drilled primarily to delineate the extent of the hydrocarbon accumulation in the thin Hildasay sands. Well 204/10a-3 was drilled close to the structural spill to prove volumes in place and test the concept of thicker sands to the north as inferred from seismic mapping and the depositional model. The well was located in order to: encounter the OWC in the thickest Hildasay sandstone unit drilled in 204/10-1; investigate the potential for deeper T40 Colsay reservoir within the NAIP volcanic rocks section; and investigate the potential for pre-NAIP Paleocene sands (essentially the same play that failed due to erosion at Lindisfarne well 204/10-2). Figure 13 is a schematic section through Cambo illustrating our understanding of the hydrocarbon and reservoir distribution prior to drilling 204/10a-3.

The 204/10a-3 well was drilled in 35 days with no disruption to operations associated with bad weather and heightened sea state. It confirmed a much thicker Hildasay sandstone interval, with

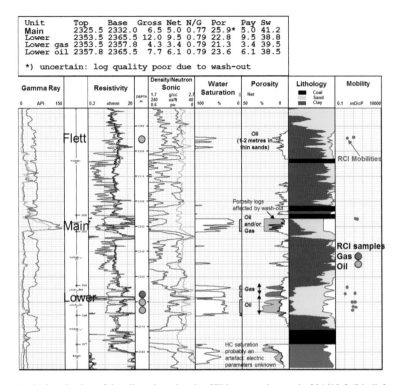

Fig. 12. Petrophysical evaluation of the oil- and gas-bearing Hildasay sandstones in 204/10-2 (Lindisfarne).

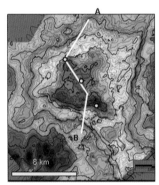

- **204/10a-3 Primary objective was to test Hildasay sandstones in a downdip location**
  - **Thicker reservoirs**
  - **Oil bearing to mapped spill point?**
- **Additional deeper targets were recognized in the intra- and /or pre-basalt Paleocene sediments.**

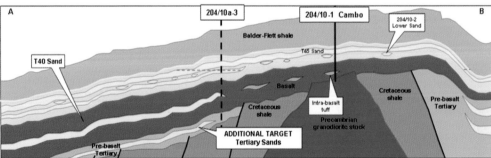

Fig. 13. Schematic section through Cambo as understood in 2009 and the location of 204/10a-3 (Cambo-3).

60 m net sand spread over four sand units. There were good oil shows at the top of the reservoir interval, but the sands were all proven water-bearing from wireline logs. Extrapolation of reservoir pressures between the penetrated hydrocarbon legs in the previous wells and the aquifer pressures in 204/10a-3 implied that significant oil volumes could lie updip. Dry hole analysis looked into all possible explanations for the presence of water-bearing reservoirs at the 204/10a-3 location. Seal breach, late tilting of the structure and the presence of thief sands on the flanks of the structure reducing the extent of dip closure were all considered possible explanations.

Figure 14 illustrates the prognosed and actual sections for the 204/10a-3 appraisal well. The Hildasay Formation was thicker and higher net-to-gross than expected. Three significant Hildasay sand units were encountered, each more than 15 m thick. Oil shows were encountered while drilling, and some residual oil has been interpreted from petrophysical evaluation. However, the mobile fluid is interpreted to be water throughout the Hildasay sandstone. The well was plugged and abandoned (P&A) without full bore/conventional core or drill stem test (DST).

The 204/10a-3 appraisal well results had some profound implications for the Cambo Hildasay prospect. Despite not encountering the prognosed oil leg, the well was in many ways a positive result.

Specifically, the well proved up over 50 m of net reservoir, dispelling concerns that there was

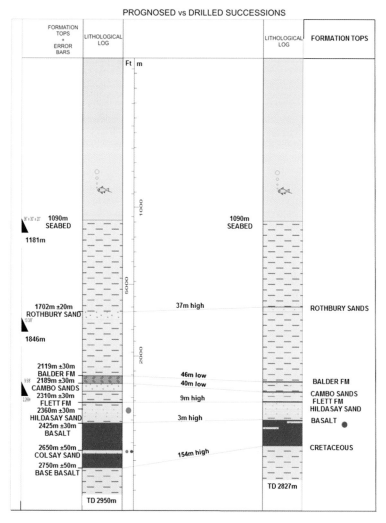

Fig. 14. Prognosed v. drilled sections for 204/10a-3 (Cambo-3).

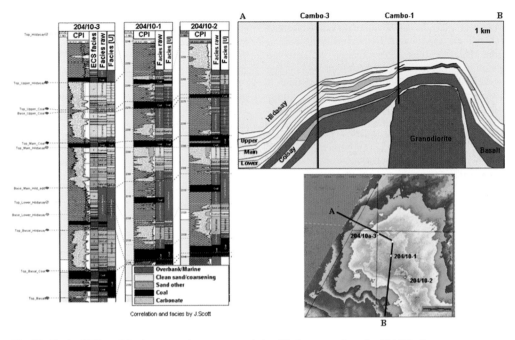

Fig. 15. Cambo Hildasay Member reservoir geometry and simplified cross-section after 204/10a-3.

not likely to be thick enough reservoir over the Cambo High to justify development. Secondly, a number of sidewall cores were recovered that provided the first evidence for high-quality reservoir, with permeability in the range of 100 mD to >1 Darcy. This led to more comfort that flow rates for production wells could be potentially high enough to justify deep-water development. Finally, successful formation pressure testing was achieved for the first time within the Hildasay water leg. Intersection of these aquifer pressures with the oil pressures from higher in the reservoir confirmed that a significant hydrocarbon column was indeed likely to be present within the Hildasay at Cambo.

After drilling the 204/10a-3 appraisal well, the status of our understanding of the hydrocarbon potential of the Cambo Structure could be summarized as follows.

- Well 204/10a-3 was an appraisal dry hole. However, the well did prove thick high-quality Hildasay reservoir, proving the viability of the reservoir play element on the structure.
- Pressure data indicated a significant oil column could be present downdip from 204/10-1, confirming seal efficacy.
- Volumes in place were calculated as being potentially of commercial interest.
- The well data also suggested >30 m of oil column in >10 m thick individual sandstone units could be present, from which significant recovery could be anticipated.
- Although there remains some controversy, the consensus view is that the structure is not filled to structural spill point and questions remain on possible re-migration or leakage of oil into nearby structures. Figure 15 summarizes the Cambo Hildasay Member reservoir units

Table 1. *The evolution of understanding of the Cambo accumulation with exploration/appraisal history*

	Trap	Seal	Reservoir	Volume	Recovery
3D Seismic	☺	?	?		
204/10-1	☺	☺?	☺?		
204/10-2	☺	☺?	☺?		
204/10a-3	☺	☺	☺	☺?	☺?
204/10a-4	☺	☺	☺	☺	☺?
204/10a-4Z	☺	☺	☺	☺	☺

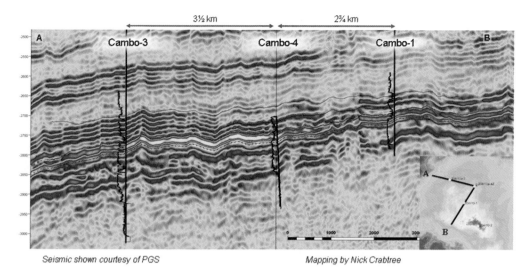

Fig. 16. Structural position of planned well 204/10a-4 (Cambo-4).

drilled in the first three wells, and provides a simplified cross-section illustrating our understanding of the reservoir geometry at the time.

One important lesson learned from 204/10a-3 is that aggressive downdip appraisal locations can be useful even if water-bearing. Indeed, as indicated

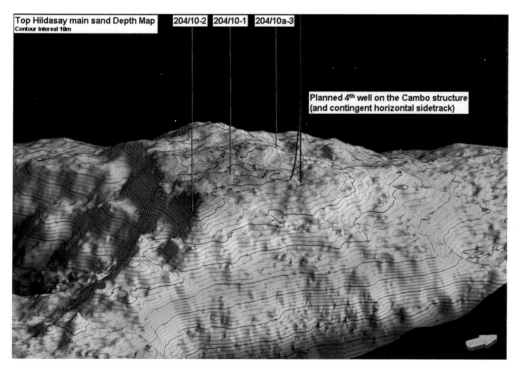

Fig. 17. The Cambo structure with existing and planned wells at 1 March 2011. For scale, 204/10a-3 and 204/10-1 are 3.5 km apart.

in Table 1, this well was the most successful in terms of moving forwards technically in the P1028 Licence Group's understanding of the subsurface, and resulted in a significant increase in expectation by the co-venturers despite its dry hole status.

As a further lesson learned, drilling the wells geometrically in line (WNW–ESE) was not ideal, as significant depth conversion uncertainty remained off this line. This depth uncertainty had profound implications for volumes in place on the northern flank.

Well 204/10a-4,4Z

Hess and the Cambo co-venturers planned and drilled the fourth well on the Cambo structure during 2011. Well 204/10a-4,4Z was designed to intersect thick Hildasay sands within the oil column, proving sufficient volumes for commercial development (Fig. 16).

The 204/10a-4 near-vertical pilot hole was planned to penetrate the Hildasay section in a structural position approximately half-way between the 204/10-1 discovery and the 204/10a-3 dry hole. The primary objective was to find thick sands in the oil column to prove sufficient in-place oil to support development, and to core the reservoir in order to define reservoir character.

In the success case of the pilot hole, a sidetrack option (204/10a-4Z) was planned that would allow a horizontal well section to be landed within the reservoir to prove reservoir continuity and development concept. There was also a plan to place a sand-face completion down-hole over the reservoir interval to allow a drill stem test (DST) to be performed. This DST was designed to both provide high-quality fluid samples and establish commercial flow. Figure 17 shows the pre-drill Top Hildasay Member reservoir depth structure map with location of pilot well and side-track.

The 204/10a-4 pilot well came close to prognosis on reservoir presence and, although the oil–water contact was encountered shallower than expected, proved significant oil in place. Two oil-bearing cores were recovered containing excellent reservoir quality sandstones. The well was considered a success and therefore the horizontal sidetrack was initiated.

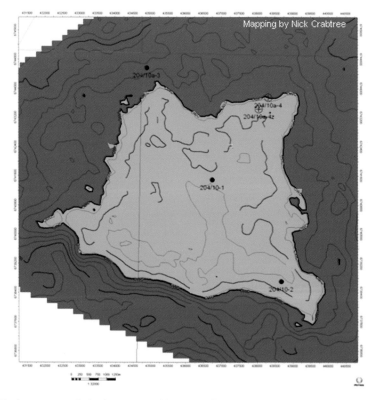

Fig. 18. The Cambo top reservoir depth structure, with outline of hydrocarbon footprint within the main Hildasay sandstone unit in green.

In 204/10a-4Z, over 500 m of continuous sand was drilled within the Hildasay at 90° confirming its lateral continuity. However, operational delays associated with the sand-face completion led to DST timing in late October–early November which was considered problematic due to unpredictable weather and heave limitations during testing. As a result, the well was suspended. The plan is to return to this well to obtain DST flow data in the future.

Despite not managing to complete the full programme within the short summer weather window west of Shetlands, the 204/10a-4,4Z well did remove a number of major subsurface risks to development including depth conversion uncertainty, the height of the oil column, permeability and reservoir continuity in this deep-water setting.

After drilling the 204/10a-4,4Z appraisal well, the status of our understanding of the hydrocarbon potential of the Cambo Structure could be summarized as follows.

- 204/10a-4 was an oil-bearing appraisal well that confirmed a significant accumulation at Cambo and significant accessible Stock Tank Oil Initially In Place (STOIIP).
- High-quality core was recovered, confirming the presence of the high-porosity high-permeability reservoir.
- The main reservoir was shown to be laterally continuous in a horizontal sidetrack 204/10a-4Z which was accurately landed in a single reservoir unit, and remained within this unit for a full 500 m section.
- Other Hildasay sand units were proven correlatable over tens–hundreds of metres between pilot and sidetrack.
- Unfortunately, an insufficient weather window remained to undertake the planned DST on the completed horizontal well-bore. We await DST flow data to confirm commercial viability. Figure 18 illustrates the interpreted extent of the oil leg in the main Hildasay sandstone unit after the drilling of 204/10a-4, 4Z.

Current status and forward plan

Table 1 indicates our current understanding of the petroleum potential of the Cambo High and its evolution with increasing data acquisition and appraisal. It has taken a number of years and significant intellectual and commercial effort to progress our understanding at Cambo to the point where we can say with some degree of certainty that there is a sizeable hydrocarbon volume present that can be recovered in a commercially viable manner.

Further evaluation of subsurface uncertainties will continue in an attempt to reduce the level of uncertainty and associated risk to manageable levels.

Efforts will now focus on finding the optimum development plan for the discovered resources. Development options include: stand-alone development; some form of combined development or export solution with nearby Rosebank Discovery; or a Hub development at Cambo with a tie back of additional accumulations. All these options are currently being assessed with a view to moving the project forward to development sanction.

Conclusions

Key conclusions from this case study include the following.

(1) Crestal locations for discovery wells are often not optimal. This is especially the case if reservoir presence is the key risk of the play type being tested. A downdip location for 204/10-1, for example, would potentially have led to the recognition of a robust volume of in-place hydrocarbon within the Hildasay reservoirs earlier in the exploration and appraisal history at Cambo. The crestal location left discovered volumes essentially unconstrained and left doubts on both the presence of effective reservoir (as it was very thin at the crest) and effective seal (as the proven hydrocarbon column after the first well was extremely limited).

(2) It is the subsurface setting that controls uncertainty in oil in-place and requirement for appraisal, not the drilling environment. An environment where wells are expensive to drill does not influence the technically required number of appraisal wells prior to making a development decision. Indeed, the standard issues of areal extent of hydrocarbons, reservoir continuity and variability, fluid character, positions of fluid contacts, and so on. will all be present and need to be addressed regardless of drilling environment and well cost. The exploration and appraisal costs required to assess these issues to an appropriate level of understanding for the play type being chased must be understood and accepted prior to entering into exploration in operationally difficult and expensive environments.

(3) If it is equally complex and laterally extensive, a volumetrically small oil field requires at least as much delineation as a large field. Indeed, more data will be required for a marginal field size in order to reduce risk of commercial failure to acceptable levels than is necessarily required for a field with more robust economics.

The exploration and appraisal story of the Cambo High has spanned the last ten years and more. It has cost in excess of £100 MM to prosecute. There have been successes such as the initial discovery at 204/10-1, the recovery of high-quality unconsolidated sandstone core from 204/10a-4 and the drilling of a continuous 500 m high-angle reservoir section in 204/10a-4Z. There have been disappointments, such as the failure of the pre-NAIP volcanic rocks targets in 204/10-2, and the cancellation of the DST at 204/10a-4Z during the 2011 operational season due to the capricious sea conditions and weather of the North Atlantic.

With hindsight, it is clear that different exploration and appraisal decisions could have resulted in more efficient data collection and more rapid progression of the Cambo project towards sanction. However, the four wells on the Cambo High have confirmed that the Hildasay reservoir play is effective there, that it contains a significant volume of accessible STOIIP and that a development solution involving high-angle or horizontal wells is viable. The Cambo Project will now concentrate on finding the optimum development solution for the discovered oil and gas resources, and on securing a fiscal environment within which exploitation of the resources can be achieved in a commercially viable manner.

The authors would like to thank the many contributors to the Hess Cambo subsurface teams, past and present, and are grateful to Hess, Chevron, OMV, DONG Energy and PGS for permission to publish.

References

CLIFFORD, P. J., O'DONOVAN, A. R., SAVORY, K. E., SMITH, G. & BARR, D. 2005. *Clair Field – managing uncertainty in the development of a waterflooded fractured reservoir*. Conference Paper at Offshore Europe, 6–9 September 2005, Aberdeen United Kingdom, Society of Petroleum Engineers, 96316.

EBDON, C. C., GRANGER, P. J., JOHNSON, H. D. & EVANS, A. M. 1995. Early tertiary evolution and sequence stratigraphy of the Faeroe-Shetland Basin: implications for hydrocarbon prospectivity. *In*: SCRUTTON, R. A., STOKER, M. S., SHIMMIELD, G. B. & TUDHOPE, A. W. (eds) *The Tectonics, Sedimentation and Palaeoceanography of the North Atlantic Region*. Geological Society, London, Special Publications, **90**, 51–69.

HERRIES, R., PODDUBIUK, R. & WILCOCKSON, P. 1999. Solan, Strathmore and the back basin play, west of Shetland. *In*: FLEET, A. J. & BOLDY, S. A. R. (eds) *Petroleum Geology of NW Europe: Proceedings of the 5th Conference*. Geological Society, London, 693–712.

LAMERS, E. & CARMICHAEL, S. M. M. 1999. The Paleocene deepwater sandstone play West of Shetland. *In*: FLEET, A. J. & BOLDY, S. A. R. (eds) *Petroleum Geology of Northwest Europe: Proceedings of the 5th Conference*. Geological Society, London, 645–659.

NICHOLS, G. J. 2005. Sedimentary evolution of the Lower Clair Group, Devonian, West of Shetland: climate and sediment supply controls on fluvial, aeolian and lacustrine deposition. *In*: DORÉ, A. G. & VINING, B. A. (eds) *Petroleum Geology: North-West Europe and Global Perspectives – Proceedings of the 6th Petroleum Geology Conference*. Geological Society, London, 957–967.

SMALLWOOD, J. R. 2004. Tertiary inversion in the Faroe-Shetland Channel and the development of major erosional scarps. *In*: DAVIES, R. J., CARTWRIGHT, J. A., STEWART, S. A., LAPPIN, M. & UNDERHILL, J. R. (eds) *3D Seismic Technology: Application to the Exploration of Sedimentary Basins*. Geological Society, London, Memoirs, **29**, 187–198.

SMALLWOOD, J. R. & KIRK, W. 2005. Paleocene exploration in the Faroe–Shetland Channel: disappointments and discoveries. *In*: DORÉ, A. G. & VINING, B. A. (eds) *Petroleum Geology: North-West Europe and Global Perspectives – Proceedings of the 6th Petroleum Geology Conference*. Geological Society, London, 977–991.

SMALLWOOD, J. R., PRESCOTT, D. & KIRK, W. 2004. Alternatives in Paleocene exploration West of Shetland: a case study. *Scottish Journal of Geology*, **40**, 131–143.

VERSTRALEN, I., HARTLEY, A. J. & HURST, A. 1995. The sedimentological record of a late Jurassic transgression: Rona Member (Kimmeridge Clay Formation equivalent), West Shetland Basin, UKCS. *In*: HARTLEY, A. J. & PROSSER, D. J. (eds) *Characterization of Deep Marine Clastic Systems*. Geological Society, London, Special Publications, **94**, 155–176.

Improved signal processing for sub-basalt imaging

N. WOODBURN*, A. HARDWICK, H. MASOOMZADEH & T. TRAVIS

TGS, Graylaw House, 21/21A Goldington Road, Bedford MK40 3JY, USA

*Corresponding author (e-mail: nick.woodburn@tgs.com)

Abstract: Sub-basalt imaging continues to provide a challenge within the West of Shetland region. Successful imaging is being achieved through the reprocessing of 2D reflection seismic data. These were acquired with conventional source array and streamer parameters and include key signal processing techniques. The first technique involves spectral processing to boost low-frequency signals at the early stages of processing. The second technique involves attenuating coherent and incoherent noise in all of the available 'time-offset' domains. Examples of the data after application of these techniques are presented and clear improvements over the original processing are demonstrated. As work on these reprocessing methods has progressed, the benefits of moving to a true broadband processing solution have become clear.

Gold Open Access: This article is published under the terms of the CC-BY 3.0 license.

The Faroe–Shetland Basin is bound to the SE by the West Shetland Platform and to the NW by the Fugloy Ridge. Voluminous Palaeogene intrusive and extrusive igneous material is found within the basin in the form of continental flood basalts, hyaloclastites, sill and dyke complexes, igneous centres, magmatic underplating and the deposition of regional tuff horizons associated with the opening of the proto North Atlantic Ocean (Naylor *et al.* 1999; Lundin & Doré 2005). The Corona Ridge approximates the limit of extrusive volcanic units which gradually thicken to the NW, exceeding 6 km onshore Faroe Islands (White *et al.* 2003). These volcanic units, present as heterogeneous high-velocity layers, continue to provide a challenge for seismic imaging of the prospective sedimentary units beneath. Lower-frequency energy in the source wavelet is more likely to penetrate through the basalt than higher frequencies as it is less attenuated by intrinsic absorption, and less scattered by the heterogeneity of the basalt reflectors (Ziolkowski *et al.* 2003). A solution to providing improved images beneath basalt flows is therefore to generate, retain and enhance as much low-frequency energy as possible.

Various methods of generating low-frequency signal have been proposed and employed in the last 10 years. While accepting that carefully parameterized acquisition can be used to provide a greater richness in low-frequency signal, Gallagher & Dromgoole (2007) and Hardwick *et al.* (2010) conclude the sub-basalt image is primarily dependent on the careful retention and enhancement of low-frequency signal at all stages of the processing.

In 2010 TGS began a program to reprocess 70 000 km of its 2D seismic database across the NW European Atlantic Margin, including 20 000 km from the West of Shetland region. These long-offset data were acquired with conventional acquisition parameters (typical gun volume *c.* 0.075 m^3, or 4600 cubic inches; source depth *c.* 7 m; cable depth *c.* 9 m). The solution to providing successful sub-basalt imaging for these data is down to careful signal processing. The key signal processing techniques are described in the following sections.

Spectral processing: low-frequency boost

After conversion to zero-phase at the beginning of data processing, the recorded source wavelet is manipulated in order to enhance the signal at the low-frequency end of the amplitude spectrum. The low-frequency components of the wavelet are edited and shaped to generate a target wavelet and appropriate zero-phase matching operators, one operator for each vintage of seismic acquisition. An example input wavelet, low-frequency boosting operator and output wavelet is shown in Figure 1. The amplitude spectrum of the operator (brown line) indicates a maximum boost of *c.* 12 dB. This maximum boost is centred in the 3–7 Hz frequency band where signal levels drop off rapidly in the input spectrum (green line), thus providing the greatest uplift where it is most required. The operator also provides a smooth increase in the 7 Hz to peak frequency range to approximately simulate a deep towed source array. This apparent spectral shaping is in alignment with some key findings made in an evaluation on the spectral output of marine airgun arrays by Parkes & Hegna (2011).

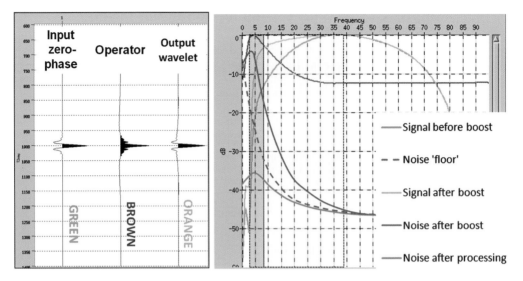

Fig. 1. Spectral analyses of a sample input zero-phase wavelet (green line), the low-frequency boosting operator (brown line) and the output wavelet (orange line). The red shaded region indicates the 3–7 Hz band where the operator provides maximum boost. The yellow shaded region indicates the 7 Hz to peak frequency band where the operator boost simulates a deeper towed source.

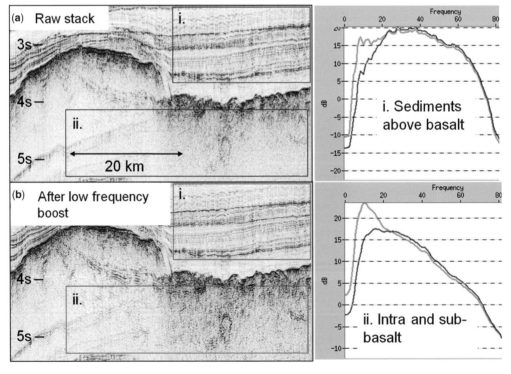

Fig. 2. Example stack section (**a**) before and (**b**) after application of the low-frequency boosting operator, with accompanying normalized spectral analyses (spectra calculated from data within the green and red boxes). Red spectra represent data before boost; green spectra after boost.

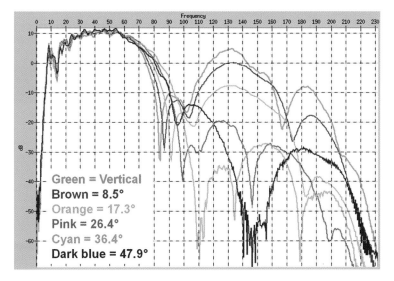

Fig. 3. Normalized spectral analyses of source wavelets derived from six separate common *p*-planes. Each common *p*-trace represents a unique emergence angle; the values are provided in the display.

They suggest there are inherent restrictions imposed on the low-frequency signal. Their assessment indicates an almost fixed decay in signal levels between *c.* 3 and 7 Hz irrespective of airgun array size, design and tow depth. Rather, modifications to source arrays will alter the low-frequency content from *c.* 7 Hz to 30–40 Hz.

A similar concept for low-frequency spectral manipulation was proposed by Masoomzadeh *et al.* (2006) using spectral whitening before the final

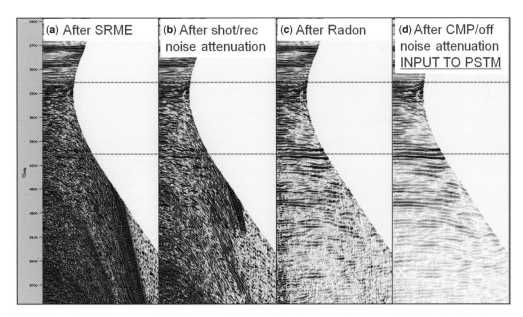

Fig. 4. Example NMO-corrected CMP after several key pre-migration processing stages: (**a**) after SRME multiple attenuation; (**b**) after noise attenuation in the shot and receiver domains; (**c**) after radon multiple attenuation; (**d**) final pre-processed gather – ready for pre-stack migration. Maximum offset = 10 km. Green dashed lines indicate top and base of basalt layer.

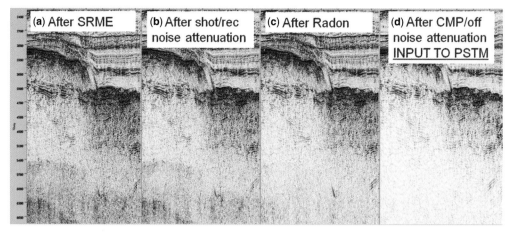

Fig. 5. Example stack image after several key pre-migration processing stages: (**a**) after SRME multiple attenuation; (**b**) after noise attenuation in the shot and receiver domains; (**c**) after radon multiple attenuation; (**d**) final pre-processed gather – ready for pre-stack migration.

stack. Moving to a pre-stack application is important, and the decision to apply the boosting operator at the beginning of the data processing sequence is considered key for the following two reasons. Firstly, as the boosting operator does not discriminate between signal and noise, the poor signal-to-noise ratio common at low frequencies is not improved after the simple process of applying the operator (see the red lines in Fig. 1). However, by applying the operator at the start of processing, the noise component assumes its true prominence relative to the flattened signal amplitude spectrum. This in turn enables the full suite of signal enhancing components in the processing sequence to

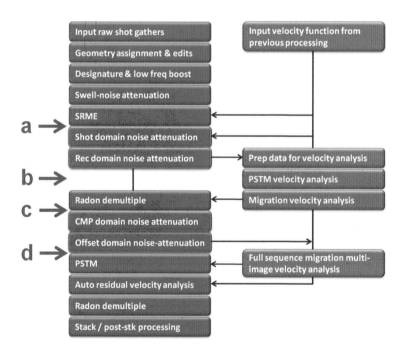

Fig. 6. Example processing flow used to reprocess 2D seismic data in the West of Shetland region. The flow chart indicates the position of the noise attenuation stages (blue boxes). The blue arrows indicate the pre-migration processing stages applied to the sample gather and stack section displayed in Figures 4 and 5, respectively.

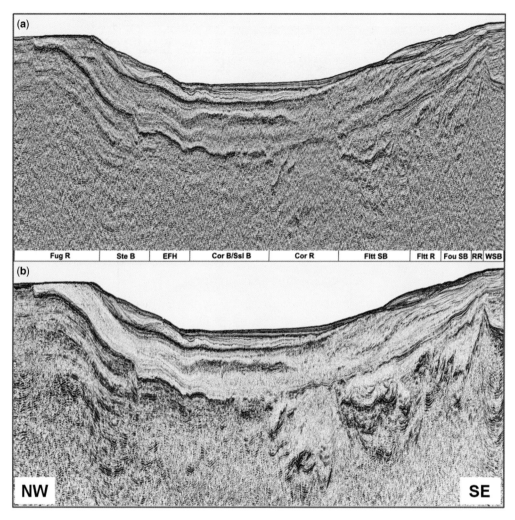

Fig. 7. Comparison of the final time-migrated section across the Faroe–Shetland Basin: (**a**) original and (**b**) reprocessed time image. Annotated above the reprocessed image are the major intra-basinal highs and basins: Fug R, Fugloy Ridge; Ste B, Steinvør Basin; EFH, East Faroe High; Cor B/Ssl B, Corona/Sissal Basin; Cor R, Corona Ridge; Fltt SB, Flett Sub-basin; Fltt R, Flett Ridge; Fou SB, Foula Sub-basin; RR, Rona Ridge; WSB, West Shetland Basin.

be tested for optimal application to the boosted low-frequency data. Secondly, seismic horizons related to the intra- and sub-basalt geology are more easily identified in low-frequency enhanced data displayed as stack images, gathers and in semblance plots. In consequence, more accurate sub-basalt velocity models can be produced throughout the processing sequence. Since many pre-migration demultiple and noise attenuation processes are guided by the primary velocity function, these algorithms can be applied to greater effect.

Figure 2 displays the results of applying the low-frequency operator to the 'raw' zero-phased data. The accompanying spectral analyses show the frequency content of tertiary sediments overlying the basalt are not compromised by this process. Furthermore, application of a single boosting operator does not affect the natural attenuation of higher frequencies through the basalt.

Some concerns have been raised regarding the stability of a single operator applied to pre-stack data. These concerns arise from the knowledge that the ghost function varies with angle of incidence. The question arises, therefore, as to how representative of the full dataset is the input wavelet, which only represents energy travelling at

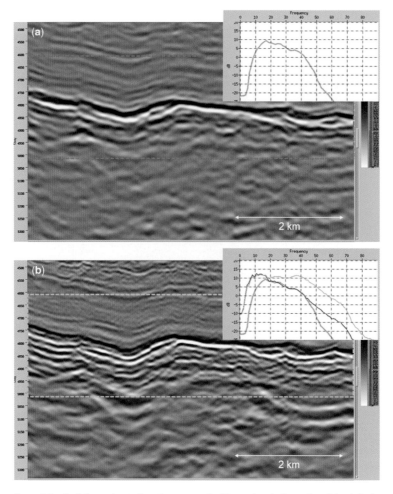

Fig. 8. Comparison of the final time-migrated section across the Flett volcanic formation: (**a**) original conventional processing and (**b**) reprocessed broadband time image. Green spectral plot is derived from the original image. Brown spectral plot after step 1: spectral processing. Orange spectral plot after step 2: inverse-Q filtering.

vertical to near-vertical incidence? As a test source wavelets were derived from the raw data transformed to plane-wave (p) domain, such that each wavelet represents a unique emergence angle. Figure 3 displays the frequency spectra of the six wavelets extracted across a range of common p-planes. As expected, the position of the first non-zero notch in the spectra moves to higher frequencies as the angle of propagation from the vertical increases. However, the frequency content below 40 Hz exhibited consistent character across all of the common p-values tested. This test suggests that a single operator convolved with the full dataset (which shapes the spectra below 40 Hz) remains a robust approach. Furthermore, it is equally valid as a method for both 2D and 3D seismic datasets.

Multi-domain noise attenuation

Several noise-attenuating processes were performed in all of the available 'time-offset' domains. Noise attenuation techniques were applied in four different domains, namely the shot, receiver, common mid-point (CMP) and common offset domains, to enhance low-frequency sub-basalt primary signal and minimize both coherent and incoherent noise. CMP gathers (Fig. 4) and stack images (Fig. 5) show the significant improvements made by the shot and receiver noise attenuation applied after surface-related multiple elimination (SRME), and the subsequent improvements made by the CMP and offset noise attenuation after Radon demultiple. Figure 6 provides a processing flow diagram for the reprocessing of 2D data from

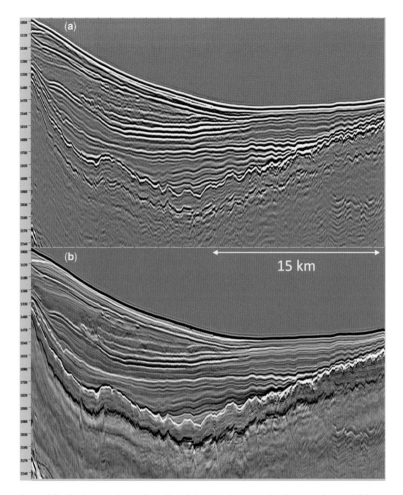

Fig. 9. Comparison of the final time-migrated section: (**a**) original conventional processing and (**b**) reprocessed broadband time image.

the West of Shetlands. Techniques employed in each of the four domains (shown as the blue boxes in Fig. 6) include: coherent noise attenuation using a time and space variant f–x apparent velocity dip filter; several iterations of an algorithm which decomposes data into frequency bands and identifies and attenuates anomalous amplitudes within those bands based on time-variant thresholds; and multiple passes of time- and space-variant f–x deconvolution, regularly operating only below the top or base basalt horizons.

Imaging results

An example reprocessed time image (Fig. 7b) transecting many of the major intra-basinal highs and sub-basins in the Faroe–Shetland Basin shows a dramatic improvement over the original version (Fig. 7a). The greatest improvements are demonstrated in the previously poorly imaged Mesozoic and Palaeozoic structure. This is provided without compromising the frequency content of the post-basalt Tertiary section.

Further work: pushing for broadband seismic from conventional streamer acquisition

The oil industry is increasingly recognizing the benefits of broadband data. Broadband seismic provides a greater richness of both robust low frequencies, ideal for sub-basalt imaging and inversion stability, and high frequencies, which aid temporal resolution. Suggested acquisition-based solutions include variable-depth streamer or slanting cable to tackle the receiver-side ghosting (Soubaras & Dowle 2010), and dual-sensor streamers combined with random-depth sources (Carlson et al. 2007; Tenghamn et al. 2007). Following the success of the methods described earlier in this paper the technique has been expanded upon to provide a processing-based broadband solution. This is highly cost effective for two reasons: it does not require any extra acquisition effort and it is applicable to the existing legacy data library acquired by conventional flat cables.

Delivering increased bandwidth (i.e. broadband) data for exploration and production (E&P) purposes has numerous advantages. However, two major factors serve to significantly reduce the useful bandwidth of the source wavelet. The first is the interference between the source pulse and reflections from the water surface ('ghosts') and the second is the fact that the Earth acts as a filter which preferentially attenuates high frequencies in the source wavefield as it travels through the Earth. In order to obtain truly 'broadband' seismic images, both factors need to be addressed. These have both been dealt with by substituting the low-frequency boosting described earlier for the following two steps.

Step 1: Broadband spectral processing

To suppress the effects of the source and receiver ghosts, the low-frequency boosting described earlier is replaced with a multi-dimensional deconvolution applied in the plane-wave (p) domain. In this domain each trace represents a common emergence angle, making it the preferred domain for de-ghosting. Using a stochastic search for the best set of parameters, a semi-deterministic stage of de-ghosting operations is applied. This is complemented by a statistical stage including a carefully designed deconvolution operation, averaging over a large number of common-slowness traces in order to address the remaining spectral defects including residual ghosts, side lobes and bubble effect. This full de-ghosting regime applied to 2D legacy data works towards replicating the benefits of broadband acquisition and processing.

Step 2: Effectively solve for the Earth's attenuation

The result of the Earth's attenuation is a preferential loss of higher frequencies during wave propagation, which induces a 'tilt' in the spectral content of deeper horizons.

This attenuation is commonly expressed as effective Q, which is a combination of the intrinsic attenuation, describing energy loss due to the propagation media, and apparent attenuation describing attenuation due to scattering, transmission, mode conversion, etc. An important feature of broadening the spectrum in step 1 is that more accurate measurements of the effective Q can be calculated from the data itself. The attenuation caused by the Earth's filter can therefore be determined and solved for on the pre-stack data. This is beneficial for deeper targets as the apparent tilt in the spectrum is minimized.

Any process that seeks to broaden the spectrum must take care to enhance the signal and not the noise. In order to maximize the whitening of the useful signal, the two steps are complemented by the multi-domain noise attenuation processes described earlier.

In the context of sub-basalt imaging, the benefit of this broadband processing technique is to provide substantial enhancement of the low-frequency signal. A further benefit is provided for imaging of intra-basalt prospects where the improved low and high frequencies combine to yield superior temporal resolution. Figure 8 provides a comparison of a portion of 2D seismic with and without our broadband processing applied. Clear benefits of higher resolution within the intrusive volcanics can be seen along with a richness of lower-frequency energy below this sequence. Figure 9 provides a comparison of a different portion of the same 2D seismic line, demonstrating the benefits of broadband seismic data in the shallower Eocene sediments.

Conclusions

Significant improvements in imaging intra- and sub-basalt geology through the reprocessing of seismic data covering the West of Shetland region is demonstrated. The two signal processing approaches key to providing these improvements are the application of a single low-frequency boosting operator at the beginning of the processing sequence, and the application of several noise attenuating processes performed in the various 'time-offset' domains. Our new broadband processing technique provides a greater richness of both robust low frequencies, ideal for sub-basalt imaging, and high frequencies, which aid temporal resolution of intra-basalt prospects.

The authors would like to thank TGS for permission to show the data examples. Our thanks and appreciation go to colleagues involved in the review of this paper.

References

CARLSON, D., SÖLLNER, W., TABTI, H., BROX, E. & WIDMAIER, M. 2007. *Increased resolution of seismic data from a dual-sensor streamer cable.* 77th Annual

International Meeting, SEG, Expanded Abstracts, San Antonio, 994–998.

GALLAGHER, J. W. & DROMGOOLE, D. 2007. Exploring below the basalt, offshore Faroes: a case history of sub-basalt imaging. *Petroleum Geoscience*, **13**, 213–225.

HARDWICK, A., TRAVIS, T., STOKES, S. & HART, M. 2010. Lows and highs: using low frequencies and improved velocity tools to image complex ridges and basement highs in the Faroe-Shetland Basin. *First Break*, **28**, 61–67.

LUNDIN, E. R. & DORÉ, A. G. 2005. NE Atlantic break-up: a re-examination of the Iceland mantle plume model and the Atlantic – Arctic linkage. *In*: DORÉ, A. G. & VINING, B. A. (eds) *Petroleum Geology: North-West Europe and Global Perspectives – 6th Petroleum Geology Conference*. Geological Society, London, 739–754.

MASOOMZADEH, H., BARTON, P. & SINGH, S. 2006. *Preservation of low frequencies in wide-angle data processing for sub-basalt imaging.* 76th SEG Annual Meeting, Extended abstracts, New Orleans, **25**, 2832.

NAYLOR, P. H., BELL, B. R., JOLLEY, D. W., DURNALL, P. & FREDSTED, R. 1999. Palaeogene magmatism in the Faeroe-Shetland Basin: influences on uplift history and sedimentation. *In*: FLEET, A. J. & BOLDY, S. A. R. (eds) *5th Petroleum Geology of Northwest Europe Conference*. Geological Society, London, 545–558.

PARKES, G. & HEGNA, S. 2011. *The low frequency output of marine air-gun arrays.* 81st Annual International Meeting, SEG, Expanded Abstracts, San Antonio, 77–81.

SOUBARAS, R. & DOWLE, R. 2010. Variable-depth streamer – a broadband marine solution. *First Break*, **28**, 89–96.

TENGHAMN, R., VAAGE, S. & BORRESEN, C. 2007. *A dual-sensor, towed marine streamer; its viable implementation and initial results.* 77th Annual International Meeting, SEG, Expanded Abstracts, San Antonio, 989–993.

WHITE, R. S., SMALLWOOD, J. R., FLIEDNER, M. M., BOSLAUGH, B., MARESH, J. & FRUEHN, J. 2003. Imaging and regional regional distribution of basalt flows in the Faeroe-Shetland Basin. *Geophysical Prospecting*, **51**, 215–231.

ZIOLKOWSKI, A., HANSSEN, P. ET AL. 2003. Use of low frequencies for sub-basalt imaging. *Geophysical Prospecting*, **51**, 169–182.

Using formation micro-imaging, wireline logs and onshore analogues to distinguish volcanic lithofacies in boreholes: examples from Palaeogene successions in the Faroe–Shetland Basin, NE Atlantic

TIM J. WATTON[1,2]*, STEVE CANNON[3], RICHARD J. BROWN[1], DOUGAL A. JERRAM[4,5] & BRENO L. WAICHEL[6]

[1]*Volcanic Margins Research Consortium, Department of Earth Sciences, Science Labs, Durham University, Durham, DH1 3LE, UK*

[2]*Present address: Statoil UK, Ltd., One Kingdom Street, Paddington, Central London, W2 6BD, UK*

[3]*DONG UK E&P Ltd, London, UK*

[4]*DougalEARTH Ltd, Solihull, UK*

[5]*Centre for Earth Evolution and Dynamics (CEED), University of Oslo, Norway*

[6]*Universidade Federal de Santa Catarina, Florianopolois, Brazil*

*Corresponding author (e-mail: timwa@statoil.com)

Abstract: Formation micro-imaging (FMI) is a tool that produces micro-resistivity images of the sidewall of the well bore. FMI logging used in conjunction with conventional well logging techniques (e.g. GR, Gamma Ray/RES, Resistivity/NPHI, Neutron Porosity/SONIC, Velocity tools) allows detailed analysis of volcanic lithofacies variation and informs a robust interpretation of volcanic sequences. This methodology is of particular use where rock core data are limited or not present. Examples are presented from the Rosebank Field in the Faroe–Shetland Basin (West of Shetland, UK continental shelf) where the re-establishment of fluvial activity between phases of effusive volcanism resulted in a complex sequence of siliciclastic sedimentary rocks and basaltic lavas. We demonstrate how high-resolution FMI images through this sequence can differentiate internal basalt lava flow features, such as vesicular zones, brecciated intervals, sediment–lava contact relationships and joint/fracture networks. If FMI data exist through volcanic packages and if assessed and calibrated properly via core, sidewall core and field analogue comparisons, it can provide additional constraints on the interpretation and classification of reservoir (siliciclastic) and non-reservoir (volcanic) rocks.

Volcanic and igneous rocks occur in many hydrocarbon-bearing sedimentary basins worldwide and pose considerable challenges in the identification and assessment of potential hydrocarbon reserves (Chen *et al.* 1999; Helland-Hansen 2009). They can reach kilometres in thickness and can overlie and mask potential reservoirs (the 'sub-basalt' problem, e.g. White *et al.* 2003; Changzhi *et al.* 2006). Volcanic rocks can be intimately interlayered with siliciclastic reservoir rocks (e.g. the 'Rosebank discovery'; Helland-Hansen 2009), which complicates the construction of accurate reservoir models. Successful petroleum exploration in such settings requires a comprehensive understanding of the emplacement mechanisms, geometries and geophysical and petrophysical properties of the volcanic and intrusive components.

The petrophysical responses of volcanic rocks encountered in boreholes have received little attention in the past; most research has been undertaken by the International Ocean Drilling Project (IODP, Planke 1994; Delius *et al.* 1995; Bücker *et al.* 1998; Delius *et al.* 1998; Planke & Cambray 1998; Brewer *et al.* 1999; Delius *et al.* 2003; Bartetzko *et al.* 2005). In the Faroe–Shetland Basin (FSB), well log identification of volcanic deposits has evolved significantly since the completion of the LOPRA-1/1A well at the Faroes Islands in 1996. LOPRA-1/1A penetrated c. 3.5 km of Palaeogene volcanic rock that also outcrops on the Faroe Islands (Heinesen *et al.* 2006). The lower 1.1 km of the borehole comprises the Lopra Formation, which consists of hyaloclastite, intrusive basaltic sills and volcaniclastic sandstones. The overlying 3.5-km-thick Beinisvørð Formation is composed of laterally extensive basalt sheet lobes (see Passey & Jolley 2009). Studies of the seismic attributes of the LOPRA-1/1A rocks have illustrated how

From: CANNON, S. J. C. & ELLIS, D. (eds) 2014. *Hydrocarbon Exploration to Exploitation West of Shetlands*. Geological Society, London, Special Publications, **397**, 173–192. First published online February 25, 2014, updated April 3, 2014, http://dx.doi.org/10.1144/SP397.7
© The Geological Society of London 2014. Publishing disclaimer: www.geolsoc.org.uk/pub_ethics

internal lithological variations within lava flows have varying petrophysical properties (Boldreel 2006; Christie et al. 2006; Andersen et al. 2009; Nelson et al. 2009). Studies elsewhere, for example in the Daqing oil field, China, have proved the effectiveness of using wireline and formation micro-image (FMI) data for interpreting felsic volcanic rocks in boreholes (Dezhi et al. 1991; Li et al. 2009; Wang et al. 2009). Other studies have shown how low-resolution data from formation micro-scanner (FMS) tools can be useful in interpreting mafic volcanic rocks (e.g. Delius et al. 1998; Brewer et al. 1999; Waagstein 2000).

Exploration in frontier volcanic margin settings is becoming more common, and some hydrocarbon fields in volcanic-affected basins are moving into development phases. There is therefore a pressing need for new, updated methodologies for interpreting subsurface data containing volcanic and igneous rocks. In this paper we examine the Palaeocene rocks of the Faroe–Shetland Basin, Faroe Islands and UK continental shelf (UKCS) that were laid down in a volcanic margin setting and are the subject of sustained petroleum exploration. They comprise a complex sequence of siliciclastic sedimentary rocks, volcaniclastic rocks (including hyaloclastite and pyroclastic rocks) and basaltic lava flows (e.g. Passey & Bell 2007; Helland-Hansen 2009; Jerram et al. 2009) associated with the opening of the Atlantic Ocean. We present new high-resolution FMI image data for volcanic rocks encountered in boreholes in the Rosebank field. We first calibrate the FMI datasets against a number of other datasets and then compare lithofacies recognized in the FMI dataset to surface outcrop examples in Iceland and Namibia. Our aim is to illustrate how volcanic lithofacies can be accurately recognized in FMI logs. We show how powerful FMI data can be in interpreting volcanic rocks when drill core is unavailable or when analyses of composite log data are ambiguous.

Uses of FMI in down-hole well analysis of volcanic rocks

The FMI is a qualitative high-resolution resistivity tool designed by Schlumberger, which produces comparative images of the sidewalls of a well bore (Ekstrom et al. 1987). Its precursor tool, the FMS, has been used extensively by IODP scientists over the last 20 years for interpreting volcanic stratigraphy. The FMS tool had only four detector pads and a lower spatial resolution than FMI. Brewer et al. (1999) showed how FMS image log evaluation could lead to the recognition of different volcanic rocks where core recovery was poor. FMS images were subjected to detailed core-log integration in order to justify their effectiveness as a formation analysis tool.

The FMI tool obtains resistivity measurements from eight sensor pads placed upon four orthogonally positioned calliper arms that run along the borehole walls (Siddiqui et al. 2004). The data coverage is dependent on the pad spacing; typically 80–100% of the hole is covered at a resolution of 5 mm or less. Processing of the FMI image yields a high-resolution vertical image of the borehole that is colour-coded for resistivity values. Static and dynamic images can be produced depending on methods used to average the down-hole resistivity values; static images are produced for the entire well, whereas dynamic images are averaged every 2 m. FMI is commonly applied to the mapping of sedimentary lithofacies in boreholes (Ekstrom et al. 1987); however, it has had only limited application to volcanic sequences (Li et al. 2009; Wang et al. 2009). The high-resolution images obtained by FMI reveal previously unrecognized diagnostic volcanic features, such as clast sizes and shapes, contacts, fractures and vesicles, which can greatly improve the accuracy of volcanic facies interpretation. We describe these features in the section 'Volcanic lithofacies identification using FMI'.

Case study area and stratigraphy

The Rosebank Field in the FSB is situated between the Faroe Islands and the Shetland Islands in the NE Atlantic Ocean (Fig. 1) in licence blocks 205/1, 213/26 and 213/27. It lies above the Corona ridge, a broad, basement-controlled anticline. The Palaeogene volcanic rocks in this study were erupted in response to the break-up of the North Atlantic and the development of the North Atlantic Igneous Province. The stratigraphy of the Rosebank Field can be simplified into reservoir units (Colsay units: siliciclastic sandstones and mudstones) and non-reservoir units (lavas, volcaniclastic siltstone, breccias and sandstones; Fig. 2). In this paper we focus on the volcanic non-reservoir units of the Rosebank Field with specific reference to well 213/27-2 which drilled through a significant thickness of volcaniclastic rocks that accumulated as an age equivalent to the uppermost Colsay unit.

Methodology

Data have been acquired from five wells and one side track (Fig. 2; Table 1). Petrophysical identification of individual volcanic lithofacies was initially determined using composite well logs (gamma ray or GR (API); p-wave velocity or V_p (km/s); shear wave velocity or V_s (km/s); density ρ (gm/cm^3); resistivity (Ohms); photoelectric

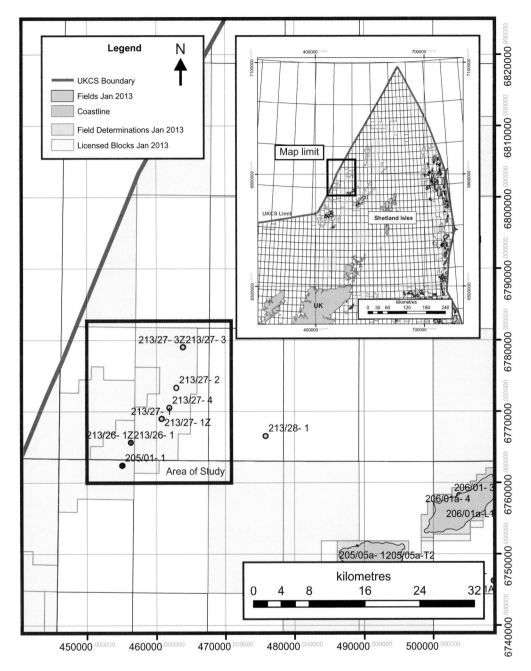

Fig. 1. Location of study areas and wells within the Faroe–Shetland Basin, UCKS. All data are in the public domain and were acquired from the UK Department of Energy and Climate Change (DECC) website. The Rosebank prospect was still not considered a field as of January 2013 by the UK government, even although it is entering the development phase.

effect or PEF (barns)) based upon methodologies outlined by Planke (1994), Brewer et al. (1999), Delius et al. (2003), Boldreel (2006), Andersen et al. (2009) and Nelson et al. (2009). In order to validate the lithofacies interpretations, FMI log data were calibrated using proven techniques for

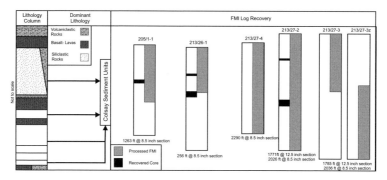

Fig. 2. A generalized stratigraphy for the Rosebank Field comprising four siliciclastic packages and five volcanic packages. Processed FMI and core recovery for each well included in the study is included to the right of the figure and summarized in Table 1.

the remote identification of volcanic rocks from rock core, sidewall core, V_p analysis and analogue examples. Table 2 outlines the perceived usefulness and accuracy of these techniques. All techniques are best used in conjunction with interpreted wireline log data to give the most accurate interpretation of available FMI data. Calibrated FMI logs were loaded into Petrel™ and tied to the other well logs at known recovery depths. This allows a side-by-side comparison of well logs and FMI logs and the production of a revised stratigraphy (methodology is outlined in Fig. 3). The FMI tool was run in 8.5- and 12.5-inch borehole diameters. All well log and FMI data were sourced from the Rosebank discovery and were kindly supplied by the Rosebank partnership (Chevron North Sea Limited UK Ltd., Statoil UK Ltd., OMV UK Ltd. and DONG E&P UK Ltd.).

Methods of FMI calibration

Recognition of individual lava flows using p-wave velocity comparisons

The first stage in interpreting volcanic lithofacies in well logs is to distinguish lithological contacts and recognize individual lava flows. This can help determine if the lavas were emplaced in a subaerial or subaqueous environment and determine the morphology of the lava flows, for example tabular or compound/braided pāhoehoe lava flows (e.g. Jerram 2002; Nelson et al. 2009). Morphologically, tabular pāhoehoe lava flows consist of individual sheet lobes that are typically metres to several tens of meteres thick (but that may reach 100 m thick) and tens–hundreds of metres (>1 km in some cases) in lateral extent (Walker 1971). Compound braided pāhoehoe lava flows are typically much thinner (typically a few metres, however overall packages may be thicker) and are composed of a series of thin stacked lobes, each up to several metres thick. Compound pāhoehoe lava flows are thought to represent small breakouts of lava either at the margin of a flow field or as volcanic activity wanes (Jerram 2002). Both tabular and compound pāhoehoe lava flows are common in the FSB (Passey & Bell 2007; Jerram et al. 2009).

Several studies have shown that V_p velocity profiles can be used to distinguish between different types of basaltic lava flow packages (Planke 1994; Boldreel 2006; Nelson et al. 2009). FMI data can also pick up important characteristics that are indicative of the different lava flow types. Figure 4 shows two 20 m intervals that display characteristic V_p response profiles for tabular lavas and for compound braided lavas. Each histogram profile matches the V_p distribution models proposed by Nelson et al. (2009) and the FMI log interpretation matches the predicted V_p distributions. In FMI logs, tabular pāhoehoe lava flows are characterized by a high conductivity and a clear differentiation between lava core, crust and base. In comparison, compound pāhoehoe lava flows exhibit a distinct bimodal resistivity response reflecting alternating core–crust signals over much

Table 1. *Well header and recovery depths. Recovered FMI log lengths and core for each well used in the study*

Well header	Core recovery (intervals)	FMI total recovery (ft)
213/27-3 and 213/27-3z	0	3821
213/27-2	2	3797
213/27-4	0	2290
213/26-1	3	256
205/1-1	1	1263

Table 2. *Summary of FMI calibration techniques used to validate volcanic lithofacies identification*

Technique	Usefulness	Accuracy
V_p comparison	The velocity histogram technique from Nelson *et al.* (2009) allows the calibration of the internal structure of lava flows to FMI images where core is not present. Can help aid the identification of lava breccia deposits in FMI logs due to their tightly distributed V_p values.	When FMI logs are compared to histogram distributions there is good correlation between observed V_p response and tabular v. compound lava flows.
Core analysis	The most direct and accurate comparison that allows direct lithological observations to be linked to FMI logs.	Usefulness is limited where core recovery is low or core is prohibitively expensive to acquire. Comparison of FMI log images to FMS logs from older studies (with extensive core recovery) yield more consistent results (Brewer *et al.* 1999).
Sidewall core/cuttings	Sidewall core can aid identification of lithological changes observed in FMI logs. Cuttings aid the separation of siliclastic and volcaniclastic rocks which can have similar visual characteristics on FMI logs.	Sidewall cores provide a more cost-effective solution than using rock core data to aid FMI calibration; needs to be closely spaced to capture changes in lithology. Cuttings greatly aid the interpretation of FMI log responses when accurate depths of recovery are recorded.
Field-based acquisition	Field-based acquisition of well log responses such as spectral gamma-ray can provide accurate interpretation of traditional composite log data. FMI images can be more accurately compared to these data.	Field-based acquisition is especially useful in aiding the interpretation of mixed siliclastic and volcaniclastic sequences where the understanding of composite log changes is limited.

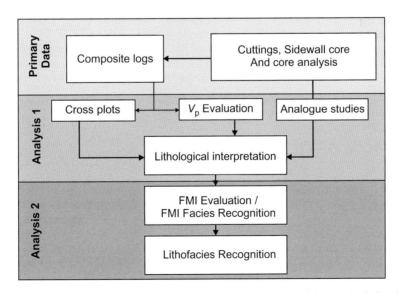

Fig. 3. A flow chart methodology of FMI-based log evaluation to determine volcanic lithofacies in the Rosebank Field.

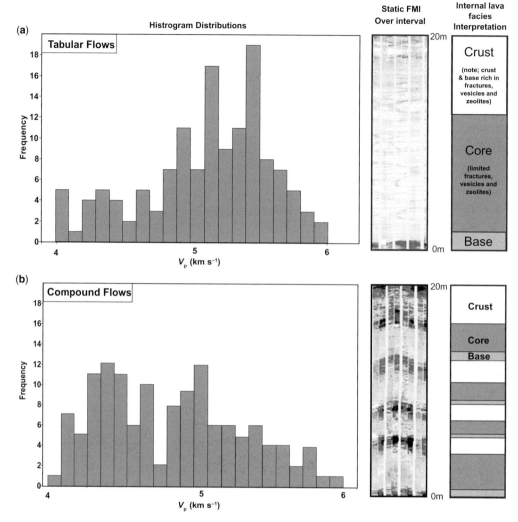

Fig. 4. An example of FMI logs from 213/27-4: (**a**) tabular flows 2890–2910 m and (**b**) compound flows 2360–2380 m. Histogram distributions (left) have been calculated for each of the packages from V_p response using the methodology of Nelson *et al.* (2009). FMI log examples (right) show bi-modal responses between resistive (dark) and conductive (light) responses. Interpretation represents core and crust relationships in a lava flow (see section 'Recognition of individual lava flows using p-wave velocity comparisons').

shorter intervals. This is a very useful technique for calibrating FMI images to lava flows where no core has been recovered, and is a more accurate method for picking flow-on-flow contacts (as compared to using conventional wireline log data such as V_p or GR).

Core analysis and analogue study

The most powerful FMI calibration technique is the tying of logs to cored intervals. Figure 5 shows two volcanic cores linked to their FMI log responses (including dynamic FMI images with resistivity values averaged every 2 m, and static FMI logs with resistivity values averaged over the entire well). The lava in the upper core shows a high degree of fracturing and has a high vesicularity (vesicles are now zeolite-filled amygdales). The latter feature is picked up in the FMI image as a distinct mottled appearance. Unfortunately, core is missing at the contact with the planar laminated unit observed in the dynamic FMI log. From core datasets we recognize this interval as a vesiculated upper crust of a tabular pāhoehoe lava flow

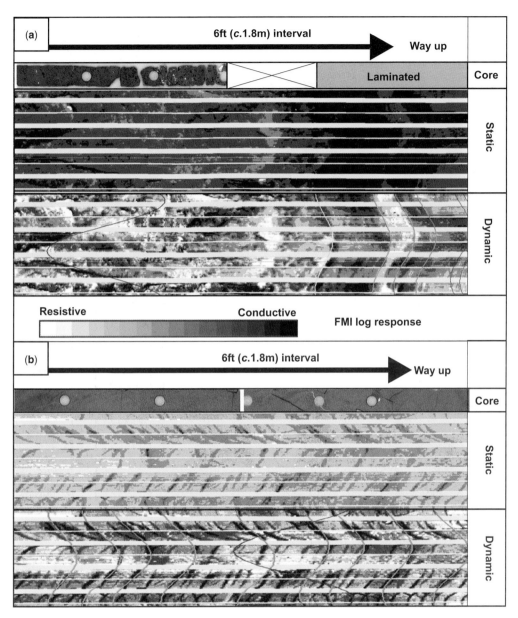

Fig. 5. (**a**) Core example from well 205/01-1 showing a highly vesicular basaltic flow top. In the FMI log we can see a distinct mottled appearance to the flow core reflecting the presence of vesicles and or amygdales. (**b**) Core shows a homogenous dense (non-vesicular) basalt with small fractures interpreted to be a flow core. FMI response is of uniform high conductivity. Horizontal fractures (green lines) are evenly spaced and probably relate to the drilling technique as they are not reflected as a volcanic feature in the recovered core. Laminations are marked with a yellow line; blue lines represent high-angle fractures.

(Fig. 5a). The second core image is less fractured, more homogenous and less vesicular (Fig. 5b). The distinct c. 45° fracture in the core has been highlighted in blue on the FMI image. Labelled in green are horizontal striations of higher resistivity that are evenly spaced and are not present in core, suggesting that they are drilling-induced features. Core observations suggest that this interval represents the dense core of a tabular basalt lava flow. Where core datasets are unavailable, for

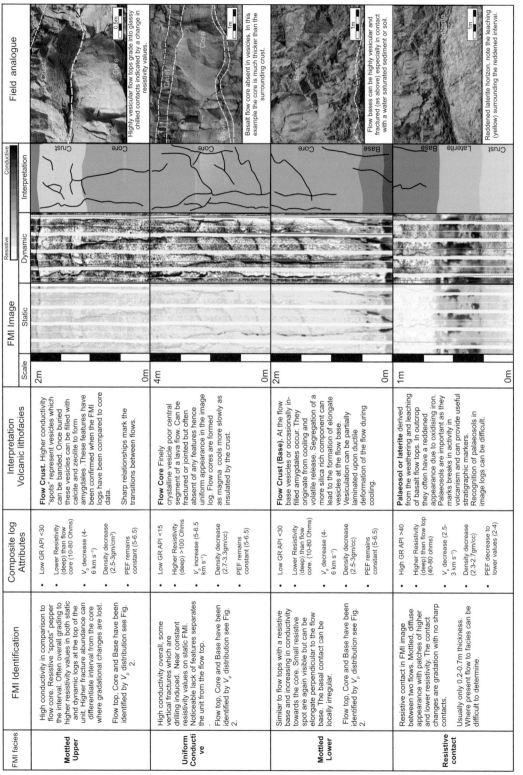

Fig. 6. FMI facies linked to volcanic lithofacies using all wells. Each example is compared to field outcrop exposures from South Iceland. Each facies is discussed in volcanic lithofacies identification using FMI

FMI facies	FMI Identification	Composite log Attributes	Interpretation Volcanic lithofacies	Scale	FMI Image (Static / Dynamic / Interpretation)	Field analogue
Conductive clast	Angular patches of conductive material situation within a resistive matrix. Contacts are often transitional from a flow core or top. Some fragments can be highly angular others more rounded. Often highly fractured by drilling.	• Low GR API <12 • Low Resistivity (deep) type I 20-40, type II 10-20 Ohms • V_p uniform (5.5-6 km s^{-1}) • Density (2.7-3.3 gm/cc) • PEF (2-5)	**Volcanic Breccia (hyaloclastite)** Difficult to distinguish from a primary volcanic breccia such as that at the base of an a'a flow. However potentially a hyaloclastite breccia due to jigsaw fit clasts and very high neutron porosity values which are indicative of water bound clays. High resistivity values between clasts are also consistent with extensive fracturing of hyaloclastite deposits.	2m – 0m	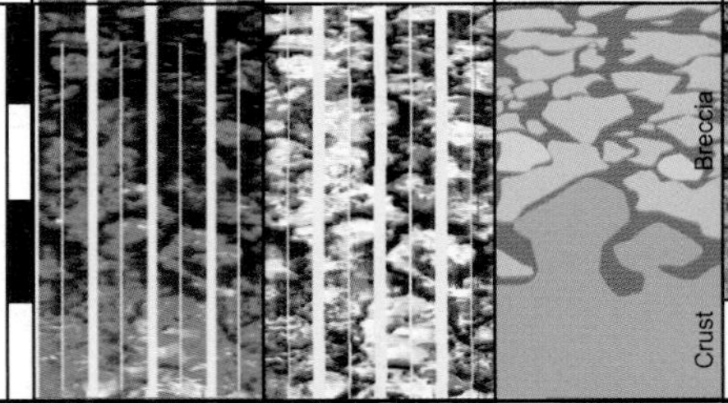	 An example of a hyaloclastite breccia. Note the highly angular clasts and brown altered matrix of palagonite clays and quenched glass.
Conductive Bulbous	Rounded to sub-rounded characteristic conductive bulb shapes occasionally ajoined but often split with a resistive layer. Potentially radial fracture in some examples are large cavities at the centre of bulbs although FMI resolution poor in this interval. Drilling through pillow lava secessions can create large cavities in the bore hole wall smearing the image log.	• Low GR API <20 • Lower Resistivity (deep) than hyaloclastite deposits (<10 ohms) • V_p (5.5-6 km s^{-1}) • Density (2.7-3.3 gm/cc) • PEF remains (2-6)	**Pillow lava** forms bulb like forms in cross section but are actually networks of small elongate lobes. Pillow lava usually have a quenched rim and radial joints developed from cooling contraction. Some pillow lava can have large cavities in the centre where volatiles have concentrated. The resistive matrix is likely to be spalled fragments of pillow lava and associated clays analogous to hyaloclastite breccias.	4m – 0m	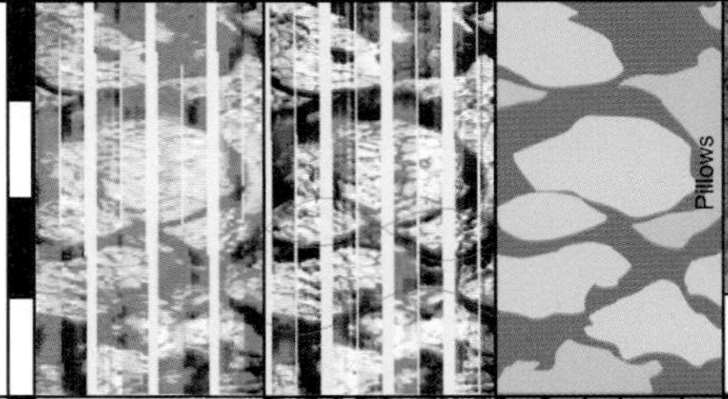	 Pillow lava example showing bulb like pillow structure in cross section. Small fragments spalled pillow lava rim make up the matrix.
Systematic Fracture	Regular spaced hexagonal or 8 sided fractures within the flow core. Can appear as 15m thick intervals. Fractures are highly resistive. Note the regularly spaced resistive striations in this example. Striations have been image analyzed and occur at even 20cm intervals. These features are caused by aggressive drill bits used to drill basal.	• Low GR API <12 • High Resistivity (deep) relative to random fracture (>90 Ohms) • V_p increase (>5-6 km s^{-1}) • Density increase (2.9-3.2 gm/cc) • PEF as flow	**Columnar jointing** forms upon water enhanced cooling of the basalt core. Joint propagate perpendicular to the cooling surface. Regularly spaced hexagonal joint patterns can develop as in this example. However as cooling is water enhanced columnar joints are often not vertical but curved (see field example) hence they cannot be used to determine dip angles of lava flows.	2m – 0m	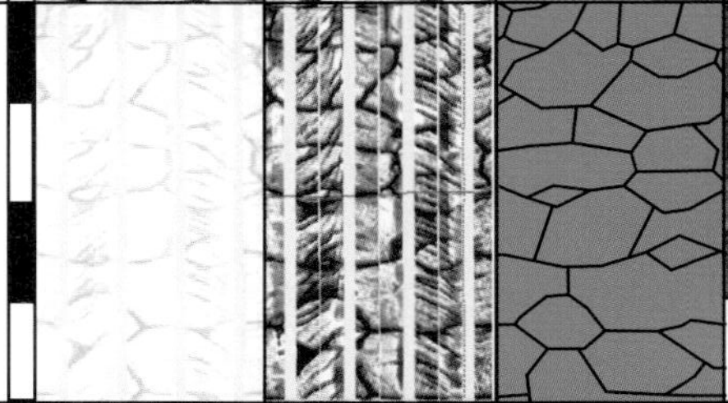	 Curvy-columnar joints. Note the regular space between each column but the highly variable direction of joint propagation.
Random fracture	Highly fractured restive features in random orientations. Easier to observe on the dynamic FMI log. Secondary sub-vertical fractures are probably drilling induced.	• Low GR API <16 • Lower Resistivity (deep) than systematic fracture (10-90 ohms) • V_p decrease (4.7-5.5 km s^{-1}) • Density decrease (2.7-3 gm/cc) • PEF as flow	**Fractured flow top (Hackley fracture)** develops due to rapid cooling which is usually water enhanced from the top of the flow. Systematic fracture is abandoned a dense fracture network is produced. Interestingly there is a significant variation in resistivity values from columnar jointed examples.	1m – 0m	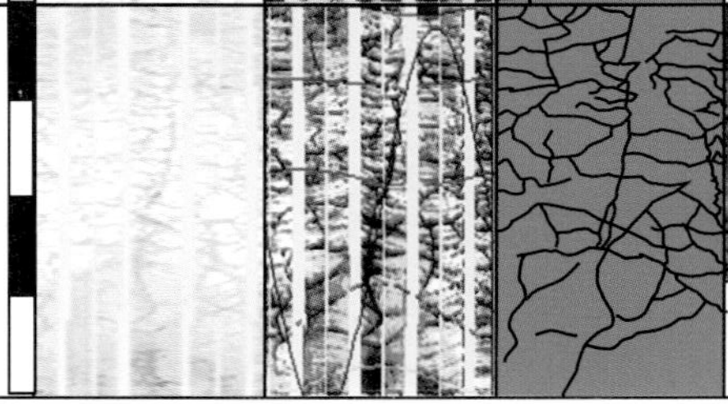	 Highly fractured flow core developed when fluvial activity reestablished itself over the top of a cooling basalt flow.

Fig. 6. *Continued.*

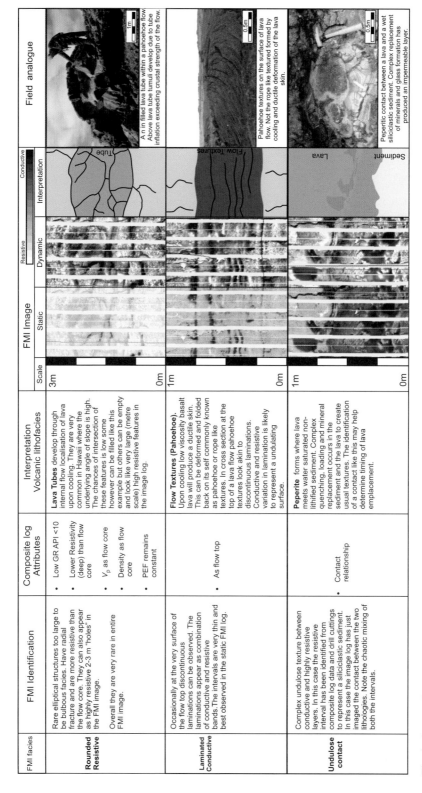

Fig. 6. Continued.

example through the conductive clast or conductive bulbous-type FMI facies (Fig. 6), then remote data can instead be used to aid interpretation and to calibrate the FMI log. In such cases we have used observations from Brewer et al. (1999) to positively identify and interpret such features in Rosebank FMI logs.

Sidewall core analysis

Sidewall cores provide another useful calibration dataset for FMI logs. For the purposes of this study three cores were selected from the uppermost Colsay unit of 213/27-2. Detailed analyses of the components of each core are shown in Figure 7. All of the cores show an increased abundance of basaltic lithic clasts relative to typical Colsay sandstone, which does not contain volcanic material. Graphs showing GR plotted against V_p for the selected interval reveal significant deviation from that of the average Colsay sandstone in 213/27-2 (Fig. 7a). Values that trend towards the average basalt are inferred to be enriched in volcaniclastic components. The FMI log reveals a conductive highly fractured zone; a coarse-grained clastic zone composed of conductive clasts and a zone of horizontal laminations (see Fig. 8). Each zone has been plotted as a point cloud in order to group FMI observations. The conductive fractured zone contains the greatest abundance of basalt lithic clasts. The basaltic component decreases as the siliciclastic component increases into the laminated zone volcaniclastic sandstone. Sidewall cores can therefore be used in conjunction with FMI logs to help identify zones that contain varying abundances of basaltic lithic fragments.

Field-based acquisition calibrating down borehole volcanic/sediment transitions

Outcrops of mixed volcaniclastic and siliciclastic sequences on the Earth's surface can provide additional constraints for FMI log calibration. Using data acquired from the Huab outliers, Namibia, we compare the spectral gamma-ray (SGR) signature of a mixed siliciclastic and volcanic succession.

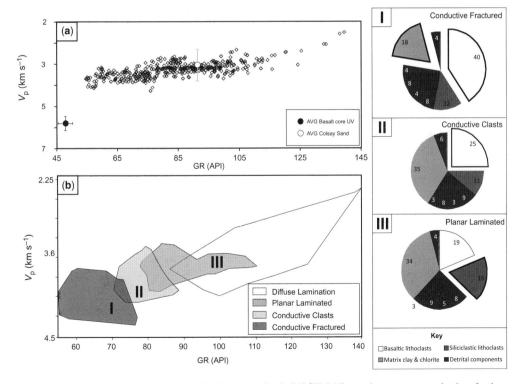

Fig. 7. V_p v. GR cross-plots of the uppermost Colsay succession in 213/27-2 (diamonds represent actual values for the sucession). (**a**) Average values for basalt core and Colsay sandstone are plotted with error bars showing well variation. (**b**) FMI facies as point clouds (see Fig. 8 for examples). I, II, III are pie charts of sidewall core components with a percentage provided in each sector.

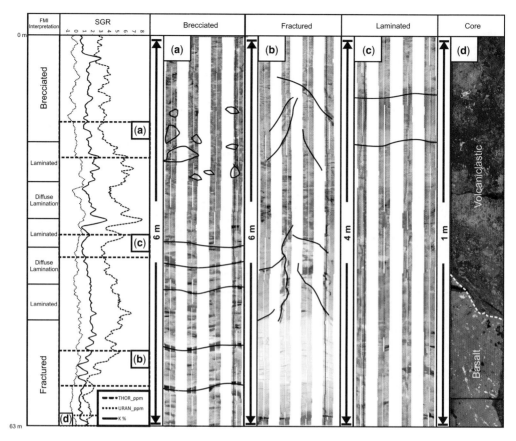

Fig. 8. Examples of FMI log responses from the uppermost Colsay interval 213/27-2. A SGR log has been provided for comparison with remotely recorded examples. (**a**) Conductive breccia interval above a laminated interval. (**b**) Fractured conductive log response. (**c**) Diffuse-laminated and laminated zones. Laminated zones are inferred to be bedding within sedimentary rocks. (**d**) Cored interval showing a mix of volcanic and basalt components. FMI log recovery has a limited pad number through this interval due to technical failure.

The Huab outliers are composed of aeolian sandstone dunes encased within basalt and basaltic andesite pāhoehoe lava flows that were emplaced during the formation of the 133–132 Ma Etendeka flood basalt province (Jerram et al. 1999, 2000). In the upper part of the Huab sequence thin sandstone intervals alternate with very fine-grained volcaniclastic rocks interbedded within lava flows with dense cores and brecciated or fractured tops. This provides an excellent opportunity to examine the transitions from lava to sediments and from clean sandstone to volcaniclastic sandstone. SGR data on the rocks were acquired using a Radiation Inc. RS-230 BGO Super-Spec handheld device at either a c. 30 cm spacing or at intervals determined by exposure. The device uses a Bismuth Germanate source to record SGR components (as U ppm, Th ppm and K%) with high precision. This data can be combined with a lithological log (Fig. 9). SGR values in the sandstone are high and decrease up into the volcaniclastic sandstone, which has similar SGR values to the basalt lava flows. The lava flow values were taken four times to validate this observation.

The Rosebank well logs show similar SGR profiles to the Namibian examples. SGR component values of the Rosebank well logs (the uppermost Colsay succession in well 213/27-2; Fig. 8) can be compared with the field logs (Fig. 9). The highest SGR values in well 213/27-2 occur in the laminated zone and are comparable to the clean sandstone in the Namibian examples. Highly fractured intervals (Fig. 8b) have values that are similar to the fractured basalt flow tops in Namibia. In well 213/27-2 the conductive clast zones show a slightly elevated K% (Fig. 8a). This

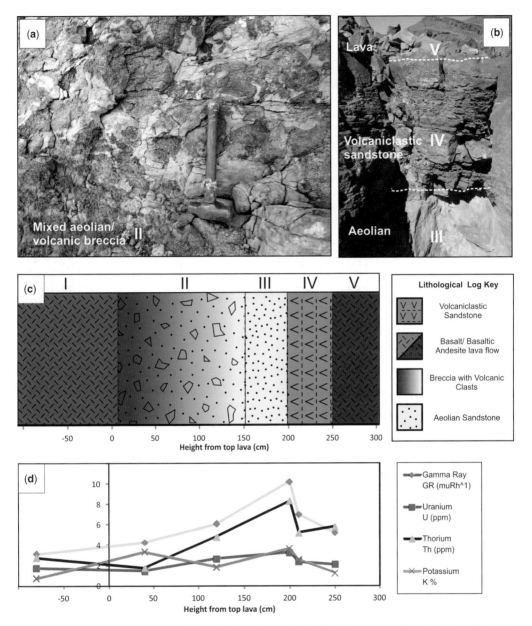

Fig. 9. A logged section with recorded SGR points along its profile. Section shows a contaminated sandstone/basalt/basaltic andesite contact from the Huab Outliers, Namibia. (**a**) Lava top breccias containing vesicular basalt clasts within an aeolian sandstone matrix, lithofacies II in schematic log C. (**b**) Photo highlighting the transition from clean aeolian sandstone to volcaniclastic sandstone and basaltic-andesite lava top, lithofacies III–V of the study section. (**c**) Schematic log highlighting the volcanic and sedimentary lithofacies distribution (labelled I–V) in the studied section, with measured thicknesses taken from the top of the lower lava. (**d**) The recorded SGR data across the study section (uranium, thorium and potassium plotted). See section 'Field-based acquisition'.

is similar to the trends seen in brecciated intervals in Namibia. The SGR trends in the Namibian and Rosebank rocks are similar due to similarities in bulk mineralogy rather than geochemical composition. A cored interval (Fig. 8d) helps constrain values of volcaniclastic material present here as a fracture in the top of a basalt flow crust.

Volcanic lithofacies identification using FMI

This study provides the opportunity to outline some of the main volcanic facies that can be identified using FMI combined with well log data and outcrop analogues, and provides a working template for future FMI interpretations of volcanic lithologies. Figure 6 outlines the volcanic lithofacies that have been identified according to their FMI data. Each of the FMI facies (interpreted from FMI log alone) include examples of both static and dynamic FMI log images. The FMI facies have been designed to be non-genetic. Diagnostic FMI characteristics and well log parameters are provided for each FMI facies. Field examples from Iceland of the various interpreted volcanic facies are linked with each FMI facies, providing a lithofacies interpretation and placing them in the context of their volcanic setting (for further information on the Icelandic examples, see Watton et al. 2013).

Mottled upper/lower to homogeneously highly conductive FMI facies

Observation. Zones exhibiting either a mottled appearance or an almost homogeneously high conductivity are the most common FMI facies in the Rosebank field (>60% of total facies volume in 205/1-1 and 213/27-4). These zones are generally 4–8 m thick but can reach thicknesses of 20 m. Mottled facies comprise c. 60% of this facies group. The mottling is defined by circular patches of c. 2–4 mm in diameter of either higher or lower conductivity than the background. Mottled and homogeneously conductive units form packages separated by contact facies. Upper and lower mottled units sandwich homogeneously conductive facies. Upper mottled units comprise up to 60% of each up-to-20-m-thick package, which is analogous to crusts in pāhoehoe lava flows. In some cases the upper part (c. 10%) can be discontinuously laminated (laminations are defined by layers of mottles). Homogeneous zones are areas of constant conductivity that are devoid of other features except for rare randomly orientated fractures or joints (see section 'Interpreting jointing and fracture systems in FMI'). The petrophysical responses of the mottled facies show significant variation in GR (15 API), V_p (1.5 km s^{-1}) and ρ (0.3 gm/cc) values in comparison with the homogenously conductive facies, although the photoelectric factor (PEF) in both remains the same.

Context. Rock core (Fig. 5) and petrophysical characteristics of the mottled and homogenously conductive units combined are consistent with the vertical variation within a single basalt lava flow (Planke 1994; Boldreel 2006; Nelson et al. 2009). The mottling is interpreted as vesicles or amygdales. In the FMI images, a single pāhoehoe lava flow is composed of a thin lower mottled unit (a poorly vesicular lower crust), a homogeneously conductive unit (a dense core) and a thick upper mottled unit (vesicular upper crust). This observation is consistent with the upper mottled facies (vesicular upper crust) comprising c. 60% of the total lava flow thickness (Nelson et al. 2009). Most basaltic lava flow fields in the FSB are pāhoehoe lavas (Passey & Jolley 2009). This observation is reinforced by the presence of an upper conductive laminated FMI facies, which shares similarities to pāhoehoe flow lobes in cross-section and consist of a series of mottle laminations.

The characteristic tri-partite internal structure of pāhoehoe lavas (lower crust, core, upper crust) results from endogenous growth during emplacement. They are emplaced slowly and advance as a series of thin, small breakout lobes. Solid crusts form quickly on these cooling lobes. The lava flow thickens through a process termed inflation whereby magma is transported to the advancing margin of the flow and cools and accretes to the base of the upper crust (e.g. Roland & Walker 1990; Hon et al. 1994). As the lava flows, gas bubbles rise upwards and the lava that accretes to the base of the crust is typically vesicular. The upper crust thickens at a faster rate than the lower crust and can grow to thicknesses of many metres (typically <60% of the flow thickness) above the flowing lava (Self et al. 1998). As the lava flow field matures, preferential pathways (lava tubes) form a distributary network channelling lava to the flow front (e.g. Keszthelyi 1995; Thordarson & Self 1998) (Figure 6, rounded resistive FMI facies). When the lava flow ceases, stagnant lava cools to form the lava core which is typically formed of degassed, dense lava. Each part (lower crust, core, upper crust) has distinctive petrophysical properties (cf. Nelson et al. 2009). The lower crust is typically thin (<10 cm) and may be poorly vesicular; the core is often non-vesicular and may exhibit widely spaced (50–100 cm) columnar joints. The upper crust is often characterized by diffuse layers of varying vesicularity. Large cavities may be present between accreted layers in the upper crust. The upper crust may additionally become fragmented into blocks and rubble during flow.

Pāhoehoe lava is distinct from other types of basaltic lava, such as a'ā lava, which characterized by basal and upper clinker breccias surrounding a dense lava core (e.g. MacDonald 1953). A'ā lavas develop due to changes in the groundmass crystal content and viscosity, induced by high volumetric flow rates and rapid convective cooling (Roland &

Walker 1990; Cashman *et al.* 1999; Kilburn 2004). In FMI logs a'ā lava flows may be recognized by two layers of breccia that sandwich a highly conductive dense lava core.

Random and systematic fractures

Observation. Both randomly oriented joints and systematically oriented joints are present in >50% of the mottled and homogenously conductive FMI facies (in 205/1-1 and 213/27-4). Random fractures are closely spaced and display high resistivity values. Generally, fracture density increases towards the top of the upper mottled facies. Dynamic FMI often needs to be used in collaboration with static FMI logs to determine fracture v. noise. Systematic fractures occur in the homogeneously conductive facies and form crude 5–8-sided polygons. The overall fracture spacing is regular (0.2–15 cm). There is a substantial difference in the petrophysical properties of the systematic joints and the random fractures; although GR values remain similar, resistivity values and V_p are >80 ohms and 1.3 km s^{-1} lower, respectively.

Context. We interpret the random fractures as hackly fractures in basalt upper crusts and the systematic fractures as columnar cooling joints in the basalt flow core. Columnar jointing varies with height in lava flows due to their layered nature and heterogeneous cooling times. Regularly spaced columnar joints develop in the stagnant flow cores. This interpretation is supported by FMI log observations; columnar joints are only present in homogeneously conductive units. Random fractures are differentiated from columnar joints on the basis of having no systematic pattern between resistive cuts in the FMI log. This interpretation is supported by the random fractures being confined to the upper mottled facies (the upper crust). Columnar joints in basalt lavas result from thermal contraction during cooling. Columnar joints propagate perpendicular to the cooling front and form approximately regularly spaced 5- and 6-sided columns (Hetényi *et al.* 2012). Joint growth is controlled by crack propagation creating plumose structures on column faces (Degraff & Aydin 1987). Water infiltration at the top of the flow can lead to radial joint sets as the isotherms in the flow core are deflected, leading to curved columnar joints as shown in Figure 6. The occurrence of radial joint patterns means that joint sets cannot be used as palaeo-dip indicators. Joint orientation is modified when water floods on top of a cooling lava flow (Saemundsson 1970; Long & Wood 1986). Hackly fracture (entablature) and Kubberberg (cube) jointing develop due to vastly increased water interaction at the top of the flow soon after emplacement (see Saemundsson 1970).

Bulbous conductive FMI facies

Observation. The bulbous conductive FMI facies occur in one well (213/27-4) as discrete packages within the lowermost volcanic unit, associated with the conductive breccia facies. It is characterized by highly conductive bulbous or oval features (>20 cm) within a higher-resistivity matrix. The conductive features display <2 mm thin fractures on their surfaces that are parallel to the outer margins. Cuspate clasts (up to 4 cm) occur at some places in the matrix in association with the bulbous shapes. The petrophysical response of this FMI facies is generally similar to that of the uniform conductive facies. The facies in question is recognized based on the FMI image rather than the composite log.

Context. This FMI facies is interpreted as pillow lavas (e.g. Brewer *et al.* 1999). Pillow lavas form when lava flows into water and reacts with it in a non-explosive manner. Pillow lavas advance in a manner similar to pāhoehoe lavas, with the lobe shape modified by the increased density of water. They typically have lower aspect ratios than subaerial pāhoehoe break-out lobes, and grow by endogenous growth of the crust that causes the stretching and fracturing of each lobe (Moore 1975). Pillows have bulbous forms in cross-section, but are tube-like in three dimensions (Walker 1992).

Conductive breccias (clast-supported, matrix-supported) FMI facies

Observation. Clast-supported conductive breccia intervals occur in a number of wells (205/1-1, 213/27-2 and 213/27-4) and cumulatively comprise *c.* 10% of the entire volcanic sequence. Clasts range in size from 2 to 15 cm in diameter. Two types of conductive breccia FMI facies are recognized. Type 1 breccias (Fig. 8) are associated with the mottled and homogenous conductivity facies, whereas type 2 breccias are associated with the conductive bulbous FMI facies (Fig. 6). They can be distinguished by variations in the matrix conductivity (high in type I and low in type II) and by their overall resistivity values (30–50 Ohms difference).

Context. Volcanic breccias can form in a number of ways, for example, from pyroclastic activity, during the emplacement of subaerial lavas (flow top breccias, a'ā lava flows), by quenching of lava upon water entry (hyaloclastite) or by sedimentary processes (e.g. debris flows). Discrimination between these different types is difficult because lateral facies variations are often required. In a 1D borehole FMI, the interpretation relies upon up- or down-section relationships. Type 1 conductive breccia facies occur adjacent to upper and lower mottled facies, suggesting they are associated with lava flows. We interpret these as the brecciated upper crusts of pāhoehoe lavas. The formation of cooling joints in crusts and fracturing associated with differential inflation of a lava lobe can result in rubbly flow tops. These can be of local extent (metres to tens-of-metres wide) or cover the entire flow field. They can be up to several metres thick and can have metre-scale relief.

Type 2 conductive breccias are associated with the conductive bulbous facies (pillow lavas). Breccias associated with pillow lavas are formed by the quenching and primary fragmentation of lava upon water contact and are termed hyaloclastite deposits (Carlisle 1963; Moore et al. 1973; Kokelaar 1986; Tribble 1991; Mattox & Mangan 1997; Umino et al. 2006). Hyaloclastite deposits can be modified by syn-eruptive resedimentation and reworking of hyaloclastite material (Watton et al. 2013). In this case, type 2 breccias are closely related to pillow lavas suggesting that they are hyaloclastite deposits derived from primary fragmentation processes. Whether this fragmentation was in a fluvial, marine or lacustrine environment is unclear.

Undulose and resistive contact FMI facies

Observation. Two types of contacts are observed in the FMI data: contacts between mottled facies units and sediments (undulose contacts) and contacts between two intervals of conductive mottled facies (resistive contacts). Undulose contacts are complex and have high-conductivity and high-resistivity components. The material in areas of high resistivity is brecciated or forms rounded blebs that typically range in size from 2 to 10 cm, but can reach 1 m in diameter. This material can be incorporated into the overlying homogeneously conductive facies. Undulose contact facies always occur at the base of lower mottled facies and Colsay units, whereas resistive contacts are associated with the upper mottled facies and occur always on the uppermost flow in the lava package in each volcanic sequence. Resistive contacts are mottled but the mottling is more diffuse, does not form coherent blobs, occurs in thinner zones (<1 m) and generally has higher resistivity values. The petrophysical characteristics of the resistive contact facies are significantly different (resistivity 40 Ohms and V_p 2 km s^{-1}).

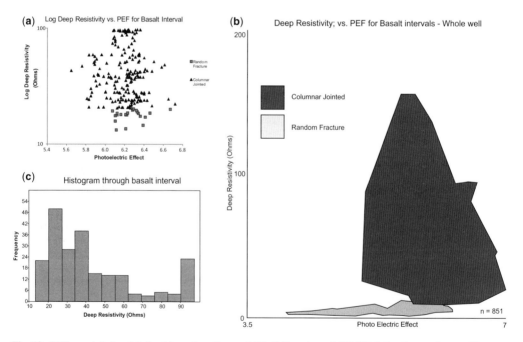

Fig. 10. PEF v. resistivity plots for (**a**) one lava flow and (**b**) all flows in well 213/27-4. Results are discussed in 'Interpreting jointing and fracture systems in FMI'. The spread and average resistivity values are higher in columnar (systematic) joint sets than in those that display random fracture. (**c**) A histogram of deep resistivity values within the basalt interval.

Context. The undulose contact facies marks the contacts between lava and sedimentary rocks of fluvial or marine origin. In FMI logs peperite intervals have been interpreted where rocks show mixed siliciclastic/volcanic log responses (higher GR and lower V_p responses), exhibit fluidal rounded textures (unlike type 1 or 2 breccias) and have clast sizes which are analogous to field observations. The distribution of conductive v. resistive material is more diffuse (unlike the sharp contacts seen in brecciated intervals). This is also consistent with vertical facies associations where proposed peperitic contacts always occur at the base of lava flow packets in contact with sedimentary deposits.

Peperites form when lava or intruding magma interacts with unconsolidated, usually saturated, sediments (e.g. Jerram & Stolhofen 2002; Skilling *et al.* 2002; Brown & Bell 2007; Waichel *et al.* 2007). They comprise an intimate mixture of igneous material and host sediment. Peperites can exhibit a range of textures dependent on host lithology, degree of lithification, lava flow viscosity, cooling rates, magma fragmentation, density contrasts and hydromagmatic activity (see McPhie *et al.* 1993). They can form quenched glassy layers which may be highly impermeable, thus limiting fluid transfer between the sediment and the lava.

We interpret the resistive contacts as palaeosols developed on the tops of lava flows. The development of both laterite (iron-rich) and bole (kaolinite-rich) soil intervals on basalt flows has been documented throughout Palaeogene volcanic successions in the UK and the Faroes (Hill *et al.* 2001; Passey & Bell 2007). Laterite, bole and thin inter-basalt sediment intervals indicate pauses in volcanic activity, and can be useful for the interpretation of palaeo-climates and palaeo-environments and the subdivision of thick sequences of stacked lavas (Jolley *et al.* 2005; Passey & Jolley 2009). The recognition of laterite or bole intervals again relies on the facies association with upper mottled facies (i.e. overlying a vesicular upper crust of a pāhoehoe lava flows). The occurrence of palaeosols can be indicative of a significant time gap between lava flow emplacement (up to 0.3 m Ma^{-1}; Pillans 1997).

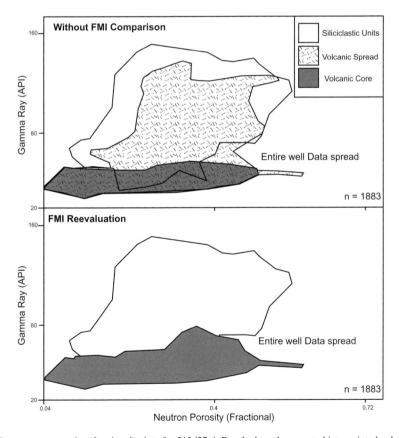

Fig. 11. GR v. neutron porosity (fractional) plots for 213/27-4. Results have been sorted into point clouds and are discussed in section 'Lithological separation using FMI data'.

Interpreting jointing and fracture systems in FMI

The FMI tool allows the recognition of lava intervals that are characterized by different joint types (regular or random) and therefore the investigation of the effects of jointing patterns on the petrophysical response. Using the FMI log it is possible to select flow lava intervals with either regular columnar jointing or random fractures (Fig. 10). If we plot the photoelectric factor values for these intervals (PEF) against deep resistivity, which does not change between lava core and crust, we can make the data spread a function of deep resistivity values and joint/fracture set characteristics. This indicates that the lava cores with columnar joints have higher resistivity values than basalt with high fracture densities within any one flow unit (Fig. 10a). This is supported by a bimodal spread in the histogram distribution (Fig. 10c). The same holds true for all basalt flows within an entire well (Fig. 10b). The regular-spaced columnar joint intervals have mean resistivity values that are almost an order of magnitude greater than those of the randomly fractured intervals. This indicates that there are significant differences in the fluid flow potential of the different joint systems, which may relate to how different orientations of joints close in response to confining pressure.

Lithological separation using FMI data

Detailed analysis of FMI images that have been calibrated to core and wireline logs can help accurately distinguish contact lithofacies (such as peperites) between siliciclastic and volcaniclastic rock types and give increased confidence to well log interpretations (Fig. 11). This cross-plot simply represents positively identified volcanic and siliciclastic rocks as well as volcanic spread of undetermined values before analysis of the FMI logs. Using FMI data in well 213/27-4 it is possible to constrain the volcanic spread (range of values for volcanic and volcaniclastic rocks) using Figure 6a–c and accurately separate volcanic from siliciclastic rocks, resulting in significantly less overlap in interpretation and thus a more accurate characterization of the reservoir intervals. Accurately classifying contacts and calibrating FMI logs using multiple techniques can also improve the geological understanding and interpretation of mixed volcanic and siliciclastic successions. Combining field-based data collection and interpretation with wireline logging and sidewall core recovery has led to an improved recognition of contaminated siliciclastic units (e.g. 213/27-2). These observations can be fed back into lithofacies interpretation, increasing confidence in the determination of reservoir and non-reservoir units. We recommend developing the use of FMI logs in combination with down-hole geochemical proxies (such as Electron Capture Spectroscopy and geochemical typing of lava flow packages) in order to provide additional constraints on the evolution and architecture of volcanic sequences.

Conclusions and recommendations

FMI log analysis of mixed volcanic and volcaniclastic rocks in the Rosebank field, FSB, has allowed the accurate characterization of distinct volcanic lithofacies. We have outlined four methods for calibrating FMI logs (p-wave velocity comparisons, wireline log collection of field-based analogues and wireline log v. sidewall core analysis), three of which can be conducted in the absence of rock core. Internal lava flow features, contact relationships and joint networks can be readily identified. Figure 6 is provided as a working template for the identification of key volcanic lithofacies using FMI, and can be extended as more FMI datasets that intersect volcanic rocks become available. This study has shown that such a tool is extremely useful in identifying potential migration pathways, seals and reservoirs in hydrocarbon-bearing basins containing mixed volcanic and siliciclastic rocks. Integration of FMI-derived lithological interpretations with seismic data should substantially increase the accuracy of interpretations of subsurface geology.

This research was funded by and partly completed while working for DONG E&P, UK Ltd. TJW's PhD was sponsored by DONG E&P as part of the Volcanic Margins Research Consortium. M. Smith (DONG Energy E&P) is thanked for discussion, help and encouragement. All well log and FMI data were sourced from the Rosebank discovery, kindly supplied by the Rosebank partnership (Chevron North Sea Limited UK, Statoil UK Limited, OMV UK Limited and DONG E&P UK Limited). S. Lumbard at OMV is also thanked for help with processing the FMI data. USCF, Brazil are thanked for the loan of the Radiation Inc. RS-230 BGO Super-Spec SGR hand-held device. K. Wright and C. Grove are thanked for discussion. N. Schofield and M. Hole are thanked for detailed comments on the manuscript. The reviews of B. Bell and S. Lumbard improved the manuscript.

References

ANDERSEN, M. S., BOLDREEL, L. O. THE SEIFABA GROUP 2009. Log responses in basalt successions in 8 wells from the Faroe-Shetland Channel – a classification scheme for interpretation of geophysical logs and case studies. In: VARMING, T., ZISKA, H. (eds) Faroe Islands Exploration Conference: Proceedings of the 2nd Conference. Annales Societatis Scientiarum Færoensis, Tórshavn, **50**, 364–391.

BARTETZKO, A., DELIUS, H. & PECHING, R. 2005. Effect of compositional and structural variations on log responses of igneous and metamorphic rocks; I, Mafic rocks. In: HARVEY, P. K., BREWER, T. S., PEZARD, P. A. & PETROV, V. A. (eds) *Petrophysical Properties of Crystalline Rocks*. Geological Society, London, Special Publications, **240**, 255–278.

BOLDREEL, L. O. 2006. Wire-line log-based stratigraphy of flood basalts from the Lopra-1/1A well, Faroe Islands. In: CHALMERS, J. A. & WAAGSTEIN, R. (eds) *Scientific Results from the Deepened Lopra-1 Borehole, Faroe Islands*. Geological Survey of Denmark and Greenland Bulletin, **9**, 7–22.

BREWER, T. S., HARVEY, P. K., HAGGAS, S., PEZARD, P. A. & GOLDBERG, D. 1999. The role of borehole images in constraining the structure of the ocean crust: case histories from the Ocean Drilling Program. In: LOVELL, M. A., WILLIAMSON, G. & HARVEY, P. K. (eds) *Borehole Imaging: Case Histories*. Geological Society, London, Special Publications, **67**, 283–294.

BROWN, D. J. & BELL, B. R. 2007. How do you grade peperites? *Journal of Volcanology and Geothermal Research*, **159**, 409–420.

BÜCKER, C. J., DELIUS, H., WOHLENBERG, J. LEG 163 SHIPBOARD SCIENTIFIC PARTY 1998. Physical signature of basaltic volcanics drilled on the northeast Atlantic volcanic rifted margin. In: LOVELL, M. A., WILLIAMSON, G., HARVEY, P. K. (eds) *Core Log Integration*. Geological Society, London, Special Publications, **136**, 363–374.

CARLISLE, D. 1963. Pillow breccias and their aquagene tuffs, Quadra Island, British Columbia. *Journal of Geology*, **71**, 48.

CASHMAN, K. V., THORNBER, C. & KAUAHIKAUA, J. P. 1999. Cooling and crystallization of lava in open channels, and the transition of Pāhoehoe Lava to'A'ā. *Bulletin of Volcanology*, **61**, 306–323.

CHANGZHI, W., LIANXING, G., ZUNZHONG, Z., ZUOWEI, R., ZHENGYAN CHEN, & WEIQIANG, L. 2006. Formation mechanisms of hydrocarbon reservoirs associated with volcanic and subvolcanic intrusive rocks: examples in Mesozoic–Cenozoic basins of eastern China. *AAPG Bulletin*, **90**, 137–147.

CHEN, Z. Y. H., YAN, J. S., LI, G., ZHANG, Z. W. & LIU, B. Z. 1999. Relationship between tertiary volcanic rocks and hydrocarbons in the Liaohe basin, People's Republic of China. *AAPG Bulletin*, **83**, 1004–1014.

CHRISTIE, P., GOLLIFER, I. & COWPER, D. 2006. Borehole seismic studies of the volcanic succession from the Lopra-1/1A borehole in the Faroe Islands, northern North Atlantic. In: CHALMERS, J. A. & WAAGSTEIN, R. (eds) *Scientific Results from the Deepened Lopra-1 Borehole, Faroe Islands*. Geological Survey of Denmark and Greenland Bulletin, **9**, 23–40.

DEGRAFF, J. M. & AYDIN, A. 1987. Surface morphology of columnar joints and its significance to mechanics and direction of joint growth. *Geological Society of America Bulletin*, **99**, 605–617.

DELIUS, H., BÜCKER, C. J. & WOHLENBERG, J. 1995. Significant log responses of basaltic lava flows and volcaniclastic sediments in ODP Hole 642E. *Scientific Drilling*, **5**, 217–226.

DELIUS, H., BÜCKER, C. & WOHLENBERG, J. 1998. Determination and characterization of volcaniclastic sediments by wireline logs: Sites 953, 955, and 956, Canary Islands. In: WEAVER, P. P. E., SCHMINCKE, H.-U., FIRTH, J. V. & DUFFIELD, W. (eds) *Proceedings of the Ocean Drilling, Scientific Results*. Ocean Drilling Program, College Station, Texas, **157**, 29–37.

DELIUS, H., BREWER, T. S. & HARVEY, P. K. 2003. Evidence for textural and alteration changes in basaltic lava flows using variations in rock magnetic properties (ODP Leg 183). *Tectonophysics*, **371**, 111–140.

DEZHI, B., ZIGANG, Q. & MINGGAO, L. 1991. *Method and Application Effects of the Identifying Igneous Lithology Based on Fuzzy Mathematics: Geological Application of the Well Logging Data*. Petroleum Industry Press, Beijing.

EKSTROM, M. P., DAHAN, C. A., CHEN, M. Y., LLOYD, P. M. & ROSSI, D. J. 1987. Formation imaging with Microelectrical Scanning Arrays. *The Log Analyst*, **May–June**, 294–306.

HEINESEN, M. V., ROSENKRANDS, L. & SØRENSEN, K. 2006. Introduction. In: CHALMERS, J. A. & WAAGSTEIN, R. (eds) *Scientific Results from the Deepened Lopra-1 Borehole, Faroe Islands*. Geological Survey of Denmark and Greenland Bulletin, **9**, 1–6.

HELLAND-HANSEN, D. 2009. Rosebank – challenges to development from a subsurface perspective. In: VARMING, T. & ZISKA, H. (eds) *Faroe Islands Exploration Conference: Proceedings of the 2nd Conference*. Annales Societatis Scientarum Faeroensis, Tórshavn, Supplementum **50**, 241–245.

HETÉNYI, G., TAISNE, B., GAREL, F., MÉDARD, É., BOSSHARD, S. & MATTSSON, H. 2012. Scales of columnar jointing in igneous rocks: field measurements and controlling factors. *Bulletin of Volcanology*, **74**, 457–482.

HILL, I. G., WORDEN, R. H. & MEIGHAN, I. G. 2001. Formation of interbasaltic laterite horizons in NE Ireland by early Tertiary weathering processes. *Proceedings of the Geologists' Association*, **112**, 339–348.

HON, K., KAUAHIKAUA, J. P., DENLINGER, R. & MACKAY, K. 1994. Emplacement and inflation of pahoehoe sheet flows: observations and measurements of active lava flows on Kilauea Volcano, Hawaii. *Geological Society of America Bulletin*, **106**, 351–370.

JERRAM, D. A. 2002. Volcanology and facies architecture of flood basalts. In: MENZIES, M. A., KLEMPERER, S. L., EBINGER, C. J. & BAKER, J. (eds) *Volcanic Rifted Margins*. Geological Society America, Boulder, CO, Special Paper, **362**, 119–132.

JERRAM, D. A. & STOLLHOFEN, H. 2002. Lava–sediment interaction in desert settings; are all peperite-like textures the result of magma–water interaction? *Journal of Volcanology and Geothermal Research*, **114**, 231–249.

JERRAM, D. A., MOUNTNEY, N., HOLZFÖRSTER, F. & STOLLHOFEN, H. 1999. Internal stratigraphic relationships in the Etendeka Group in the Huab Basin, NW Namibia: understanding the onset of flood volcanism. *Journal of Geodynamics*, **28**, 393–418.

JERRAM, D. A., MOUNTNEY, N., HOWELL, J. A., LONG, D. & STOLLHOFEN, H. 2000. Death of a sand sea: an active aeolian erg systematically buried by the Etendeka flood basalts of NW Namibia. *Journal of the Geological Society, London,* **157**, 513–516.

JERRAM, D. A., SINGLE, R. T., HOBBS, R. W. & NELSON, C. E. 2009. Understanding the offshore flood basalt sequence using onshore volcanic facies analogues: an example from the Faroe-Shetland basin. *Geological Magazine,* **146**, 353–367.

JOLLEY, D. W., MORTON, A. & PRINCE, I. 2005. Volcanogenic impact on phytogeography and sediment dispersal patterns in the NE Atlantic. *In:* JOLLEY, D. W., MORTON, A. & PRINCE, I. (eds) *Atlantic Margins: New Insights, Regional Synthesis and Large-Scale Tectonics.* Geological Society, London, Petroleum Geology Conference Series, **6**, 969–975.

KESZTHELYI, L. 1995. A preliminary thermal budget for lava tubes. *Journal of Geophysical Research,* **100**, 411–420.

KILBURN, C. 2004. Fracturing as a quantitative indicator of lava flow dynamics. *Journal of Volcanology and Geothermal Research,* **132**, 209–224.

KOKELAAR, P. 1986. Magma-water interactions in subaqueous and emergent basaltic Volcanism. *Bulletin of Volcanology,* **48**, 275–289.

LI, N., TAO, H. G. & LIU, C. P. 2009. Theory, method and application of acidic volcanics well log interpretation. *Petroleum Exploration and Development,* **36**, 683–692.

LONG, P. E. & WOOD, B. J. 1986. Structures, textures, and cooling histories of Columbia River Basalt flows. *Geological Society of America Bulletin,* **97**, 1144–1155.

MACDONALD, G. A. 1953. Pāhoehoe, a'ā and block lava. *Amercian Journal of Science,* **215**, 169–191.

MATTOX, T. N. & MANGAN, M. T. 1997. Littoral hydrovolcanic explosions: a case study of lava-seawater interaction at Kilauea Volcano. *Journal of Volcanology and Geothermal Research,* **75**, 1–17.

MCPHIE, J., DOYLE, M. & ALLEN, R. 1993. *Volcanic Textures: a guide to the interpretation of textures in volcanic rocks.* Centre for Ore Deposits and Exploration Studies, University of Tasmania Press.

MOORE, J. G. 1975. Mechanism of formation of pillow lava. *American Scientist,* **63**, 269–277.

MOORE, J. G., PHILLIPS, R. L., GRIGG, R. W., PETERSON, D. W. & SWANSON, D. A. 1973. Flow of lava into the sea, 1969–1971, Kilauea Volcano, Hawaii. *Geological Society of America Bulletin,* **84**, 537–546.

NELSON, C. E., JERRAM, D. A. & HOBBS, R. W. 2009. Flood basalt facies from borehole data: implications for prospectivity and volcanology in volcanic rifted margins. *Petroleum Geoscience,* **15**, 313–324.

PASSEY, S. R. & BELL, B. R. 2007. Morphologies and emplacement mechanisms of the lava flows of the Faroe Islands Basalt Group, Faroe Islands, NE Atlantic Ocean. *Bulletin of Volcanology,* **70**, 139–156.

PASSEY, S. R. & JOLLEY, D. W. 2009. A revised lithostratigraphic nomenclature for the Palaeogene Faroe Islands Basalt Group, NE Atlantic Ocean. *Earth and Environmental Science Transactions of the Royal Society of Edinburgh,* **99**, 127–158.

PILLANS, B. 1997. Soil development at snail's pace: evidence from a 6 Ma soil chronosequence on basalt in north Queensland, Australia. *Geoderma,* **80**, 117–128.

PLANKE, S. 1994. Geophysical response of flood basalts from analysis of wireline logs: ODP Site 642, Voring volcanic margin. *Journal of Geophysical Research,* **99**, 9279–9296.

PLANKE, S. & CAMBRAY, H. 1998. Seismic properties of flood basalts from Hole 917A downhole data, southeast Greenland volcanic margin. *In:* LARSEN, H. C., SAUNDERS, A. D. & CLIFT, P. D. (eds) *Proceedings of the Ocean Drilling Program, Scientific Results,* 152. Ocean Drilling Program, College Station, TX, 453–462.

ROLAND, S. K. & WALKER, G. P. L. 1990. Pahoehoe and aa in Hawaii: volumetric flow rate controls the lava structure. *Bulletin of Volcanology,* **52**, 615–628.

SAEMUNDSSON, K. 1970. Interglacial lava flows in the lowlands of southern Iceland and the problem of two-tiered columnar jointing. *Jökull,* **20**, 62–77.

SELF, S., KESZTHELYI, L. & THORDARSON, T. 1998. The importance of pahoehoe. *Annual Review of Earth and Planetary Sciences,* **26**, 81–110.

SIDDIQUI, S., HUGHES, G., SADLER, K. R. & KAMEES, A. 2004. Density-based imaging to supplement FMI images for sedimentary facies modelling. *Saudi Aramco Journal of Technology,* **Spring**, 60–68.

SKILLING, I. P., WHITE, J. D. L. & MCPHIE, J. 2002. Peperite: a review of magma–sediment mingling. *Journal of Volcanology and Geothermal Research,* **114**, 1–17.

THORDARSON, T. & SELF, S. 1998. The Roza Member, Columbia River Basalt Group: a gigantic pāhoehoe lava flow field formed by endogenous processes? *Journal of Geophysical Research,* **103**, 411–27, http://dx.doi.org/10.1029/98JB01355.

TRIBBLE, G. W. 1991. Underwater observations of active lava flows from Kilauea volcano, Hawaii, *Geology,* **19**, 633–636.

UMINO, S., NONAKA, M. & KAUAHIKAUA, J. 2006. Emplacement of subaerial pahoehoe lava sheet flows into water: 1990K(u)over-barupaianaha flow of Kilauea volcano at Kaim(u)over-bar Bay, Hawai'i. *Bulletin of Volcanology,* **69**, 125–139.

WAAGSTEIN, R. 2000. Formation micro-scanner image logging of a basalt lava – hyaloclastite sequence from the Lopra-1/1A well, Faroe Islands. *Danmarks og Grønlands Undersogelse Rapport,* 2000/7.

WAICHEL, B., DE LIMA, E. F., SOMMER, C. A. & LUBACHESKY, R. 2007. Peperite formed by lava flows over sediments: an example from the central Paraná Continental Flood Basalts, Brazil. *Journal of Volcanology and Geothermal Research,* **159**, 343–354.

WALKER, G. P. L. 1971. Compound and simple lava flows and flood basalts. *Bulletin Volcanology,* **35**, 579–590.

WALKER, G. P. L. 1992. Morphometric study of pillow size spectrum among pillow lavas. *Bulletin of Volcanology,* **54**, 459–474.

WANG, Z., JIN, L., GUAN, Q., YANG, L. & WU, J. 2009. Comprehensive evaluation of volcanic reservoir based upon FMI and ECS. *Journal of Southwest Petroleum University (Science & Technology Edition),* **5**, 5–10.

WATTON, T. J., JERRAM, D. A., THORDARSON, T. & DAVIES, R. J. 2013. Three-dimensional lithofacies variations in hyaloclastite deposits. *Journal of Volcanology and Geothermal Research,* **250**, 19–33.

WHITE, R. S., SMALLWOOD, J. R., FLIEDNER, M. M., BOSLAUGH, B., MARESH, J. & FRUEHN, J. 2003. Imaging and regional distribution of basalt flows in the Faroe-Shetland Basin. *Geophysical Prospecting,* **51**, 215–231.

A fresh approach to ditch cutting analysis as an aid to exploration in areas affected by large igneous province (LIP) volcanism

J. M. MILLETT*, M. J. HOLE & D. W. JOLLEY

Geology and Petroleum Geology, School of Geosciences, University of Aberdeen, Meston Building, Aberdeen, AB24 3UE, UK

*Corresponding author (e-mail: j.millett@abdn.ac.uk)

Abstract: Where hydrocarbon exploration targets occur within basins affected by large igneous province (LIP) sequences, an understanding of the volcanic stratigraphy is essential in compiling accurate basin models at all scales. Ditch cutting samples are one of the most commonly available sources of data yielding information in the context of LIP stratigraphy such as phenocryst load, degree of secondary precipitation and lava composition, largely unattainable by remote sensing. Where core data is limited or absent, cuttings provide the only means of accessing such data along with valuable inference of volcanic facies development and down-hole conditions. Interpretations based on cuttings data are widely used in industry and, as such, a repeatable and well-defined methodology for the analysis and designation of volcanic facies from cuttings is an important requirement for regional and individual play modelling. Such an approach has not been common practice to date. We propose a system of basic percentage-based cuttings analysis and ternary classification specifically tailored to LIP sequences, and argue for the benefits of a coherent and transparent basin-wide approach. The classification system is further developed into a log-style output for easy integration and comparison with other down-hole geophysical and biostratigraphic data.

Exploration in areas affected by large igneous provinces (LIPs) such as much of the offshore West of Shetland North Atlantic region has many associated difficulties. In the West of Shetland Basin (WSB) region these may include deep water, difficulties in seismic imaging, regular bad weather, difficult formation properties and a relative lack of regional down-well data with which to predict the extent and volume of volcanic rocks (Archer et al. 2005; Close et al. 2005; Gallagher & Dromgoole 2007). With additional difficulties come additional costs and, as such, it is a matter of no small significance to gain the absolute maximum in information from completed wells. In this contribution we specifically approach the analysis of ditch cuttings generated through LIP sequences in an attempt to promote a standardized analytical procedure. At present a transparent and repeatable system for analysis is not apparent and limits the usability and correlative potential of cuttings-based data on LIP stratigraphy to date. We propose a simple non-genetic percentage-based cutting classification system from which an assessment of the degree of error inherent in cuttings data can be made. The classification system also provides a platform from which subsequent genetic inference can be made where appropriate. Further, as ditch cuttings data are continuous in good well conditions, a subsequent stratigraphic output is proposed whereby easy comparison between other down-hole data sources and, importantly, other wells can be made. This development has been largely undertaken using data from, and in the context of, offshore exploration for hydrocarbons. The results and methodology are however deemed equally applicable to onshore drilling for irrigation or hydrothermal developments where volcanic stratigraphy is encountered and correlation is important.

Background

One of the major issues that have inhibited exploration in areas affected by LIP sequences is the difficulty in acquiring reliable seismic data from beneath basalt cover (Roberts et al. 2005; Gallagher & Dromgoole 2007). In recent years detailed investigations into the geophysical signatures of different flood basalt facies from down-well logging tools have been undertaken (Planke 1994; Bartetzko et al. 2001; Helm-Clark et al. 2004; Boldreel 2006; Nelson et al. 2009). These and other studies have significantly improved the potential for interpretation of LIP sequences from both well log and subsequently seismic data; however, there are still many aspects of LIP stratigraphy that cannot be resolved purely by geophysical means. These include inter-formational chemical variations (Kerr 1995; Larsen et al. 1999; Jolley et al. 2012), phenocryst populations (Hald & Waagstein 1984; Passey &

From: CANNON, S. J. C. & ELLIS, D. (eds) 2014. *Hydrocarbon Exploration to Exploitation West of Shetlands.*
Geological Society, London, Special Publications, **397**, 193–207.
First published online February 21, 2014, http://dx.doi.org/10.1144/SP397.2
© The Geological Society of London 2014. Publishing disclaimer: www.geolsoc.org.uk/pub_ethics

Jolley 2009), degree and type of secondary mineral precipitation (Walker 1970; Jørgensen 1984, 2006) and interbed ecosystem development (Jolley 1997). Not only do these factors have implications for building a robust stratigraphy that is integral to maximized correlation and basin modelling, but some may also have large implications for drilling and associated complications. These data may commonly be deduced from core and sidewall cores; however, coverage is commonly sparse in industry-driven wells due to the greater expense involved. Where core is unavailable, cuttings provide the only means of accessing such lithological-based data albeit with a much greater requirement for data screening and error quantification prior to interpretation (Georgi et al. 1993). Ditch cuttings therefore comprise one of the most valuable and easily available ground truth datasets available for nearly all wells completed in a given exploration region. Development in the analysis and interpretation of cuttings from wells penetrating LIPs does not appear to have undergone the same amount of systematic scientific attention as the geophysical responses. We therefore feel they require a transparent, repeatable and accurate analytical workflow with which to enhance their usability in regional modelling and formation evaluation. Ditch cutting data can be used to feed lithology data into geophysical well log interpretations and, by a combination of methods, can be used to calibrate information on mixing, lag time and cavings. This in turn can affect the accuracy of other cuttings-derived data such as biostratigraphic analysis and bulk geochemistry sampling.

Current procedures

A number of references to the analysis of volcanic-derived cuttings exist in the literature. Hald & Waagstein (1984) describe assigning flow boundaries in the Faroe Islands Lopra 1 onshore borehole by the recognition and occurrence of oxidized flow top cuttings at the surface. Brister & McIntosh (2004) describe sampling the central portions of large ignimbrites for geochronology from combined cutting petrography and log assessments. Many studies mention the use of cuttings through LIP deposits as part of an integrated approach to well analysis, but do not expand on the subject (Morton et al. 1988; Ellis et al. 2002; Jolley & Bell 2002; Passey 2004; Archer et al. 2005). In many cases this results from the availability of more accurately constrained data in the form of core or sidewall core coverage; the cuttings are therefore commonly only utilized as supplementary data. A large amount of industry-driven wells that encounter LIP deposits do not have expansive core coverage and may therefore have to rely largely on cuttings data for lithological inference. Volcanic LIP cuttings are also not commonly encountered by most industry mud loggers and, as such, a detailed approach to the sometimes subtle variations involved is not part of standard training. As these wells comprise the largest amount of current data and coverage in the WSB region, a robust system for cuttings analysis and assessment is an integral requirement for effective formation evaluation in the exploration and appraisal spheres.

From our experience of dealing with cuttings-based volcanic stratigraphy reports from offshore wells, there appears to be very little to no indication as to the process used for assessing the reported stratigraphic packages. We clarify at this point that we are discussing specific volcanic stratigraphy reports and not the results of standard contractor mud logging. If the process of defining a volcanic package from cuttings was very simple and without room for error, then it may be argued that such a justification is not required; however, this is commonly not the case and is certainly not the case all of the time. Where cuttings are collected from the drill site shale-shaker it is rare to have a single lithology comprising the full 100% of the sample; more often than not a degree of mixing between separate and sometimes genetically quite different lithologies is observed. To positively identify and define a unit of volcanic rock based on ditch cuttings without acknowledging the quantitative percentages from which the designation has been made leaves a large potential for error. Firstly error may exist in the designation itself but secondly, and more importantly, error may also then be propagated into the external assessment of that designation. Without defining the limits or basis of an interpretation, there is no way for an external reader or regional geologist to know what parameters separate analysts are using for designation. The net result is that when attempting to undertake regional correlations of volcanic stratigraphy, involving cuttings analysis from different companies using different analysts, there is as far as we can see no level playing field from which to verify a meaningful attempt. If recourse to the volcanic cuttings is necessary in certain and indeed many instances, their accurate designation and description in a coherent and an as far as is possible repeatable manner should be a bare minimum to aid interpretive modelling.

LIP facies from cuttings

Prior to discussing or proposing a methodology for cuttings analysis, it is first important to define the main objectives that its design should address.

What is the most accurate and informative way to define and classify an assemblage of cuttings collected over a known interval of a well drilled through a LIP deposit? The problem is three-fold. Firstly, it requires a method of systematically analysing the components of a cuttings assemblage. Secondly, it requires a non-genetic classification scheme within which to allocate the cutting type or types present which allows later inference to LIP stratigraphy. And lastly, it requires a means of assessing where possible the potential errors involved in bit to bag cuttings recovery and assemblage.

Ditch cutting generation

Any paper treating the analysis of ditch cuttings must acknowledge at least briefly some of the inherent problems attributable to cuttings and their assemblage which may obscure their accuracy in stratigraphy. Problems relating to cuttings data generally result from discrepancies in the required in-sequence transfer of drill cuttings through the medium of the drilling fluid flowing up the well annulus to the rig site (Brister & McIntosh 2004). Turbulence in drilling fluid during annulus transit can cause carried cuttings to mix with those from lower or higher in the annulus, causing cuttings from different stratigraphic horizons to emerge at the surface not representing their assigned depth lithology. The same turbulence can cause erosion of weakly consolidated units from anywhere within an un-cased interval causing contamination of the at-bit lithology, commonly termed cavings. The well inclination also has a large effect on cuttings transportation (Sifferman & Becker 1992). Also of consideration is the size and shape of cuttings produced by drilling; larger-sized or higher-density cuttings may undergo a degree of gravity settling depending on the drilling fluid properties (Hussaini & Azar 1983; Belavadi & Chukwu 1994; Al-Kayiem et al. 2010). The type of drill bit, rotation speed, penetration rate and formation physical properties all affect the size of coherent cuttings generated. The physical properties of the generated cuttings along with the length of annulus, drilling fluid circulation speed and turbulence all affect subsequent cutting degeneration *en route* to the sample bag. The last important concern when assessing cuttings is the minimum time it takes for the mud stream and first cuttings to reach the surface from the drill bit, usually termed the lag time. This calculation is dependent on a number of factors such as annulus volume, circulation flow rate and fluid losses. The calculation of lag can be fairly accurate, but the deeper a well penetrates the longer the travel time for the drilling fluid and cuttings and thus the greater potential for error. Most of these factors can be monitored by a combination of careful sample screening, by comparison of cutting first occurrences in relation to well-defined log signatures, by the use of tracers and by careful examination of the drilling and mud engineer reports. We propose that few of the problems just outlined should affect the description and reporting of cuttings assemblages; instead, they should be treated with the subsequent interpretation of the cuttings assemblages.

Sample preparation

In dealing with the analytical method of cuttings assemblage assessment, the basic sample preparation outlined in Table 1 is given as a guide where contractor dried cuttings are not used. It is deemed excessively time consuming and expensive to point count each individual cuttings assemblage, given the large number of samples per well penetration. Instead a visual assessment of each individual assemblage is deemed appropriate given constant reference to percentage charts such as in Terry & Chilingar (1955) or equivalent and access to a high-powered ring-light binocular microscope. Where cuttings are found which cannot be identified by the binocular microscope alone, we suggest the production of a polished slide suitable for both

Table 1. *Sample preparation workflow for cuttings received straight from the rig-site shale-shaker*

Stages	Notes
1. Sample washing	Sample taken from shale-shaker should be washed with tap water (or mild detergent if hydrocarbons are present) using a 0.178–0.5 mm mesh sieve depending on the drill bit used and cuttings average size and quality.
2. Sample drying	Sample should then be fully dried in an oven at <100 °C (usually overnight).
3. Sample selection	A 2.5 mL measuring spoon of representative sample omitting cuttings larger than around 2 mm as cavings should be selected (this size may vary depending on the modal cutting size of the well and the drill bit used).
4. Sample display	The sample should then be spread over a clear plastic tray with a percentage grid underlay for aided estimation ready for analysis.

petrographic and scanning electron microscope (SEM) backscatter analysis if required.

Facies and classification scheme

Before discussing how and what to record in LIP cuttings, it is first important to deal with the last two objectives mentioned above. These points, namely a cutting-based classification scheme for expected LIP deposits and a method of assessing errors, will be dealt with together.

The broad facies variations observed in LIPs can be defined as: (1) tabular-classic flow facies; (2) compound-braided flow facies; (3) hyaloclastite facies; (4) intrusive facies; and (5) inter-basaltic facies, after the scheme of Jerram (2002). In fine detail, large variations and complexities exist in LIP deposits (Single & Jerram 2004; Ross et al. 2005; Passey & Bell 2007; Bryan et al. 2010; Watton et al. 2013); however, as a general guideline most can be broadly attributed to one of these categories (Fig. 1), particularly in the North Atlantic Igneous Province (NAIP). We propose that of these five facies the physical attributes of ditch cuttings, produced by drill bit penetration, can be allocated between three broad end-members prior to further more-detailed designation if possible. We define the end-members of the classification scheme as: (1) volcanic glass; (2) crystalline/scoriaceous; and (3) epiclastic/bole (Fig. 2). With a distinct definition of all these end-member parameters (dealt with in the following sections in more detail) it is assessed that an individual cutting from the vast majority of major LIP deposits should fit into the ternary classification (Fig. 2). At this point all we are asserting is that the vast majority of LIP lithologies at a cuttings assemblage scale will fit into one of these basic non-genetic categories, and not that the category defines any facies association by itself.

The percentages that designate each classification field in Figure 2 have been arbitrarily assigned in order to help classify the cuttings analyses in light

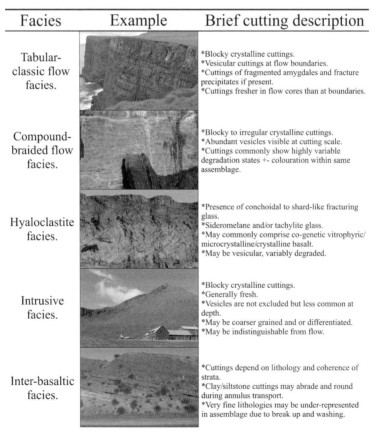

Fig. 1. LIP facies after the scheme of Jerram (2002). All examples from the Faroe Islands except the Hyaloclastite and Inter-basaltic facies which are from the Columbia River Basalt province. In the third column is a very broad description of some of the common attributes of each facies at cuttings scale.

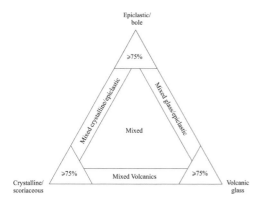

Fig. 2. Ternary classification scheme for LIP-derived cuttings.

of their potential accuracy issues. It is assessed that 75% and above of one end-member constitutes enough of a majority to define the analyses as exclusively that end-member in most cases. If the highest abundance is below 75% then the assemblage must be defined as mixed to a lesser or greater degree, depending on whether two or only one of the other constituents comprise 26% or more of the sample. If the sample is defined as mixed (the largest field in Fig. 2), then using it as an interpretive datum without careful examination should be avoided. The three mixed classification fields around the edge of the ternary diagram may all be important for different facies.

End-member classification

In this section we provide a detailed description of the three classification end-members including details on all the inclusions and exclusions of each. Classification schemes for volcanic rocks and in particular volcaniclastic rocks have been a matter of contentious debate for decades (Fisher 1966; Fisher & Schmincke 1984; White & Houghton 2006). What is common to the majority of systems is a reliance on detailed field and petrographic descriptions including, in many cases, direct observations of formational process. When dealing with cuttings there is obviously much less coherent background information such as facies association and this, along with the drilling fragmentation process and cuttings size, significantly complicates the picture to be deciphered. Any approach to cuttings analysis therefore requires an initial oversimplification of more in-depth classification systems from which detailed observations can then be used to attempt, where possible, to infer greater detail. This initial simplification is the designed purpose of the three end-members presented here, specifically

not any theoretical development on standard classification schemes for volcanic rocks or facies. It is worth remembering at this point that a sample interval of 10 ft which is fairly standard for an industry well sample collection can only resolve (assuming perfect bit to surface transfer) lithological units that are that thickness or greater without subsequent interpretation of the assemblage. Any units below the 10 ft interval resolution will require inference, which should be transparently based upon a detailed description of the grounds for interpretation.

Scoriaceous/crystalline. Crystalline is defined on the ternary as any cutting that shows a holocrystalline groundmass including any phenocrysts, amygdales or fracture precipitates that may be present within it. The fracture surface under magnification is rough to irregular due to the presence of intergrown crystals (Fig. 3). Such cuttings with any degree of degradation are still classified as crystalline as long as remnant groundmass crystal structure can be observed. Scoriaceous material is defined as highly vesicular micro-crystalline to glassy basalt. These may be identified as fragments of cuspate striated cuttings that may display a surface microtexture similar to pāhoehoe lavas or as variably degraded highly vesicular cuttings (Fig. 3). Scoriaceous material may be found at the frothy upper margin of sub-aerial lava flows formed by vesicle migration and accumulation or as vent proximal spatter formed by the rapid expansion of volatiles at the surface (Cas & Wright 1987). In cuttings it is generally not possible to differentiate the two types unless individual pyroclast morphologies are maintained and have a uniformly degraded outer rim.

From this broad definition we would expect the majority of cuttings produced from sub-aerial lavas including compound, simple, pāhoehoe, 'a'ā and eruption proximal spatter lithologies to fall into this category (Head & Wilson 1989; Self et al. 1997; Jerram 2002). Intrusive lithologies such as sills or dykes will also predominantly fit into this category.

Volcanic glass. Glass is defined here as any volcanic cuttings that have a massive non-crystalline structure under magnification. A range exists from transparent to opaque shiny lustre typical glassy (sideromelane) material to more dull/soapy lustre dark (tachylitic) glass. Glass can be distinguished from micro-crystalline basalt by its lack of any identifiable matrix crystals under high binocular magnification and by the usual presence of a conchoidal to smooth fracture appearance. Glass can be porphyritic (vitrophyric), containing any size of crystals formed prior to or during eruption; the important point in these instances is the glass

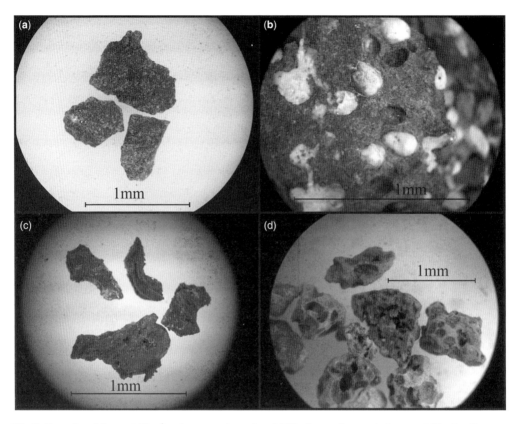

Fig. 3. Examples of the crystalline/scoriaceous end-member. (**a**) Blocky angular morphology crystalline basalt cuttings. (**b**) Vesicular, variably degraded basalt from a compound lava flow, vesicles mostly filled with secondary zeolite; however, some appear hollow as amygdales have been knocked out during drilling. (**c**) Broken striated scoriaceous shards (micro-vesicular) formed by Hawaiian-style eruption spatter, easily reworked if loose. (**d**) Highly vesicular degraded basalt from scoriaceous flow top; unfilled vesicles can indicate potential permeable loss zones. All examples are from West of Shetland wells except for (c) which is from Hawaii.

texture of the groundmass (Williams *et al.* 1958). Also included in this category are any fragmental glass shards included in a glassy matrix as well as any cuttings displaying basalt sediment interaction (Fig. 4). Basalt sediment interaction at the cutting scale may be recognized by a micro-scale mingling association between the two lithologies or glass fused against sediment. The term fused is used where the cuttings fracture in a massive way across the glass sediment boundary. The term peperite may be used in a genetic role during subsequent interpretation (White *et al.* 2000).

From this broad definition we expect lithologies produced from the interaction of basaltic magma/lava with water/wet sediment to be encompassed. These will include hyaloclastite sequences, pillow complexes and peperite. Because of the cooling-related continuum that exists between glass and crystalline basalt, it must be noted that many deposits of the above facies may comprise considerable quantities of crystalline basalt and/or sediment at the cuttings scale (Moore 1975; Schmincke *et al.* 1979; Skilling *et al.* 2002). It is therefore implied that the presence of glass in the mixed volcanic field in Figure 2 may be diagnostically more important; however, careful analyses of the assemblage and cuttings associations is always required in each separate instance.

Epiclastic/boles. The epiclastic/bole apex of the classification system comprises perhaps the largest simplification of all the end-members. The end-member encompasses all the products of clastic sedimentology also including any lithified volcani-clastic rocks (Ross *et al.* 2005), coals, carbonates or preserved soil horizons. The justification behind this very broad end-member grouping is that in terms of volume, all the above-mentioned deposits comprise by far the smallest component of most LIP sequences. This does however depend on the

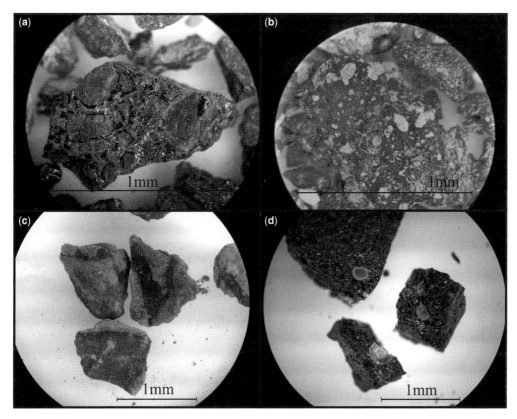

Fig. 4. Examples of the Volcanic glass end-member. (**a**) Fragmental zoned glass shards in a dull glass matrix. (**b**) Dull lustre tachylite glass with small euhedral olivine crystals and small filled vesicles. (**c**) Degraded glassy basalt in a mingling association with dark fine-grained sediment. (**d**) Sparsely vesicular massive fresh glass from pillow rind. (a) and (b) are from West of Shetland wells, (c) is from the Inner Hebrides, Scotland and (d) is from the Columbia River Basalt Province, USA.

proximity of the penetration location to the centre or periphery of any particular province. Where a significant thickness of interbed is encountered, a separate analysis using standard sedimentology method is advised. What *is* important in a simple LIP classification is the presence, broad description and location of inter-basaltic facies that indicate hiatuses in volcanism. This is why soil horizons, coals and epiclastic rocks (after Fisher 1966), whether siliciclastic or volcaniclastic, are deemed to be produced during volcanic hiatuses at the given location which in turn is an important factor in LIP stratigraphy.

What this equates to in terms of analysis is the designation of any cuttings that have no demonstrably primary volcanic features to this end-member. In cuttings, epiclastic rocks can only be assigned where more than one grain or particle is associated, by matrix or cementation, within a single cutting. The percentage of volcanic cuttings compared to non-volcanic cuttings cannot be used to infer a volcaniclastic origin of the assemblage. Only when the two populations can be demonstrably associated within a single cutting, or alternatively the character of the volcanic cuttings is such that they can be confidently inferred to have been reworked, can such an inference be made due to potential assembly mixing. Cuttings from soil horizons and poorly consolidated muds and silts may be common; however, depending on the thickness of a particular bed and the degree of drill bit degradation, they may occur at underrepresented percentages due to loss of fines in the drilling mud. Depending on the climatic conditions (temperature/humidity/rainfall) at the time of formation, soils may appear brick red in cuttings as seen in the upper Beinisvørð Formation interbeds that were deposited during the Paleocene–Eocene Thermal Maximum (Hald & Waagstein 1984; Jolley *et al.* 2002; Passey & Jolley 2009). In many cases the amount of information available from the microscopic analysis of mud and silt-grade material may be limited due to grain size. In these cases, noting the intervals of largest fines percentage is probably most useful in the identification of intervals for biostratigraphic targeting. Siliciclastic cuttings can usually be determined by the abundance of quartz; however, determining the percentages of volcanic v. non-volcanic constituents at silt grade and below is difficult (if not impossible) based solely on microscopic work. X-ray diffraction (XRD) analysis is one obvious option for further study in such cases where the significant intervals are encountered.

An important omission to this group which may at first seem illogical is that which is briefly mentioned earlier whereby an individual clast or fragment of either crystalline or glassy volcanic material, regardless of the degree of rounding, is always entered into its respective end-member class. The assemblage must then be attributed a reworked nature in the accompanying notes. The rationale behind this decision is simple; where a reworked volcaniclastic unit comprises loose or disaggregated volcanic clasts, and if larger clasts are broken by the drilling process, then there is no way to distinguish them from fresh volcanic fragments broken exclusively by the drilling process.

As long as a qualification on the amount of reworking evidence that is present, for example, minor/pervasive, is given along with the analyses, then the likelihood of error and incorrect classification is decreased.

Simple log output

An important end product from any cuttings analysis should be the production of a stratigraphic log output to display the primary results which can then be easily integrated with all other down-hole data. From the ternary classification and associated percentages, it is possible to produce a simple output in Microsoft Excel (or equivalent software) denoting all the classification fields in an association, aiding subsequent interpretation of the stratigraphy. The output format along with sample data entry is presented in Figure 5. The visual log acts as a basic framework from which to hang the lithological descriptions not displayed by the basic output. These should take the form of accompanying annotations and interval summaries. Any individual data point from the log output is directly

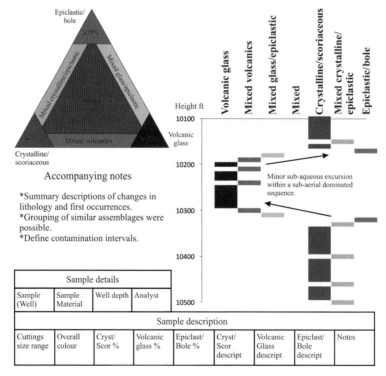

Fig. 5. Basic stratigraphic output from ternary diagram. Each data point in the log output has an accompanying row in the source spreadsheet within which the detailed description and source end-member percentages are held. The tabulated headings for the basic analysis are entered at the bottom of the figure. A spreadsheet containing the basic format and log output set-up is available from the authors on request.

linked back to an individual analysis of the cuttings returned for that specific 10 ft interval (Fig. 5).

Idealized volcanic stratigraphy from cuttings

The methodology outlined above has been utilized for a number of wells in the West of Shetland area, the majority of which are not released. For the purposes of this contribution, we have compiled a conceptual model-based stratigraphy and associated cutting responses to display the broad range of stratigraphy encountered (Fig. 6). In Figure 6 the separate end-member percentages are also presented in stratigraphic scatter plots which give an important display of more subtle variations with height than the basic ternary fields (Fig. 5). The stratigraphy has been compiled based on a number of common LIP components but does not represent any real continuous sequence. Depending on the location of penetration within a LIP province, any combination of the separate components at various scales may be encountered.

The cutting percentages in Figure 6 have been generated to display an idealized response to lithology with only minimal lag and mixing present. We briefly discuss some of the aspects of the idealized responses. Firstly, regarding the sub-aerial lava sequences of compound and tabular type, there is little to no variation in the generic cutting signatures apart from increases in epiclastiic/bole cuttings at well-developed flow boundaries. It is not always possible to distinguish between the two types from cuttings, depending on the thickness of the intervals and cutting quality. As some general guidelines, there should be a greater abundance of non-vesicular basalt in larger tabular flow sequences (Passey & Jolley 2009), a larger abundance of smaller vesicles throughout a compound sequence at cutting scale and also a generally larger amount of degraded basalt in compound sequences, at least in Palaeogene-aged flows (Noe-Nygaard & Rasmussen 1968).

Two hydrovolcanic rocks are depicted in Figure 6, a pillow and a hyaloclastite sequence. Again, differentiating these two facies from cuttings may be impossible as a full transition exists between the two. The percentage of true glass may be expected to decrease with an increasing density of pillows in comparison to more hydroclastic-dominated sequences as well as a relative decrease in the percentage of fragmental glass (Fig. 4). Minor glass and glass fused with sediment may be observable at intervals of interaction between lava and wet sediment (Fig. 4).

In volcaniclastic sediments there may be a mixture of the end-member assemblages depending on the clast source. The example given at the base of Figure 6 c. 10 200 ft, is that expected from a pro-volcanic delta deposit. What is important is the recognition of any aspects of reworking in the cuttings of glass and crystalline material for example, any rounding above that is likely from annulus and sieving abrasion which depends strongly on the hardness of the cuttings.

Where annulus mixing is increased, the boundary transitions in percentage will generally become less defined; in particular, the peak abundances relating to thinner lithology intervals will be deflated making the potential for smaller-scale boundary identification much less likely. In Figure 7, the same lithology log is produced but with a number of commonly associated problems known from drilling added into the cuttings responses. These include losses in circulation within permeable flow boundary zones and associated contamination, casing point contamination, uncased hole mixing, drill bit change mixing and increased caving and erosion within softer unconsolidated intervals (Fig. 7). The cutting response is again conceptually created to represent some of the likely reductions in data quality, including loss of returns. The responses may in reality be affected to a larger or lesser degree depending on formation properties and various drilling factors. The point is that the cuttings quality may be influenced by many factors related to both lithology and drilling, and all must be checked as thoroughly as possible prior to and during analysis. The loss in circulation at around 8400 ft is inferred to relate to the highly porous and permeable zone at the top of a simple flood basalt flow that has not undergone significant secondary mineralization (Freeze & Cherry 1979; Petford 2003; McGrail et al. 2006). The similar facies lower in the sequence are inferred to have secondary minerals filling all the vesicular pore-space and thus do not represent potential loss zones. These data are easily picked up from cuttings (Fig. 3) and should be accurately monitored during real-time drilling to indicate potential loss zones if present. Fracture porosity and permeability may only be inferred from cuttings in the negative case where fracture precipitates are observed in individual cuttings. The mineral filling state of fractures will commonly mirror those of vesicle voids where the fractures formed prior to mineralization; post-mineralization fractures are also possible however, the severity of which depends largely on the tectonic history of the region post-LIP development.

Lastly, within a volcanic sequence it may be hard to identify a cutting population that can be unequivocally related back to a log signature, especially if there are many repeated lithology intervals. Wherever possible, units such as coal occurrences that generally show a distinctive log character

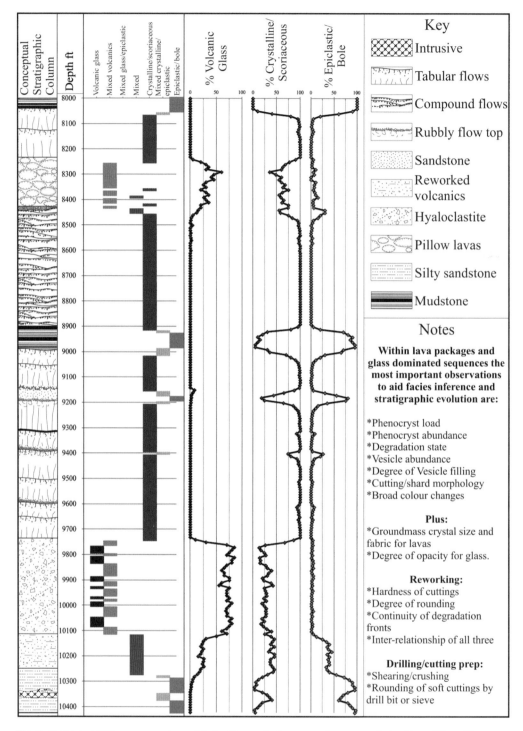

Fig. 6. Conceptual LIP sequence and associated cutting analysis response for near-perfect well conditions. The percentage of each end-member is included to show variations below the resolution of the basic ternary classification.

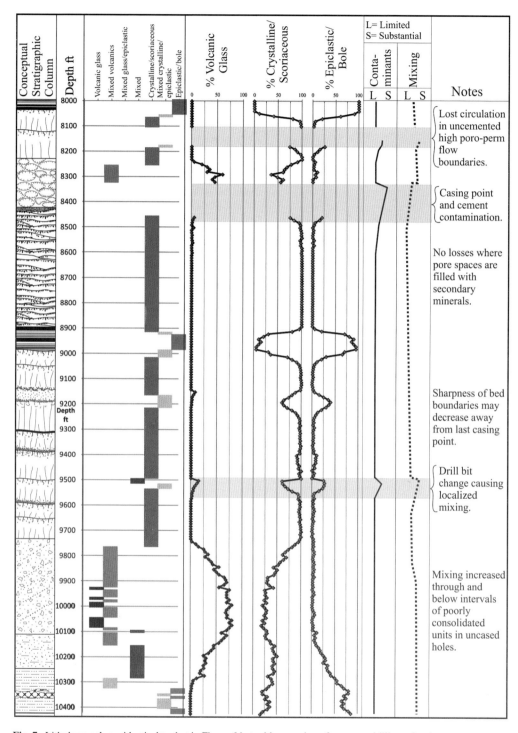

Fig. 7. Lithology column identical to that in Figure 6 but with a number of common drilling-related accuracy constraints added to the theoretical cuttings response.

should be used to calibrate the degree of mixing and lag between the log signatures and the cuttings. The transitions into and out of volcanic sections should also be used where possible to this effect where the log signature is very clear. This transition is not always clearly defined, however; combination must therefore be undertaken with caution where low confidence exists in either dataset.

Beyond initial cutting analysis

The degree of mixing and character of cuttings assemblages through LIP successions has important implications when assessing cutting-derived data. The results of a detailed microscopic analysis combined with log calibration can be fed into mixing assessments for biostratigraphical data. There also exists a large amount of data suggesting that the geochemistry of LIP volcanic sequences are stratigraphically zoned (Noe-Nygaard & Rasmussen 1968; Larsen 1989; Kerr 1995; Jolley et al. 2012) and can also be correlated over both small and large distances (Fitton et al. 1998; Larsen et al. 1999). The potential then exists for using cuttings as a basis for geochemical analysis of volcanic rocks to aid stratigraphy. Having the cuttings stratigraphy and errors within it well defined prior to sample picking where core data are unavailable is very important for such correlation attempts. If a sample is picked from a mixed zone or bulk sampled prior to assessing mixing, then the results may potentially be useless or (worse) misleading, meaning wasted time and expense.

Lastly, what has been presented is an idealized scheme appropriate for scenarios where the cutting quality is such that detailed inference is deemed possible. Where turbine polycrystalline diamond compact (PDC) drill bits are used or drilling problems and associated contaminants dilute the recovered cuttings to a large degree, useful inference may not always be possible. Where such conditions prevail, only the level of detail that is available should be reported; for instance, in a sequence of high contamination it is very unlikely that individual flow boundaries will be picked out from cuttings. This may later be inferred from the log responses, but the grounds for interpretation should not be clouded (Fig. 8) as this may propagate into the literature and mislead subsequent lithological work/sampling.

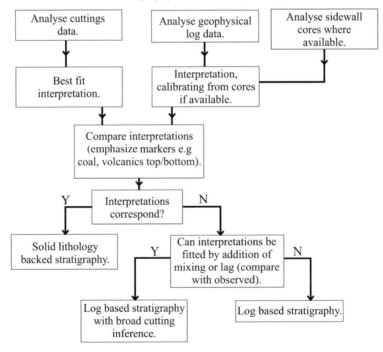

Fig. 8. Basic workflow for approaching a LIP sequence well penetration. The final interpretation should be defined according to the data that was used in its production. Fully defining the basis of an interpretation is important for assigning confidence levels to model input parameters.

Conclusions

In this contribution we have outlined some of the current obstacles to correlation within LIP-affected industry well penetrations where cuttings data are used. These relate to the analysis, reporting and characterization of cutting assemblages. We have proposed an analytical workflow and classification scheme that is non-genetic and that characterizes cutting assemblages in a way that allows easy and quick comparison between different LIP sequences. A log-style output is also proposed from which easy comparison with the raw analysis and other down-well data can be made. Where cutting quality is very poor, the level of cutting-based inference may be very low and should be reported as such. The detailed reporting of cuttings data through LIP sequences allows inference of mixing and lag times with comparison to geophysical log signatures, which in turn aids sample selection for biostratigraphy and geochemistry. Until a systematic and repeatable approach to cuttings analysis on a province-wide basis within LIP-affected basins is implemented, the full correlative potential of cutting-derived well data cannot be met. The results of this study will hopefully be of significance in approaching developments in volcanic stratigraphy and well-site contractors for any volcanic-affected drilling operations.

Comments from an anonymous reviewer are gratefully acknowledged in improving the manuscript. We would like to acknowledge members of the Chevron WoS team for helpful discussions and access to data. Stuart Archer, Simon Passey and Nick Schofield are thanked for fruitful discussions. John Still and Fiona Thomson at the University of Aberdeen are thanked for invaluable help in sample preparation and analysis. Finally Keiren Wall is thanked for discussions and analysis undertaken during concept development.

References

AL-KAYIEM, H. H., ZAKI, N. M., ASYRAF, M. Z. & ELFEEL, M. E. 2010. Simulation of the cuttings cleaning during the drilling operation. *American Journal of Applied Sciences*, **7**, 800–806.

ARCHER, S. G., BERGMAN, S. C., ILIFFE, J., MURPHY, C. M. & THORNTON, M. 2005. Palaeogene igneous rocks reveal new insights into the geodynamic evolution and petroleum potential of the Rockall Trough, NE Atlantic Margin. *Basin Research*, **17**, 171–201.

BARTETZKO, A., PEZARD, P., GOLDBERG, D., SUN, Y. & BECKER, K. 2001. Volcanic stratigraphy of DSDP/ODP Hole 395A: an interpretation using well-logging data. *Marine Geophysical Researches*, **22**, 111–127.

BELAVADI, M. N. & CHUKWU, G. A. 1994. Experimental study of the parameters affecting cutting transportation in a vertical wellbore annulus. *In*: *Society of Petroleum Engineers Western Regional Meeting*. Long Beach, California, 1st Jan 1994, http://dx.doi.org/10.2118/27880-MS

BOLDREEL, L. O. 2006. Wireline log-based stratigraphy of flood basalts from the Lopra-1/1Awell, Faroe Islands. *In*: CHALMERS, J. A. & WAAGSTEIN, R. (eds) *Scientific Results from the Deepened Lopra-1 borehole, Faroe Islands*. Geological Survey of Denmark and Greenland Bulletin, Copenhagen, **9**, 7–22.

BRISTER, B. S. & MCINTOSH, W. C. 2004. Identification and correlation of Oligocene ignimbrites in well bores, Alamosa Basin (northern San Luis Basin), Colorado, by single crystal laser fusion ^{40}Ar/^{39}Ar geochronology of well cuttings. *New Mexico Bureau of Geology & Mineral Resources, Bulletin*, **160**, 281–296.

BRYAN, S. E., UKSTINS PEATE, I. *ET AL*. 2010. The largest volcanic eruptions of earth. *Earth Science Reviews*, **102**, 207–229.

CAS, R. A. F. & WRIGHT, J. V. 1987. *Volcanic Successions: Modern and Ancient*. Allen & Unwin, London.

CLOSE, F., CONROY, D., GREIG, A., MORIN, A., FLINT, G. & SEALE, R. 2005. Successful drilling of basalt in a West of Shetland deepwater discovery. *In*: Society of Petroleum Engineers, Offshore Europe Oil and Gas Conference and Exhibition Proceedings, Aberdeen, http://dx.doi.org/10.2118/96575-MS

ELLIS, D., BELL, B. R., JOLLEY, D. W. & O'CALLAGHAN, M. 2002. The stratigraphy, environment of eruption and age of the Faroes Lava Group, NE Atlantic Ocean. *In*: JOLLEY, D. W. & BELL, B. R. (eds) *The North Atlantic Igneous Province: Stratigraphy, Tectonic, Volcanic and Magmatic Processes*. Geological Society, London, Special Publications, **197**, 253–269.

FISHER, R. V. 1966. Rocks composed of volcanic fragments and their classification. *Earth-Science Reviews*, **1**, 287–298.

FISHER, R. V. & SCHMINCKE, H.-U. 1984. *Pyroclastic Rocks*. Springer-Verlag, Berlin.

FITTON, J. G., SAUNDERS, A. D., LARSEN, L. M., HARDARSON, B. S. & NORRY, M. J. 1998. Volcanic rocks from the southeast Greenland margin at 63°N: composition, petrogenesis, and mantle sources. *In*: SAUNDERS, A. D., LARSEN, H. C., WISE, S. & ALLEN, J. R. (eds) *Proceedings of the Ocean Drilling Program, Scientific Results*, *152*. Ocean Drilling Program, College Station, TX, 331–350.

FREEZE, R. A. & CHERRY, J. A. 1979. *Groundwater*. Prentice-Hall, Englewood Cliffs, NJ.

GALLAGHER, J. W. & DROMGOOLE, P. W. 2007. Exploring below the basalt, offshore Faroes: a case history of sub-basalt imaging. *Petroleum Geoscience*, **13**, 213–225.

GEORGI, D. T., HARVILLE, D. G. & ROBERTSON, H. A. 1993. Advances in cuttings collection and analysis. *SPWLA 34th Annual Logging Symposium, Calgary*, 1–20.

HALD, N. & WAAGSTEIN, R. 1984. Lithology and chemistry of a 2-km sequence of Lower Tertiary tholeiitic lavas drilled on Suðuroy, Faroe Islands (Lopra-1). *In*: BERTHELSEN, O., NOE-NYGAARD, A. & RASMUSSEN, J. (eds) *The Deep Drilling Project 1980–1981 in the Faeroe Islands*. Annales Societatis Scientiarum Færoensis, Tórshavn, Supplementum IX, 15–38.

HEAD, J. W. & WILSON, L. 1989. Basaltic pyroclastic eruptions: influence of gas-release patterns and volumes fluxes on fountain structure, and the formation of cinder cones, rootless flows, lava ponds and lava flows. *Journal of Volcanology and Geothermal Research*, **37**, 261–271.

HELM-CLARK, C., RODGERS, D. & SMITH, R. 2004. Borehole geophysical techniques to define stratigraphy, alteration and aquifers in basalt. *Journal of Applied Geophysics*, **55**, 3–38.

HUSSAINI, S. M. & AZAR, J. J. 1983. Experimental study of drilled cuttings transport using common drilling mud. *Society of Petroleum Engineers Journal*, **23**, 11–20, http://dx.doi.org/10.2118/10674-PA.

JERRAM, D. A. 2002. Volcanology and facies architecture of flood basalts. *In*: MENZIES, M. A., KLEMPERER, S. L., EBINGER, C. J. & BAKER, J. (eds) *Volcanic Rifted Margins*. Special Paper, Geological Society of America, Boulder, **362**, 119–132.

JOLLEY, D. W. 1997. Palaeosurface palynofloras of the Skye Lava Field, and the age of the British Tertiary Volcanic Province. *In*: WIDDOWSON, M. (ed.) *Palaeosurfaces; Recognition, Reconstruction and Interpretation*. Geological Society, London, Special Publications, **120**, 67–94.

JOLLEY, D. W. & BELL, B. R. 2002. Genesis and age of the Erlend volcano, NE Atlantic Margin. *In*: JOLLEY, D. W. & BELL, B. R. (eds) *The North Atlantic Igneous Province: Stratigraphy, Tectonic, Volcanic and Magmatic Processes*. Geological Society, London, Special Publications, **197**, 95–109.

JOLLEY, D. W., CLARKE, B. & KELLEY, S. 2002. Paleogene time scale miscalibration: evidence from the dating of the North Atlantic igneous province. *Geology*, **30**, 7–10.

JOLLEY, D. W., PASSEY, S. R., HOLE, M. J. & MILLETT, J. M. 2012. Large scale magmatic pulses drive plant ecosystem dynamics. *Journal of the Geological Society*, **169**, 703–711.

JØRGENSEN, O. 1984. Zeolite zones in the basaltic lavas of the Faeroe Islands. A quantitive description of the secondary minerals in the deep wells of Vestmanna-1 and Lopra-1. *In*: BERTHELSEN, O., NOE-NYGAARD, A. & RASMUSSEN, J. (eds) *The Deep Drilling Project 1980–1981 in the Faeroe Islands*. Annales Societatis Scientiarum Færoensis, Torshavn, Supplementum IX, 71–92.

JØRGENSEN, O. 2006. The regional distribution of zeolites in the basalts of the Faroe Islands and the significance of zeolites as palaeotemperature indicators. *In*: CHALMERS, J. A. & WAAGSTEIN, R. (eds) *Scientific Results from the Deepened Lopra-1 Borehole, Faroe Islands*. Geological Survey of Denmark and Greenland Bulletin, Copenhagen, **9**, 123–56.

KERR, A. C. 1995. The geochemical stratigraphy, field relations and temporal variation of the Mull-Morvern Tertiary lava succession. *Transactions of the Royal Society Edinburgh; Earth Sciences*, **86**, 35–47.

LARSEN, L. M., WATT, W. S. & WATT, M. 1989. Geology and petrology of the Lower Tertiary plateau basalts of the Scoresby Sund region, East Greenland. *Geology of Greenland Survey Bulletin*, **157**, 1–164.

LARSEN, L. M., WAAGSTEIN, R., PEDERSEN, A. K. & STOREY, M. 1999. Trans-Atlantic correlation of the Palaeogene volcanic successions in the Faeroe Islands and East Greenland. *Journal of the Geological Society, London*, **156**, 1081–1095.

MCGRAIL, B. P., SCHAEF, H. T., HO, A. M., CHIEN, Y. J., DOOLEY, J. J. & DAVIDSON, C. L. 2006. Potential for carbon dioxide sequestration in flood basalts. *Journal of Geophysical Research B: Solid Earth*, **111**, article number B12201, 1–13.

MOORE, J. G. 1975. Mechanism of formation of pillow lava. *American Scientist*, **63**, 269–277.

MORTON, A. C., DIXON, J. E., FITTON, J. G., MACINTYRE, R. M., SMYTHE, D. K. & TAYLOR, P. N. 1988. Early Tertiary volcanic rocks in the Well 163/6-1A, Rockall Trough. *In*: MORTON, A. C. & PARSON, L. M. (eds) *Early Tertiary Volcanism and the Opening of the NE Atlantic*. Geological Society, London, Special Publications, **39**, 293–308.

NELSON, C. E., JERRAM, D. A. & HOBBS, R. W. 2009. Flood basalt facies from borehole data: implications for prospectivity and volcanology in volcanic rifted margins. *Petroleum Geoscience*, **15**, 313–324.

NOE-NYGAARD, A. & RASMUSSEN, J. 1968. Petrology of a 3000 metre sequence of basaltic lavas in the Faeroe Islands. *Lithos*, **1**, 286–304.

PASSEY, S. R. 2004. *The Volcanic and Sedimentary Evolution of the Faeroe Plateau Lava Group, Faeroe Islands and the Faeroe-Shetland Basin, NE Atlantic*. Unpublished PhD thesis. University of Glasgow.

PASSEY, S. R. & BELL, B. R. 2007. Morphologies and emplacement mechanisms of the lava flows of the Faroe Islands Basalt Group, Faroe Islands, NE Atlantic Ocean. *Bulletin of Volcanology*, **70**, 139–56.

PASSEY, S. R. & JOLLEY, D. W. 2009. A revised lithostratigraphic nomenclature for the Palaeogene Faroe Islands Basalt Group, NE Atlantic Ocean. *Earth and Environmental Science Transactions of the Royal Society of Edinburgh*, **99**, 127–158.

PETFORD, N. 2003. Controls on primary porosity and permeability development in igneous rocks. *Geological Society, London, Special Publications*, **214**, 93–107.

PLANKE, S. 1994. Geophysical response of flood basalts from analysis of wire line logs: Ocean Drilling Program Site 642, Voring volcanic margin. *Journal of Geophysical Research*, **99**, 9279–9296.

ROBERTS, A. W., WHITE, R. S., LUNNON, Z. C., CHRISTIE, P. A. F., SPITZER, R. & TEAM, I. 2005. Imaging magmatic rocks on the Faroes Margin. *In*: DORÉ, A. G. & VINING, B. A. (eds) *Petroleum Geology: North-West Europe and Global Perspectives*. Proceedings of the 6th Petroleum Geology Conference. Geological Society, London, 755–766.

ROSS, P.-S., UKSTINS PEATE, I., MCCLINTOCK, M. K., XU, Y. G., SKILLING, I. P., WHITE, J. D. L. & HOUGHTON, B. F. 2005. Mafic volcaniclastic deposits in flood basalt provinces: a review. *Journal of Volcanology and Geothermal Research*, **145**, 281–314.

SCHMINCKE, H.-U., ROBINSON, P. T., OHNMACHT, W. & FLOWER, M. J. F. 1979. Basaltic hyaloclastites from Hole 396B, DSDP Leg 46. *In*: DMITRIEV, L., HEIRTZLER, J. ET AL. (eds) *Initial Reports of the Deep Sea Drilling Project*. US Govt. Printing Office, Washington, **46**, 341–346.

SELF, S., THORARSON, T. & KESZTHELYI, L. 1997. Emplacement of continental flood basalt lava flows. *In*: MAHONEY, J. J. & COFFIN, M. L. (eds) *Large Igneous Provinces: Continental, Oceanic, and Planetary Flood*

Volcanism. AGU, Washington, Geophysical Monograph, **100**, 381–410.

SIFFERMAN, T. R. & BECKER, T. E. 1992. Hole cleaning in full-scale inclined wellbores. *Society of Petroleum Engineers Drilling Engineering*, **7**, 115–120, http://dx.doi.org/10.2118/20422-PA.

SINGLE, R. T. & JERRAM, D. A. 2004. The 3D facies architecture of flood basalt provinces and their internal heterogeneity: examples from the Palaeogene Skye Lava Field. *Journal of the Geological Society, London*, **161**, 911–926.

SKILLING, I. P., WHITE, J. D. L. & MCPHIE, J. 2002. Peperite: a review of magma–sediment mingling. *Journal of Volcanology and Geothermal Research*, **114**, 1–17.

TERRY, R. D. & CHILINGAR, G. V. 1955. Summary of 'concerning some additional aids in studying sedimentary formations' by M. S. Shvetsov. *Journal of Sedimentary Petrology*, **25**, 229–234.

WALKER, G. P. L. 1970. The distribution of amygdale minerals in Mull and Morvern (Western Scotland). *In*: MURTY, T. V. V. G. R. K. & RAO, S. S. (eds) *Studies in Earth Sciences*. West Commemoration Volume, Today & Tomorrow's Printers & Publishers, Faridabad, India, 181–194.

WATTON, T. J., JERRAM, D. A., THORDARSON, T. & DAVIES, R. J. 2013. Three-dimensional lithofacies variations in hyaloclastite deposits. *Journal of Volcanology and Geothermal Research*, **250**, 19–33.

WHITE, J. D. L. & HOUGHTON, B. F. 2006. Primary Volcaniclastic Rocks. *Geology*, **34**, 677–680.

WHITE, J. D. L., MCPHIE, J. & SKILLING, I. 2000. Peperite: a useful genetic term. *Bulletin of Volcanology*, **62**, 65–66.

WILLIAMS, H., TURNER, F. J. & GILBERT, C. M. 1958. *Petrography: An Introduction to the Study of Rocks in Thin Sections*. W. H. Feeman and Company, San Francisco.

Detrital zircon age constraints on basement history on the margins of the northern Rockall Basin

ANDREW MORTON[1,2]*, DIRK FREI[3], MARTYN STOKER[4] & DAVID ELLIS[5]

[1]*HM Research Associates, 2 Clive Road, Balsall Common, CV7 7DW, England, UK*

[2]*CASP, University of Cambridge, 181a Huntingdon Road, Cambridge CB3 0DH, England, UK*

[3]*Central Analytical Facility, Stellenbosch University, Chamber of Mines Building, Matieland 7602, South Africa*

[4]*British Geological Survey, Murchison House, West Mains Road, Edinburgh EH9 3LA, Scotland, UK*

[5]*Statoil (UK) Ltd., 1 Kingdom Street, London W2 6BD, England, UK*

*Corresponding author (e-mail: heavyminerals@hotmail.co.uk)

Abstract: Detrital zircon dating has proven to be an effective way to constrain ages of submerged basement terranes on the margins of the northern Rockall Basin, a region where direct evidence of crustal affinities is scarce or absent. Zircons have been dated from sandstones of Paleocene–Oligocene age known to have been derived from the east (Hebridean Platform) and west (Rockall and George Bligh highs). The results show that the Hebridean Platform is a westward extension of the Lewisian Complex, with Archaean and Palaeoproterozoic ages that can be directly correlated with events identified in the Outer Hebrides and NW Scotland. The detrital zircons derived from the Hebridean Platform also provide evidence for a Mesoproterozoic thermal event and two phases of intrusions in the Palaeozoic. The Rockall High consists of a Palaeoproterozoic terrane dated as *c.* 1760–1800 Ma, similar to ages previously determined from both basement samples and detrital sediment. The data also provide evidence for the subsequent intrusion of alkaline igneous rocks in the Paleocene–Eocene. The George Bligh High represents an Archaean terrane heavily affected by Palaeoproterozoic tectonothermal events, and was also the site of intrusion of alkaline igneous rocks during Paleocene time.

The northern Rockall Basin lies to the west of the Hebridean margin of the British Isles and east of a number of structural and bathymetric highs ranging from the Rockall High in the south to the George Bligh High and the Rosemary Bank Seamount in the north (Fig. 1). The present bathymetric expression of the Rockall Basin represents a large sediment-starved highly extended rift basin. Although the timing of rifting remains somewhat ambiguous, the general consensus is that the Rockall Basin is probably a Cretaceous rift, with earlier pulses of rifting during Permo-Triassic and Jurassic times (cf. Ritchie *et al.* 2013a). The region remains a frontier area for hydrocarbon exploration, with comparatively few wells having been drilled despite many years of activity which began with the stratigraphic test well 163/6-1A. Consequently, comparatively little is known either of the stratigraphic development within the basinal areas or of the age of the basement rocks on the flanks of the rift.

The basement on the Hebridean margin is considered to represent a westwards extension of the Precambrian Lewisian Complex of the Outer Hebrides and mainland NW Scotland, on the basis of borehole and seabed samples together with geophysical observations (Jones *et al.* 1986; Stoker *et al.* 1993; Ritchie *et al.* 2013b). On the eastern margin of the Rockall Plateau, the only direct evidence concerning basement ages comes from samples recovered from the Rockall High during diving expeditions (Roberts *et al.* 1973) and, more recently, from boreholes drilled on the Hatton High by the British Geological Survey (Hitchen 2004). Sm–Nd model (T_{dm}) ages from these samples lie within the range 1.89–2.14 Ga (Morton & Taylor 1991; Hitchen 2004). U–Pb zircon age data have been acquired from some of these samples and, although the results have not been formally published, they have yielded results similar to that of the Annagh Gneiss of north Mayo (Ireland) which has a U–Pb zircon crystallization age

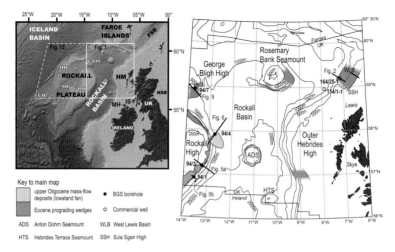

Fig. 1. Location map showing positions of borehole and well sites described in this paper, the distribution of Eocene prograding wedges on the eastern and western margins of the northern Rockall Basin and of upper Oligocene mass-flow deposits, as well as the lines of geoseismic/seismic sections illustrated in Figures 2, 5, 6 and 9. Bathymetric values in metres. Inset map shows area covered by main map, as well as expanded area (incorporating the Rockall Plateau) that is illustrated in Figure 12. Abbreviations on inset map: EH, Edoras High; FSB, Faroe–Shetland Basin; GBH, George Bligh High; HB, Hatton Basin; HH, Hatton High; HM, Hebridean Margin; IS, Islay; MH, Malin Head (adjacent to Inishtrahull); NSB, North Sea Basin; RH, Rockall High. Adapted from McInroy et al. (2006) and Stoker et al. (2012). Structural terminology from Ritchie et al. (2013a).

of c. 1750 Ma (Daly et al. 1995; Scanlon & Daly 2001). These limited data suggest that the Rockall High metamorphic basement comprises a juvenile Palaeoproterozoic terrane (Morton & Taylor 1991; Dickin 1992).

Crystalline basement rocks have not been recovered from the other bathymetric highs in the northern Rockall Basin area (George Bligh High and Rosemary Bank Seamount). However, Palaeogene basalts recovered from the George Bligh High display evidence of contamination by Archaean crust (Hitchen et al. 1997), suggesting that the boundary between the Archaean and Palaeoproterozoic terranes lies between the Rockall High and the George Bligh High, as proposed by Dickin (1992) and Dickin & Durant (2002). A borehole drilled on the Rosemary Bank Seamount also proved the existence of basalts, in this case dated as Late Cretaceous, but these rocks do not display any evidence of crustal contamination (Morton et al. 1995).

In this paper, we present an alternative approach for constraining the crustal history of the basement rocks on the flanks of the northern Rockall Basin by utilizing detrital zircon ages of sediment known to be shed from the basin margins. The same approach was previously adopted in order to determine ages of basement rocks comprising the Edoras and Hatton highs (Fig. 1) on the western flank of the Rockall Plateau (Morton et al. 2009). The data have been acquired from Palaeogene sandstones recovered from two hydrocarbon exploration wells (154/1-1 and 164/25-1) and three British Geological Survey (BGS) boreholes (94/1, 94/4 and 94/7). The locations of these sites are shown in Figure 1. The sandstones recovered in 154/1-1 and 164/25-1 were derived from the Hebridean margin, whereas those recovered in BGS boreholes 94/1, 94/4 and 94/7 were derived from the Rockall and George Bligh highs. The zircon age data reported here therefore provide constraints on the nature of the crystalline basement forming the Rockall High and George Bligh High and the westward extension of the Hebridean margin outboard of the Outer Hebrides High, thereby adding considerably to our knowledge of the crustal history of the northern Rockall Basin region.

Zircon U–Pb analysis

All U–Pb age data were obtained at the Central Analytical Facility, Stellenbosch University, by laser-ablation single-collector magnetic-sectorfield inductively coupled plasma mass-spectrometry (LA-SF-ICP-MS) employing a Thermo Finnigan Element2 mass spectrometer coupled to a New-Wave UP213 laser ablation system. All age data presented here were obtained by single spot analyses with a spot diameter of 30 μm and a crater depth of c. 15–20 μm, corresponding to an ablated zircon mass of c. 150–200 ng. The methods employed for analysis and data processing are described in detail by Gerdes & Zeh (2006) and Frei & Gerdes (2009). For quality control, the Plešovice (Sláma et al. 2008) and M127 (Nasdala et al. 2008; Mattinson 2010) zircon reference materials were analysed and the results were consistently in excellent agreement with the published isotope-dilution thermal ionization mass spectrometry (ID-TIMS) ages. Full analytical details and the results for all quality control materials analysed are reported in Table 1. Plotting of concordia diagrams was performed using Isoplot/Ex 3.0 (Ludwig 2003) and probability–density plots were generated using AgeDisplay (Sircombe 2004). For zircons younger than 800 Ma the radiogenic $^{206}Pb/^{238}U$ age has been used for the probability density plots, whereas the $^{207}Pb/^{206}Pb$ age has been used for grains older than 800 Ma.

Table 1. *LA-SF-ICP-MS U–Th–Pb dating methodology at CAF, Stellenbosch University*

Laboratory and sample preparation	
Laboratory name	Central Analytical Facility, Stellenbosch University
Sample type/mineral	Detrital zircons
Sample preparation	Conventional mineral separation, 1 inch resin mount, 1 μm polish to finish
Imaging	CL, LEO 1430 VP, 10 nA, 15 mm working distance
Laser ablation system	
Make, model and type	ESI/New Wave Research, UP213, Nd:YAG
Ablation cell and volume	Custom build low volume cell, volume c. 3 cm^3
Laser wavelength	213 nm
Pulse width	3 ns
Fluence	2.5 J cm^{-2}
Repetition rate	10 Hz
Spot size	30 μm
Sampling mode/pattern	30 μm single spot analyses
Carrier gas	100% He, Ar make-up gas combined using a T-connector close to sample cell
Pre-ablation laser warm-up (background collection)	40 s
Ablation duration	20 s
Wash-out delay	30 s
Cell carrier gas flow	0.3 L min^{-1} He
ICP-MS instrument	
Make, model and type	Thermo Finnigan Element2 single collector HR-SF-ICP-MS
Sample introduction	Via conventional tubing
RF power	1100 W
Make-up gas flow	1.0 L min^{-1} Ar
Detection system	Single collector secondary electron multiplier
Masses measured	202, 204, 206, 207, 208, 232, 233, 235, 238
Integration time per peak	4 ms
Total integration time per reading	c. 1 s
Sensitivity	20 000 cps/ppm Pb
Dead time	16 ns
Data processing	
Gas blank	40 s on-peak
Calibration strategy	GJ-1 used as primary reference material, Plešovice and M127 used as secondary reference material (Quality Control)
Reference material info	M127 (Nasdala et al. 2008; Mattinson 2010); Plešovice (Sláma et al. 2008); GJ-1 (Jackson et al. 2004)
Data processing package used	In-house spreadsheet data processing using intercept method for laser-induced elemental fractionation (LIEF) correction
Mass discrimination	Standard-sample bracketing with $^{207}Pb/^{206}Pb$ and $^{206}Pb/^{238}U$ normalized to reference material GJ-1
Common-Pb correction, composition and uncertainty	204-method, Stacey & Kramers (1975) composition at the projected age of the mineral, 5% uncertainty assigned
Uncertainty level and propagation	Ages are quoted at 2σ absolute, propagation is by quadratic addition. Reproducibility and age uncertainty of reference material and common-Pb composition uncertainty are propagated
Quality control/validation	Plešovice: Wtd ave $^{206}Pb/^{238}U$ age = 337 ± 4 (2SD, MSWD = 0.2) M127: Wtd ave $^{206}Pb/^{238}U$ age = 520 ± 5 (2SD, MSWD = 0.8)
Other information	Detailed method description reported by Frei & Gerdes (2009)

Hebridean margin

Detrital zircon age data have been acquired from two Paleocene sandstone samples from wells adjacent to the Hebridean margin, one from well 154/1-1 and one from well 164/25-1. Both wells are located on the eastern flank of the Rockall Basin, although 164/25-1 was drilled in the West Lewis Basin which is separated from the Rockall Basin by the West Lewis High (Fig. 2). The West Lewis High is part of a complex of basement structural highs and Late Palaeozoic–Mesozoic basins on the Hebridean margin (Ritchie et al. 2013a). The Palaeogene sequences in both wells were derived from the Hebridean margin, which has been a constant source of sediment throughout the Cenozoic (Stoker 2013) (Fig. 2).

U–Pb isotopic compositions and ages of detrital zircons in the Paleocene sandstones from wells 154/1-1 (2865.0 m) and 164/25-1 (2651.0 m) are shown in Figures 3 and 4. Most of the zircons in the sandstone from 154/1-1 have Archaean ages within the range 2700–2900 Ma, together with a subsidiary group in the 1700–1900 Ma range and five young zircons dated as c. 280–320 Ma. There are also small numbers of older Archaean zircons ranging back to c. 2970 Ma, and younger Archaean grains down to c. 2510 Ma. The detrital zircon age spectrum in 164/25-1 is similar, with the majority having Archaean ages within the range 2700–2900 Ma, a subsidiary group of zircons in the 1700–2000 Ma range and two young zircons dated as 296 Ma and 433 Ma. The oldest-dated Archaean zircon is c. 3060 Ma and the youngest is c. 2530 Ma.

Rockall High

The eastern margin of Rockall High is fringed by a series of sediment wedges of Early–Late Eocene age that prograde eastwards and north-eastwards into the Rockall Basin (Figs 1 & 5). A detailed analysis of BGS borehole 94/3, which penetrated 210 m into the East Rockall Wedge (Fig. 5), proved that the material that forms these wedges was derived locally from the Rockall High (Stoker et al. 2012). BGS borehole 94/1 also sampled the East Rockall Wedge, albeit the collapsed outer part of the wedge (Fig. 5), and a sample from this borehole was chosen to provide constraints on the nature of the Rockall High source. In addition, an upper Oligocene mass-flow deposit, which accumulated as part of a lowstand fan on the eastern flank of the Rockall Basin (Fig. 6), was similarly derived from the top of Rockall High (Stoker et al. 2001; McInroy et al. 2006). BGS borehole 94/4 recovered sediment from this deposit, and zircons from a sandstone sample were included in the analytical programme.

U–Pb isotopic compositions and ages of detrital zircons in the Eocene sandstone from BGS borehole 94/1 (58.8 m) and the Oligocene sandstone from BGS borehole 94/4 (41.0 m) are shown in Figures 7 and 8. In both cases, the great majority of the

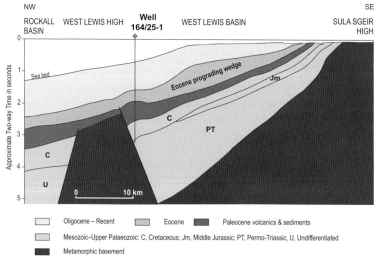

Fig. 2. Schematic geoseismic profile showing location of well 164/25-1 in the West Lewis Basin and its relationship to the Eocene prograding wedge. Profile is based on information derived from several sources including Tate et al. (1999), Isakson et al. (2000), British Geological Survey (2007) and Ritchie et al. (2013a). See Figure 1 for location of profile.

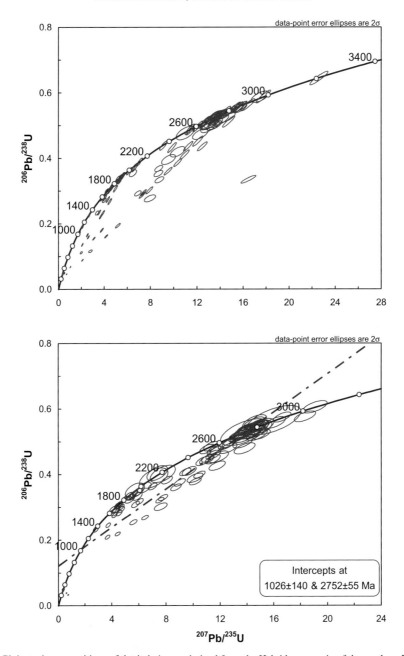

Fig. 3. U–Pb isotopic compositions of detrital zircons derived from the Hebridean margin of the northern Rockall Basin, plotted on Wetherill concordia diagrams.

zircons have compositions that are 90–110% concordant. In borehole 94/1, the zircon spectrum displays a single large peak at c. 1800 Ma together with a small number of older zircons (four in the 2030–2170 Ma range and one at c. 2900 Ma). Due to the small sample size and relatively low proportion of detrital zircons, it was possible to make only 48 analyses from this sample; there is therefore a possibility that some minor age components have not been recognized. This is not the case for

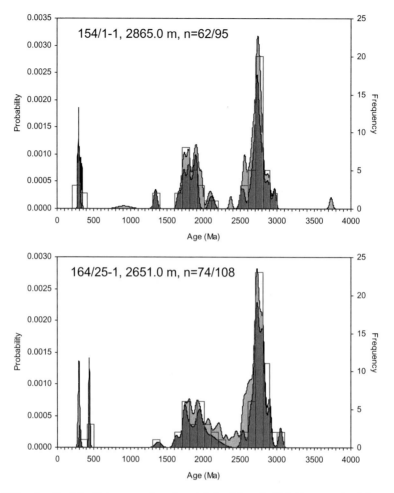

Fig. 4. Probability–density plots of zircon ages in sediment derived from the Hebridean margin of the northern Rockall Basin. Dark grey areas are zircons with 90–110% concordance; light grey areas are zircons with >10% discordance.

borehole 94/4, where zircons proved to be more abundant and >100 analyses were possible. Virtually all the zircons in borehole 94/4 form a single group of age 1670–1850 Ma, peaking at c. 1760 Ma, the only exceptions being a small Paleocene–Eocene group dated as 53–60 Ma comprising two concordant zircons and two with similar ages but with >10% discordance.

George Bligh High

A prograding wedge has been identified on the south-eastern margin of the George Bligh High (Fig. 1). On seismic data, the form of the wedge is defined by the top basalt reflector (underlying the wedge) and the sea bed. Internal reflections, although weak, show a consistent dip away from the top of the high towards the Rockall Basin (Fig. 9). This wedge was penetrated by BGS borehole 94/7 and proved to comprise bioclastic carbonate-rich sandstones of Middle–Late Eocene age (McInroy et al. 2006; Stoker et al. 2012).

Analysis of the sample from borehole 94/7 (11.2–11.8 m) was hampered by the relatively small amount of siliciclastic material due to its predominantly bioclastic composition.

In total, only 33 zircons were available for analysis, of which 21 have isotopic compositions that are 90–100% concordant (Fig. 10). The majority of the concordant grains have young ages, with 14 zircons falling in the 55–64 Ma range. The young zircons display an almost Gaussian distribution (Fig. 11) with a peak at c. 57 Ma, although there is also a hint of a slightly older population.

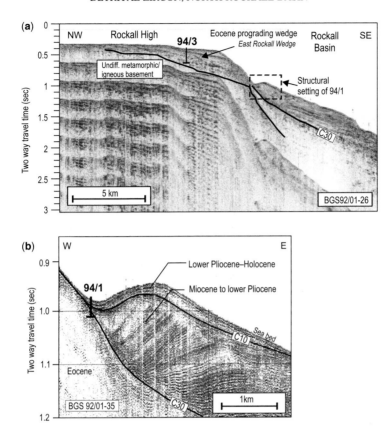

Fig. 5. Illustrative seismic sections through the Eocene prograding sedimentary wedge on the eastern margin of the Rockall High, sampled by BGS boreholes (**a**) 94/3 and (**b**) 94/1. Borehole 94/1 sampled the collapsed outer margin of the wedge, following rapid Late Eocene deepening of the Rockall Basin; the borehole site was subsequently buried beneath onlapping, upslope-accreting, deep-water Neogene strata. See Figure 1 for location of sections. Adapted from Stoker *et al.* (2001, 2012) and McInroy *et al.* (2006). *Abbreviations*: C10, Early Pliocene unconformity; C30, Late Eocene unconformity.

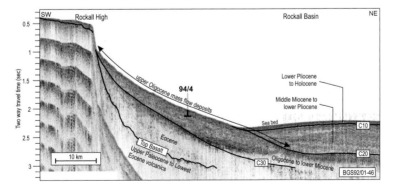

Fig. 6. Illustrative seismic line across the upper Oligocene mass-flow deposit (lowstand fan) on the NE slope of the Rockall High, sampled by BGS borehole 94/4. See Figure 1 for location of section. Adapted from Stoker *et al.* (2001) and McInroy *et al.* (2006). *Abbreviations*: C10, Early Pliocene unconformity; C20, late Early/early Middle Miocene unconformity; C30, Late Eocene unconformity.

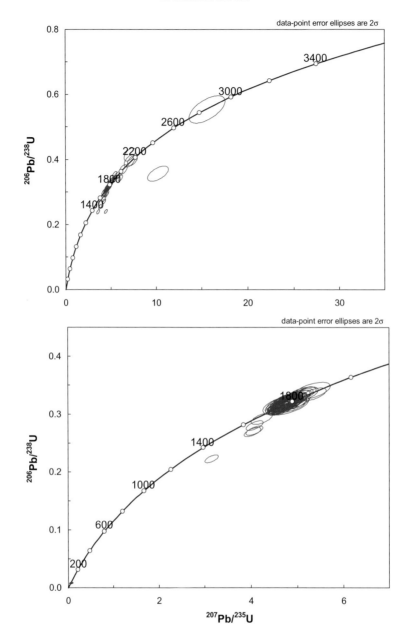

Fig. 7. U–Pb isotopic compositions of detrital zircons derived from the Rockall High, plotted on Wetherill concordia diagrams.

In addition, a significant proportion of the population comprises Proterozoic and Archaean grains (Fig. 11). The concordant zircons appear to define two groups, one of Palaeoproterozoic age (1760–1930 Ma) and one of Neoarchaean–Mesoarchaean age (2760–2930 Ma), in approximately equal proportions.

Discussion

Hebridean margin

The zircon spectra from wells 154/1-1 and 164/25-1, which mainly comprise a large Archaean group of age 2700–2900 Ma and a subsidiary

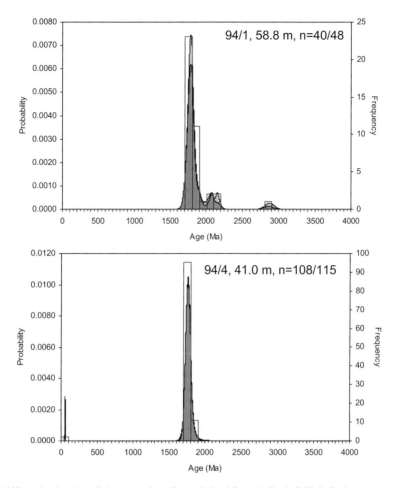

Fig. 8. Probability–density plots of zircon ages in sediment derived from the Rockall High. Dark grey areas are zircons with 90–110% concordance; light grey areas are zircons with >10% discordance.

Palaeoproterozoic group in the 1700–1900 Ma range, are consistent with derivation from crystalline basement similar to that found on the Outer Hebrides massif which consists of three of the terranes that form the Lewisian complex of NW Scotland (Fig. 12). The most northerly (Tarbert Terrane) comprises tonalite-trondhjemite-granodiorite (TTG) gneisses dated as 2800–3125 Ma, with metamorphic reworking and granite intrusion at c. 1675 Ma (Kinny et al. 2005). The

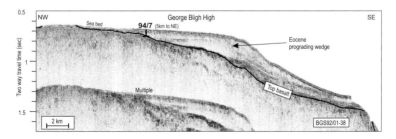

Fig. 9. Illustrative seismic line across the Eocene prograding sedimentary wedge on the SE George Bligh High that was sampled by BGS borehole 94/7. See Figure 1 for location of section. Adapted from McInroy et al. (2006).

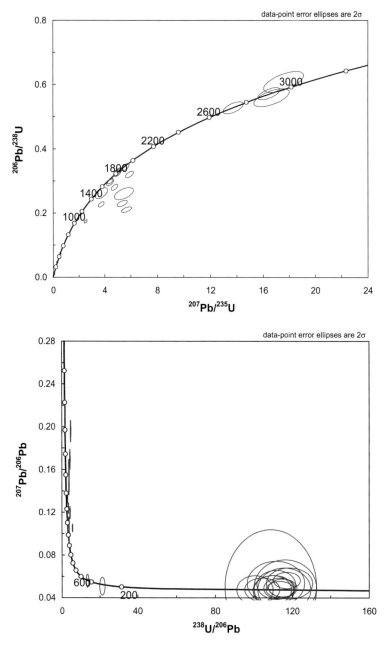

Fig. 10. U–Pb isotopic compositions of detrital zircons derived from the George Bligh High, plotted on Wetherill (upper) and Tera-Wasserburg (lower) concordia diagrams.

Roineabhal Terrane located in Harris comprises metasediments and metamafic rocks, with TTG gneisses dated as 1880 Ma (Kinny et al. 2005). The Uist block further to the south has not been as precisely dated but comprises Archaean protoliths with Proterozoic reworking (Kinny et al. 2005), similar to the Tarbert Terrane. The limited amount of data from the offshore part of the Outer Hebrides High are consistent with observations from the onshore region, where both Archaean (c. 2713,

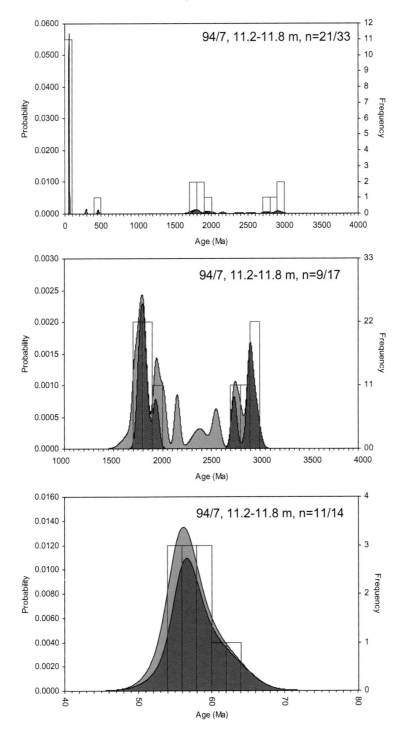

Fig. 11. Probability–density plots of zircon ages in sediment derived from the George Bligh High. Upper diagram shows all zircons, middle diagram shows the Precambrian zircons and the lower plot shows the Paleocene zircons. Dark grey areas are zircons with 90–110% concordance; light grey areas are zircons with >10% discordance.

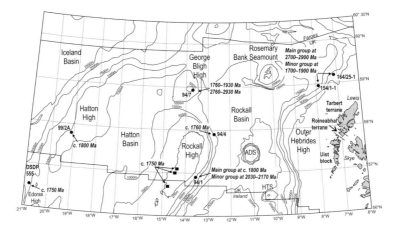

Fig. 12. Compilation of zircon U–Pb data from the northern Rockall Basin and Rockall Plateau (Rockall High–Hatton Basin–Hatton High) areas. Locations identified as black circles are detrital age data. Locations shown as black squares are dive sites from Rockall High with zircon ages inferred from Daly *et al.* (1995). Data from DSDP Site 555 and BGS borehole 99/2A are from Morton *et al.* (2009). Tarbert Terrane, Roineabhal Terrane and Uist block are from Kinny *et al.* (2005). Area of map is located in Figure 1. *Abbreviations*: ADS, Anton Dohrn Seamount; HTS, Hebrides Terrace Seamount.

c. 2767 and c. 2838 Ma) and Palaeoproterozoic (c. 1633, c. 1745 and c. 1791 Ma) ages have been recorded (Ritchie *et al.* 2013b).

The latest Carboniferous–earliest Permian zircons in the samples from 154/1-1 and 164/25-1 probably represent derivation from lamprophyre dykes of this age that occur across NW Scotland, such as the Loch Roag dyke on Lewis (Upton *et al.* 1999). The zircon dated as 433 Ma in 164/25-1 is likely to have been derived from the middle Silurian syenite dated as c. 427 Ma recovered from the Flannan High west of Lewis (Ritchie *et al.* 2013b).

The discordant zircon compositions in the sample from 164/25-1 can be interpreted as being due to Late Mesoproterozoic disturbance. A regression line fitted through the entire dataset (Fig. 3) intersects the concordia line at 1026 ± 140 Ma and 2752 ± 55 Ma. A thermal event at c. 1100 Ma was identified on Lewis and North Harris by Cliff & Rex (1989) on the basis of variations in Rb–Sr biotite ages. This thermal event was attributed by Imber *et al.* (2002) to tectonic burial of the footwall gneisses during ductile thrusting along the Outer Hebrides Fault Zone. It is possible that the discordance observed within the 164/25-1 zircon population also reflects this event.

Rockall High

The zircon age data from BGS boreholes 94/1 and 94/4 derived from Rockall High show a marked difference in character compared with sediment shed from the Hebridean margin. The abundance of c. 1800 Ma zircons in borehole 94/1 is closely comparable with data previously acquired from an Albian sandstone recovered from BGS borehole 99/2A on the Hatton High (Fig. 12), which indicated the existence of c. 1800 Ma crust on the Rockall Plateau (Morton *et al.* 2009). The presence of a small number of older Palaeoproterozoic zircons and a very minor c. 2900 Ma component indicates that the basement dated as c. 1800 Ma on the Rockall High may not be entirely juvenile, and may have a prehistory as far back as the Mesoarchaean. Hf isotopic compositions of zircons in the Albian sandstones from borehole 99/2A indicated a significant Archaean prehistory for the c. 1800 Ma zircons on the Hatton High (Morton *et al.* 2009).

The dominant zircon group in borehole 94/4 is dated as c. 1760 Ma, slightly younger than the main group in borehole 94/1. This indicates the existence of geochronological heterogeneities within the Rockall High basement. This heterogeneity is consistent with observations from the Hatton High and Edoras High (Morton *et al.* 2009), with zircons derived from the Edoras High being dated as c. 1750 Ma and those from the Hatton High at c. 1800 Ma. The zircons in borehole 94/4 have ages that compare well with those acquired from Rockall High basement samples, which yielded zircon U–Pb crystallization ages similar to that of the c. 1750 Ma Annagh Gneiss of north Mayo (Daly *et al.* 1995; Scanlon & Daly 2001).

The data from boreholes 94/1 and 94/4 therefore confirm that the crust underlying the Rockall High and Hatton High represents a

Palaeoproterozoic terrane, apparently similar to gneisses found in Islay and Inishtrahull (Fig. 1), which are dated as c. 1780 Ma (Marcantonio et al. 1988; Daly et al. 1991). There are increasing indications however, from both Hf and U–Pb isotopic data, that the crust in this region has a prolonged history dating back to the Archaean.

The small group of Palaeocene–Eocene zircons in borehole 94/4 is interpreted as representing input from Palaeogene magmatic rocks on the Rockall High. The associated heavy mineral assemblage includes sodic pyroxene (aegirine) and sodic amphibole (riebeckite-arfvedsonite), which together form >20% of the population. These minerals are characteristic of an alkaline igneous provenance and are known to occur in the Early Eocene Rockall Granite (Harrison 1975). The Paleocene–Eocene zircon dates from borehole 94/4 (53–60 Ma) are consistent with input from the Rockall Granite, or from equivalent alkaline intrusions elsewhere on the Rockall High.

George Bligh High

Given the small numbers of grains involved, caution is needed when interpreting the data from borehole 94/7. Nevertheless, it appears that the George Bligh High basement has both Palaeoproterozoic and Archaean crustal components. The Palaeoproterozoic zircons have closely comparable ages to those found in sediments derived from the Rockall High and Hatton High (Fig. 8), and suggest that the George Bligh High formed part of the extensive Palaeoproterozoic terrane that characterizes the Rockall–Hatton area to the south and west together with its onshore representation in Islay and Inishtrahull (Fig. 1). However, the presence of c. 2760–2930 Ma zircons indicates the basement underlying the George Bligh High also had a significant Archaean component.

The heavy mineral assemblage in borehole 94/7 includes a high proportion of minerals with alkaline igneous provenance (aegirine, aenigmatite, riebeckite-arfvedsonite and sodic-calcic amphibole), which together form nearly 50% of the population. This indicates the presence of alkaline magmatic rocks on the George Bligh High, similar to those that occur on the Rockall High, in addition to the basalts in which BGS borehole 94/7 terminated (Hitchen et al. 1997). The Paleocene zircon age of c. 57 Ma is interpreted as representing the age of this alkaline magmatic event.

Conclusions

Detrital zircon ages from Palaeogene first-cycle sediments in the northern Rockall Basin have provided important constraints on the nature and distribution of basement terrains, as well as dating penecontemporaneous magmatic events in the region. Sediments shed westwards from the Hebridean Platform contain zircons of Archaean and Palaeoproterozoic age, consistent with derivation from rocks with Lewisian affinities. They also contain a minor Carboniferous–Permian group, which is likely to be derived from lamprophyre dykes such the Loch Roag dyke on Lewis. One sample also contains a middle Silurian zircon, probably derived from the Flannan High. The pattern of discordance in one sample suggests that a Mesoproterozoic thermal event affected a part of the source region, supporting previous evidence for a Grenville-age event identified on the Outer Hebrides.

The zircon spectra in the two samples derived from the Rockall High are dominated by Palaeoproterozoic grains, one being dominated by a group at c. 1760 Ma and the other by a group at c. 1800 Ma. These results confirm previous data, both from direct dating of basement and from detrital zircon studies, which indicated that the crust underlying the Rockall High and Hatton High represents a Palaeoproterozoic terrane. The samples also contain a small number of Paleocene–Eocene zircons, which together with heavy mineral evidence for input from alkaline igneous sources can be tied back to the Rockall Granite or equivalent rocks in the Rockall High region.

Detrital zircons in sediment shed from George Bligh High have provided the first direct indications of the age of the crust underlying this area. Although only a small number of zircons were recovered from the analysed sample, the data positively identify both Palaeoproterozoic and Archaean zircons, indicating that the crust beneath George Bligh High represents an Archaean terrane heavily affected by Palaeoproterozoic tectonothermal events. In addition, the zircon and associated heavy minerals indicate George Bligh High was the site of Paleocene alkaline magmatism.

We are grateful to the Rockall Consortium (ENI, Shell, Statoil and DECC) for supporting BGS activities to the west of Britain, in particular with regard to the coring programme that yielded the samples discussed in this paper. The paper benefited from the constructive comments of J. Sláma and D. Ritchie. The contribution of MSS is included with the permission of the Executive Director, BGS (NERC).

References

BRITISH GEOLOGICAL SURVEY 2007. *North Rockall Basin. Sheet 58°N 13°W. Bedrock Geology. 1:500 000 Scale Geology Series*. British Geological Survey, London.

CLIFF, R. A. & REX, D. C. 1989. Evidence for a 'Grenville' event in the Lewisian of the northern Outer Hebrides. *Journal of the Geological Society, London*, **146**, 921–924.

DALY, J. S., MUIR, R. J. & CLIFF, R. A. 1991. A precise U–Pb age of the Inishtrahull syenitic gneiss, County Donegal, Ireland. *Journal of the Geological Society, London*, **148**, 639–642.

DALY, J. S., HEAMAN, L. M., FITZGERALD, R. C., MENUGE, J. F., BREWER, T. S. & MORTON, A. C. 1995. Age and crustal evolution of crystalline basement in western Ireland and Rockall. *In*: CROKER, P. F. & SHANNON, P. M. (eds) *The Petroleum Geology of Ireland's Offshore Basins*. Geological Society, London, Special Publications, **93**, 433–434.

DICKIN, A. P. 1992. Evidence for an Early Proterozoic crustal province in the North Atlantic region. *Journal of the Geological Society, London*, **149**, 483–486.

DICKIN, A. P. & DURANT, G. P. 2002. The Blackstones Bank igneous complex: geochemistry and crustal context of a submerged Tertiary igneous centre in the Scottish Hebrides. *Geological Magazine*, **139**, 199–207.

FREI, D. & GERDES, A. 2009. Precise and accurate in situ U–Pb dating of zircon with high sample throughput by automated LA-SF-ICP-MS. *Chemical Geology*, **261**, 261–270.

GERDES, A. & ZEH, A. 2006. Combined U–Pb and Hf isotope LA-(MC)-ICP-MS analyses of detrital zircons: Comparison with SHRIMP and new constraints for the provenance and age of an Armorican metasediment in Central Germany. *Earth and Planetary Science Letters*, **249**, 47–61.

HARRISON, R. K. (ed.) 1975. *Expeditions to Rockall 1971–1972*. Report for the Institute of Geological Sciences, HMSO, London, **75/1**.

HITCHEN, K. 2004. The geology of the UK Hatton-Rockall margin. *Marine and Petroleum Geology*, **21**, 993–1012.

HITCHEN, K., MORTON, A. C., MEARNS, E. W., WHITEHOUSE, M. & STOKER, M. S. 1997. Geological implications from geochemical and isotopic studies of Late Cretaceous and early Tertiary igneous rocks around the northern Rockall Trough. *Journal of the Geological Society, London*, **154**, 517–521.

IMBER, J., STRACHAN, R. A., HOLDSWORTH, R. E. & BUTLER, C. A. 2002. The initiation and early tectonic significance of the Outer Hebrides Fault Zone, Scotland. *Geological Magazine*, **139**, 609–619.

ISAKSON, G. H., WILKINSON, D. R. & HITCHEN, K. 2000. Geochemistry of organic-rich Cretaceous and Jurassic mudstones in the West Lewis and West Flannan basins, offshore north-west Scotland: implications for source rock presence in the north-east Rockall Trough. *Marine and Petroleum Geology*, **17**, 27–42.

JACKSON, S., PEARSON, N., GRIFFIN, W. & BELOUSOVA, E. 2004. The application of laser ablation – inductively coupled plasma – mass spectrometry to in situ U–Pb zircon geochronology. *Chemical Geology*, **211**, 47–69.

JONES, E. J. W., PERRY, R. G. & WILD, J. L. 1986. Geology of the Hebridean margin of the Rockall Trough. *Proceedings of the Royal Society of Edinburgh, Section B, Biological Sciences*, **88**, 27–51.

KINNY, P. D., FRIEND, C. R. L. & LOVE, G. J. 2005. Proposal for a terrane-based nomenclature for the Lewisian Gneiss Complex of NW Scotland. *Journal of the Geological Society, London*, **162**, 175–186.

LUDWIG, K. R. 2003. *Isoplot/Ex 3.00. A Geochronological Toolkit for Microsoft Excel*. Berkeley Geochronological Center, Berkeley, CA, Special Publications, **4**.

MARCANTONIO, F., DICKIN, A. P., MCNUTT, R. H. & HEAMAN, L. M. 1988. A 1800-million-year-old Proterozoic gneiss terrane in Islay with implications for the crustal structure and evolution of Britain. *Nature*, **335**, 620–624.

MATTINSON, J. M. 2010. Analysis of the relative decay constants of ^{235}U and ^{238}U by multi-step CA-TIMS measurements of closed-system natural zircon samples. *Chemical Geology*, **275**, 186–198.

MCINROY, D. B., HITCHEN, K. & STOKER, M. S. 2006. Potential Eocene and Oligocene stratigraphic traps of the Rockall Plateau, NE Atlantic. *In*: ALLEN, M. R., GOFFEY, G. P., MORGAN, R. K. & WALKER, I. M. (eds) *The Deliberate Search for the Subtle Trap*. Geological Society, London, Special Publications, **254**, 247–266.

MORTON, A. C. & TAYLOR, P. N. 1991. Geochemical and isotopic constraints on the nature and age of basement rocks from Rockall Bank, NE Atlantic. *Journal of the Geological Society, London*, **147**, 631–634.

MORTON, A. C., HITCHEN, K., RITCHIE, J. D., HINE, N. M., WHITEHOUSE, M. & CARTER, S. G. 1995. Late Cretaceous basalts from Rosemary Bank, northern Rockall Trough. *Journal of the Geological Society, London*, **152**, 947–952.

MORTON, A. C., HITCHEN, K., FANNING, C. M., YAXLEY, G., JOHNSON, H. & RITCHIE, J. D. 2009. Detrital zircon age constraints on the age of the basement on Hatton Bank and Edoras Bank, NE Atlantic. *Journal of the Geological Society, London*, **166**, 137–146.

NASDALA, L., HOFMEISTER, W. ET AL. 2008. Zircon M257 – a homogeneous natural reference material for the ion microprobe U–Pb analysis of zircon. *Geostandards and Geoanalytical Research*, **32**, 247–265.

RITCHIE, J., JOHNSON, H., KIMBELL, G. & QUINN, M. 2013a. Structure. *In*: HITCHEN, K., JOHNSON, H. & GATLIFF, R. W. (eds) *Geology of the Rockall Basin and Adjacent Areas*. British Geological Survey, London, Research Report **RR/12/03**, 10–46.

RITCHIE, J., NOBLE, S., DARBYSHIRE, F., MILLAR, I. & CHAMBERS, L. 2013b. Precambrian basement. *In*: HITCHEN, K., JOHNSON, H. & GATLIFF, R. W. (eds) *Geology of the Rockall Basin and Adjacent Areas*. British Geological Survey, London, Research Report **RR/12/03**, 47–53.

ROBERTS, D. G., ARDUS, D. A. & DEARNLEY, R. 1973. Precambrian rocks drilled on Rockall Bank. *Nature, Physical Science*, **244**, 21–23.

SCANLON, R. & DALY, J. S. 2001. Basement architecture of the rifted Northeast Atlantic margin: evidence from a combined geochronology, fission-track and potential field study. *Geological Society of America Annual Meeting*, Boston, Massachusetts, Abstracts with Programs, paper 64-0.

SIRCOMBE, K. N. 2004. AgeDisplay: an EXCEL workbook to evaluate and display univariate geochronological data using binned frequency histograms and probability density distributions. *Computers & Geosciences*, **30**, 21–31.

SLÁMA, J., KOŠLER, J. ET AL. 2008. Plešovice zircon – a new natural reference material for U–Pb and Hf isotopic microanalysis. *Chemical Geology*, **249**, 1–35.

STACEY, J. S. & KRAMERS, J. D. 1975. Approximation of terrestrial lead isotope evolution by a two-stage model. *Earth and Planetary Science Letters*, **26**, 207–221.

STOKER, M. S. 2013. Cenozoic sedimentary rocks. *In*: HITCHEN, K., JOHNSON, H. & GATLIFF, R. W. (eds) *Geology of the Rockall Basin and Adjacent Areas*. British Geological Survey, London, Research Report **RR/12/03**, 96–136.

STOKER, M. S., HITCHEN, K. & GRAHAM, C. C. 1993. *United Kingdom Offshore Regional Report: The Geology of the Hebrides and West Shetland Shelves and the Adjacent Deep-Water Areas*. HMSO, London, for the British Geological Survey.

STOKER, M. S., VAN WEERING, T. C. E. & SVAERDBORG, T. 2001. A Mid- to Late Cenozoic tectonostratigraphic framework for the Rockall Trough. *In*: SHANNON, P. M., HAUGHTON, P. D. W. & CORCORAN, D. V. (eds) *The Petroleum Exploration of Ireland's Offshore Basins*. Geological Society, London, Special Publications, **188**, 411–438.

STOKER, M. S., KIMBELL, G. S., MCINROY, D. B. & MORTON, A. C. 2012. Eocene post-rift tectonostratigraphy of the Rockall Plateau, Atlantic margin of MW Britain: linking early spreading tectonics and passive margin response. *Marine and Petroleum Geology*, **30**, 98–125.

TATE, M. P., DODD, C. D. & GRANT, N. T. 1999. The Northeast Rockall Basin and its significance in the evolution of the Rockall–Faeroes/East Greenland rift system. *In*: FLEET, A. J. & BOLDY, S. A. R. (eds) *Petroleum Geology of Northwest Europe: Proceedings of the 5th Conference*. Geological Society, London, 391–406.

UPTON, B. G. J., HINTON, R. W., ASPEN, P., FINCH, A. & VALLEY, J. W. 1999. Megacrysts and associated xenoliths: evidence for migration of geochemically enriched melts in the upper mantle beneath Scotland. *Journal of Petrology*, **40**, 935–956.

Index

Page numbers in *italics* refer to Figures. Page numbers in **bold** refer to Tables.

a'ā lava 186–187
Aegir Ridge *12*, 13
  rifting 14, 15, 22
African Plume *see* Large Low Shear Velocity Province
Alligin Field, discovery 2, *60*, *61*
Amos Field, discovery 2, **60**, *61*
amplitude anomalies 63, *64*, 65, 70, 74
  Assynt prospect *64*, *73*, 74–77
  Glenlivet Field *63*, 64, 65
  North Fleet prospect *75*, *76*, 77–78
  Schiehallion Field 65
  Tormore Field 65
  Tornado Field *64*
apatite fission track analysis 23
Arkle Field, discovery **60**, *61*
Assynt prospect, amplitude anomalies *64*, *73*, 74–77
AVO analysis 65, 69–70, 73–74
  Foinaven Field *71*, *72*
  Glenlivet Field *63*, 64, 65, 135, *136*
  Laggan Field 65, *73*, 107, *108*
  North Fleet prospect *76*, 77–78
  Schiehallion Field 65
  Tornado Field 64

Baffin Bay spreading ridge 14
Balder Formation *34*, 44, *61*, 65
basalt
  Aegir Ridge 13
  GIFR 16
  Palaeocene, micro-imaging 173–190
  Palaeocene-Eocene 20–21
  Palaeogene 3, 5
  sub-basalt imaging 163–170
  'sub-basalt problem' 173, 193
  see also Mid-Ocean Ridge Basalts
basement
  as exploration play 81–103, *84*
  Lewisian, Rona Ridge 82, 85–103
  see also Lancaster Prospect
Bedlington Field, discovery *60*, *61*
Beinisvørð Formation 173
Benbecula discovery 3
Bering land bridge 17
biogeography *see* migration
Bouguer anomaly, Vøring Spur 17
breccia, FMI *181*, 187–188
broadband seismic data 169–170
Brown Field and Southern North Sea Allowance 6
Bunnehaven Field
  discovery 2, *60*, *61*, 63
  reservoir properties 63

calcite nodules, Laggan Field 107, 116–117
Cambo Field 2, *146*
  commercial challenges 5
  discovery 3, *60*, *61*, 63
  exploration 146–147, *148*
  four-way dip closure 145–162
  pre-drill evaluation 147–149, *150*, *159*
  reservoir properties 65
  seal integrity 148–149, 157
  stratigraphy 147
  Well 204/10-1 149–151, *152*, *159*
  Well 204/10-2 (Lindisfarne) 151–155, *156*, *159*
  Well 204/10a-3 155–160
  Well 204/10a-4,4Z *159*, 160–161
Cambo High, hydrocarbon potential 148–162
Cambo Sandstone Member 145
  potential reservoir 148, 149, 150, 152
cherry, cornelian, migration 17
chlorite
  Laggan Field 62, 66, 116, 117
  Tormore Field 64
Clair Field
  basement reservoir 82
  discovery 2, 3
Colsay Sandstone Member *21*, *61*, 63, 65, 145
  hydrocarbons 147, 174
commercial challenges 5, 6
continent-ocean boundary 13, *14*, 15
controlled source electromagnetic survey (CSEM) 3, 70
core analysis
  FMI 178–183
    sidewall 183
Cornus, migration 17
Corona Basin *167*
Corona Ridge *21*, *34*, 163, *167*
  hydrocarbon discovery 59, **60**, 63
creaming curves, UKCS 3, *4*
Cuillin Field, discovery **60**, *61*

DHI (direct hydrocarbon indicator) anomalies 70
diapirism, mud/shale 51–52
discoveries, WoS 2, 3
ditch cutting analysis
  LIPs 193–205
    end-member classification 197–200
    facies classification 196–197
    generation 195
    idealized volcanic stratigraphy 201–204
    sample preparation 195
    simple log output 200–201
drilling, basement reservoir, Lancaster Prospect 92–94
dual rift model 21–22
  sediment provenance studies 23–24
dykes
  age *34*, *35*, 36–37
  emplacement 35–37, *39*, 45

East Faroe High *167*
East Rockall Wedge, zircon U-Pb analysis 212, *215*
East Solan Basin 90–91
Edradour discovery 2, 3

Eocene
  Atlantic margin development 18–19, *21*
  inversion 23
  see also Palaeocene-Eocene
European Union, response to Macondo oil spill 5
exploration
  basement reservoirs 81–103
  commercial challenges 5, 6
  political challenges 5–7
  technical challenges 3, 6
    large igneous provinces 193
  WoS
    history 1, 3, 59–64
    post-drill analysis 70–78
    pre-drill evaluation, Cambo Field 147–149

Faroe Conduit 23
Faroe Islands, LIP facies classification 196
Faroe-Shetland Basin 11–26, *12*
  geological setting *34*, 35–37
  Geostreamer 3D seismic imaging 147
  hot fluid flow 23
  igneous intrusions 33–55
  sediment provenance studies 23–24
  stratigraphy *34*
  sub-basalt imaging 163–170
  volcanic margin rocks, Rosebank prospect 174–190
  see also Rosebank Field
Faroes-Kangerlussauq lineaments 18, *19*
Fault Zones 89–90
  Lancaster Field 95–96, 98–99, 100, *101*
faults
  Glenlivet Field 133
  Lewisian Basement 85
    conceptual model 88–90, 99
    Lancaster Field 86–87, 95–96, 100, *101*, *102*
field allowances 5, 6
Flett Formation
  broadband processing *168*, 170
  hydrocarbon discovery *61*, 63
Flett Ridge *34*, *167*
Flett Sub-basin *34*, *131*, *167*
  hydrocarbon discovery 59, **60**, 63, 70
  hydrothermal vents 52
  post-drill analysis 70, 77
  seal presence 67, *68*
  seismic data *36*, *37*, *38*, *44*
  sill emplacement *44*, 47, 50
    and Vaila turbidite deposition 45, *51*, 53–54
  source rocks 69
  trap model 64–65
fluid flow
  in fracture networks 85, 87
  hot, Faroe-Shetland Basin 23
Foinaven Field
  discovery 2, 3, *60*, *61*, 62–63
  post-drill analysis 70, *71*, *72*, 74
  reservoir properties *62*, *63*, 65, *66*
  seal presence 68
  source rocks 69
formation micro-imaging (FMI)
  volcanic rock 174–190
    core analysis and analogues 178–179, 183
    down-hole well analysis 174
    field-based acquisition 183–185
    interpreting joint and fracture systems 190
    lithofacies identification *180–182*, 186–189
    lithological separation 190
    methods of calibration 176–185
    sidewall core analysis 183
Foula Sub-basin *60*, *167*
  kitchen area *83*, 85, 90, 91, *92*
four-way dip closures, White Zone, Cambo High 145–162
fracture networks
  fluid flow 85, 87, 88, 90
  FMI *181*, 187, 190
  Lancaster Prospect 88–90, 100
  Lewisian Basement 85–86
  see also faults; joints; shear fractures; sheet fractures
Front-End Egineering Design (FEED), Rosebank prospect 6, 8
Fugloy Ridge 163, *167*
FUKA (Frigg UK Association) Pipeline 123, *124*

geoid anomaly, North Atlantic Ocean 25–26
George Bligh High 210
  zircon U-Pb analysis 214, 216, *217*, *218*, *219*, 221
Geostreamer 3D seismic imaging 147
Glenlivet Field 131–143, *132*
  AVO 135, *136*
  conceptual model 143
  depositional model 137–141
    lower zone 138–139, *141*
    middle zone 139–141, *142*
    upper zone 141, *142*
  discovery 2, 3, *60*, *61*, 131
  porosity/permeability 131
  reservoir properties *63*, *64*, *65*, 131, 134–137
  reservoir thickness 134
  seismic interpretation 134–137
  shale 131
  structural and deposition evolution 133–134
  Vaila turbidites 131
  zone boundaries 141, 143
Great Britain, Atlantic margin development 18–19
Greenland, Palaeocene-Eocene succession 19–21
Greenland-Faroes Ridge 11
Greenland-Iceland-Faroe Ridge 11, *12*, 13
  composition 16
  developmental model 17–18
Gross Rock Volume, Lancaster prospect 89, 90, 101

Hatton Basin 19, *20*
Hatton High 19
  folding 23
  zircon U-Pb analysis 220
Hebridean margin, zircon U-Pb analysis 212, *213*, *214*, 216–218, 220
Hildasay Sandstone Member *61*, 63, 65, 145
  hydrocarbon 147, 150, 152, 154–158, 160–161
Huab outliers, Namibia, FMI 183–184
hydrocarbons
  exploration WoS
    history 1, 3, 59–64
    post-drill analysis 70–78
    pre-drill analysis 147–149

reservoirs
  basement 81–103
  presence 65
  quality 65–67
  Vaila turbidites 52–53
 seals 67–68
 source rocks 68–69
 trap models 64–65
hydrothermal alteration, Rona Ridge basement 86, 87
hydrothermal vents 52

Iceland
 age and role 24–25, 26
 mantle plume 11, 17, 24
igneous intrusions
 and turbidite deposition 33–55
 see also dykes; sills
Igtertiva basalts 21
inflation anticlines *38–44*, 45, 50
infrastructure 2, 8
 commercial considerations 5, 6
Inner Fault Zone 89–90
integrated asset model
 WoS 126–129
  additional field tie-in case study *127*, 129
  software 127–129
  subsea gas compression case study *128*, 129
inversion, North Atlantic province 23
inversion anticlines 50–51

Jan Mayen Fracture Zone *12*, 13, *14*
 Bouguer anomaly 17
 developmental model 17, *18*
Jan Mayen micro-continent 14–15, *16*
joints
 FMI *181*, 187, 190
 regional, Lewisian Basement 85
Judd Sub-basin
 hydrocarbon discovery 59, **60**, 62–63, 65, 70
 post-drill analysis 70, 74
 reservoir properties 65, 67
 seal presence 67
 source rocks 69

Kangerlussauq *12*, *14*, *20*, *21*
 sediment provenance 23
Kangerlussauq lineament 18, *19*
Kettla Tuff *34*, 44, *49*
 seal presence *52*, 53, 63, 67, *70*
Kimmeridge Clay Formation 69, 85, 148
Kolbeinsey Ridge *12*
 rifting 14, 15, 24

Labrador Sea extinct Ridge *12*, *14*
Lagavulin prospect 5
Laggan gas condensate field *108*
 3D geological modelling 107–121
 channel feature 113, *115*
 damage zones 111
 discovery 2, 3, *60*, *61*, 62, 63, 107
 facies modelling 117
 fault extensions 110, 111
 lineaments 110, 111
 petrophysical modelling 117–120

3D 119–120
connate water saturation correction 118–119
depth correction 117
Klinkenberg correction 118
permeability and petrophysical logs 119
porosity/permeability 107, *109*, *110*, 116, 117, 120
post-drill analysis 70, *72*, *73*, 74
reservoir modelling 111–113, *114*
reservoir properties 62, 65, 66, 107, *109*, 116–117
 heterogeneities 116–117
 quality 107
 thickness 107, *112*, 113, *114*, 115–116
sedimentological model 113, 115
seismic data 107–108, *109*, *110*, *111*, 113
seismic faults 108
structural elements 108, 110–111
structural model building 111, 113
 base reservoir construction *112*, 113, *114*
 fault modelling 111
 top reservoir construction 111, 113
water saturation modelling 120
well test matching 120
Laggan-Tormore gas condensate development 3
 Eclipse models 123, *125*, 126, 128
 HYSYS models *125*, *126*, 128
 integrated modelling 123–129
 reservoir modelling 123, *125*, *126*
 Resolve software 128–129
 WoS integrated asset model 126–129
Laggan-Tormore pipeline 5, 8, 123, *124*
Lamba Formation *34*, 44, 64
Lancaster Prospect 2, 3, 81, *82*, *83*, 84
 appraisal wells 102
 establishing resource potential 99–103
 exploration well planning 86–93
 funding 84
 operational plan 93–99
  drill stem tests 93, 95–96, *97*
 petroleum system 85–86
  drilling 92–94
  fracture network 85–86, 88
   conceptual model 88–90, 99
  oil-down-to 90, 100, 101
  porosity/permeability 86, 88, 90, 100, *101*
  source rock 85
  source and type of oil 91
  tectonic history and charge 91, *93*
  trap and seal 85
 stress fields 87–88
land bridges 17, 22
large igneous provinces
 ditch cutting analysis 193–205
  idealized volcanic stratigraphy 201–204
  simple log output 200–201
 end-member classification
  crystalline/scoriaceous 197, *198*
  epiclastic/boles 198–200
  volcanic glass 197–198, *199*, 201
 facies classification 196–197
Large Low Shear Velocity Province 18
lava flows, FMI 176, 178, 179, *182*, 186–187
Laxford Field 131, *133*, 137
 discovery 2, 3, *60*, *61*

Lewisian basement reservoir
　　Lancaster Prospect 82, 85–103
　　　　conceptual model 88–90
　　　　porosity/permeability 86, 88, 90
Lopra Formation 173
LOPRA-1/1a well 173
Loyal Field *2*, 8
　　discovery *60*, *61*, 63
　　seal presence 68

Macondo 2010 oil spill, political aftermath 5, 6
magmatism, Palaeocene 35
magneto-chrons, North Atlantic 12, 13, *14*, *15*
mantle plume
　　Greenland 11–12, 17, 24
　　geoid anomaly 25–26
melting, decompressional 15–16, 17
Mid-Ocean Ridge Basalts, Iceland 24
migration, Thulian land bridge 16–17
MMO drilling fluid 92, 94, *95*
Mohns Ridge *12*
monoethylene glycol (MEG) pipeline 123, *124*
Muckle Basin *34*
　　seismic data 37, *38*, *39–43*
　　Vaila turbidite deposition 45, *51*
　　　　reservoir quality *50*
　　　　and sill intrusion *48*, *49*, *50*, 53–54

noise attenuation *165*, *166*, 168–169
North Atlantic Igneous Province 11
　　Cambo High 148, 154, 155
　　facies classification 196
　　igneous intrusions *34*
　　Palaeocene igneous activity 18, 174
North Atlantic Ocean
　　decompressional melting 15–16, 17
　　Eocene development 18–19
　　evolution 14–16
　　geoid anomaly 25–26
　　hot fluid flow 23
　　magneto-chrons 12, 13, *14*, *15*
　　oceanic crust 12–15, 17
　　opening 11–12, 18
　　plate tectonics 11, 12–16
　　topography *12*
　　triple junction 14
North Fleet prospect, amplitude anomalies *75*, *76*, 77–78

oak pollen, migration 17
oceanic crust, North Atlantic Ocean 12–15, 17
offlaps 51
Oligo-Miocene, inversion 23
onlaps 51
Otterburn potential reservoir 149, *150*
Outer Fault Zone 89
overcrusting, Vøring Spur 18

pāhoehoe lava flows, FMI 176, 178, *182*, 186
Palaeocene
　　exploration WoS
　　　　history 59–64
　　　　post-drill analysis 70–78
　　　　trap definition 64–65

magmatism 35
palaeotopography 45, 47, *48*, *49*, 50–51
post-sill depositional model 53–54
'T-scheme' 145
see also palaeotopography, Palaeocene
Palaeocene-Eocene
　　dual rift model 21–22
　　sediment provenance studies 23–24
　　succession 19–21
palaeotopography
　　Palaeocene 45, 47, *48*, *49*, 50–51
　　　　diapiric mud models 51–52
peperite *182*, 189
pillow lava *181*, 187
Pilot Whale Anticline 51
plate tectonics, North Atlantic 11, 12–16
political challenges 5–7
Porcupine Basin 19, *20*
post-drill analysis 70–78
pre-drill analysis 147–149
Pseudomatrix 89–90, 96
　　Lancaster Prospect 95–96, 98–99, 100, *101*

reservoir modelling, Laggan-Tormore 123, *125*, *126*
reservoirs
　　basement 81–103, *84*
　　presence 65
　　quality 65–67
　　seals 67–68
　　see also hydrocarbons, reservoirs
resistive contacts, FMI *182*, 188
Reykjanes Ridge *12*, 13
　　rifting 14, 15, 22, 24
rifting
　　dual rift model 21–22
　　North Atlantic 13–16, 18
Rockall Basin 19, *20*
　　northern 209–210
　　　　zircon U-Pb analysis 210–221
Rockall Granite 221
Rockall High 210
　　zircon U-Pb analysis 212–214, *215*, *216*, 220–221
Rona Ridge *34*, *167*
　　Lancaster Prospect *83*, 84–85
　　Lewisian Basement reservoir 81, *82*, 85–103
　　　　exploration well planning 86–93
　　　　hydrothermal alteration 86
　　　　petroleum system 85–86
Rosebank Field *175*
　　basalt 3, 5, 173
　　commercial challenges 5
　　discovery 2, 3, *60*, *61*
　　FEED project 6, 8
　　formation micro-imaging
　　　　calibration methods 176–185
　　　　jointing and fracture systems 190
　　　　lithological separation 190
　　　　volcanic lithofacies identification 186–189
　　reservoir properties 63, 65
　　stratigraphy 174, *176*
Rosemary Bank Seamount *20*, 210
rotation, regional stress field 22–23
Rothbury sandstone potential reservoir 148, 149, *150*, *151*, 152

safety, post-Macondo 5
sand, Laggan Field 107, *109*, 113, 115–116, 117
Schiehallion Field
  discovery 2, 3, *60*, *61*, 63
  redevelopment 8
  reservoir properties 63, 65
  seal presence 68
seals
  Lancaster Prospect 85
  presence and effectiveness 67–68
sediment provenance studies, dual rift model 23–24
shale, Laggan Field 107, *109*, 116
shear fractures, Lewisian Basement 85
sheet fractures, Lewisian Basement 85
Shetland Gas Plant (SGP) 123, *124*
Shetland Islands Regional Gas Export (SIRGE) pipeline 8, 123
Shetland Platform *34*
sills
  emplacement 35–37, 45
    age 33, *34*, 35, 36–37, 45–47
    depositional model 53–54
    exploitation risk 52–53
    inflation anticlines *38–44*
    inversion anticlines 50–51
  and turbidite deposition 33
    palaeotopography 45, 47, *48*, *49*
    seismic geometries *36*, 44–45
Solan, discovery 2, 3
source rocks 68–69
  Lancaster Prospect 85
spectral processing, sub-basalt imaging 163–168
spreading *see* rifting
SRME (surface-related multiple elimination) *165*, *166*, 168
Steinvør Basin *167*
Strathmore, discovery 2, 3
stress fields, Lancaster Prospect 87–88
sub-basalt imaging 163–170
  broadband seismic data 169–170
  noise attenuation 168–169
  spectral processing, low-frequency boosting 163–168
'sub-basalt problem' 173, 193
Suilven Field
  discovery 2, 3, *60*, *61*, 63
  source rocks 69
Sullom Formation *34*, 44, *48*, 53
Sullom Voe oil Terminal (SVT) 8, 123, *124*

tax, UK 2011 Budget 5–6
technical challenges 3, 5
Teilhardina magnoliana 17
Thulian land bridge 17, 22
Tobermory discovery 2, 3
Tormore Field
  discovery 2, 3, *60*, *61*
  reservoir properties 64, 65

  see also Laggan-Tormore gas condensate development;Laggan-Tormore pipeline
Tornado Field
  commercial challenges 5
  discovery 2, 3, *60*, *61*
  reservoir properties 64
  seal presence 68
  seismic amplitude anomaly 3
  source rocks 69
Torridon Field
  discovery 2, *60*, *61*
  reservoir properties 66–67
traps 64–65, 70
  amplitude anomalies *64*, 65
  Glenlivet Field 131, 133–134
  Lancaster Prospect 85
turbidites
  deposition
    igneous intrusions 33–55
    seismic geometries *36*, 44–45
  Glenlivet Field 131, 133–134, 137–143
  Laggan Field 107, *109*

UK government, 2011 tax increase 5–6
UKCS, creaming curves 3, *4*
undulose contacts, FMI *182*, 188–189

Vaila turbidite sandstone formation *34*, *38*, 44
  amplitude anomalies 65, *75*
  Foinaven Field *61*, 62–63
  Glenlivet Field 64, 131
  hydrocarbon reservoirs 52–53, *61*, 62, 63
    quality 65–67
  Laggan Field 59, 62, 66
  Loyal Field 63
  palaeotopographical control *40–43*, 45, *48*, 53–54
  Schiehallion Field 63
  seal presence 67, 68
  Suilven Field 63
Victory discovery 2, 3
volcanic glass, end-member classification 197–198, *199*, 201
volcanic rock, FMI 174–190
Vøring Spur 16, 17, *18*

West Lewis Basin, zircon U-Pb analysis 212
West Shetland Basin *167*
Westray Ridge *34*, 65
Whirlwind basement discovery 81, *82*
White Zone
  four-way dip closures, Cambo High 145–162
  seismic data 149, *150*, *153*, *155*
Wyville Thompson Ridge 19, *20*

zircon U-Pb analysis
  Greenland 23
  northern Rockall Basin 210–221